Engineers' and Surveyors'

INSTRUMENTS of PRECISION

MADE BY

C. L. BERGER & SONS

HAND-BOOK AND ILLUSTRATED CATALOGUE

OF THE

Engineers' and Surveyors'

INSTRUMENTS of PRECISION

MADE BY

C. L. BERGER & SONS,

1900

Written and Edited by C.L. Berger

THE ASTRAGAL PRESS
Mendham, New Jersey

Library of Congress Catalog Card Number: 92-75614
International Standard Book Number: 1-879335-33-6

Printed in the United States of America

Published by:
THE ASTRAGAL PRESS
5 Cold Hill Road, Suite 12
Mendham, New Jersey 07945-0239

INTRODUCTION

The 1900 *Hand-book and Illustrated Catalogue of the Engineers' and Surveyors' INSTRUMENTS OF PRECISION made by C.L. Berger & Sons* was the first catalogue issued by this famous Boston instrument maker after the company was founded on October 18, 1898. As such it is a rare and wonderful collectors' piece, representing an excellent reference work for students and collectors of American scientific instruments.

Christian Louis Berger was born in Stuttgart, Germany, on September 26, 1842. By the time he was 24 years old he had apprenticed for four years with German instrument maker Christian Seeger, had studied mechanical technology for one year at the Royal Polytechnic Institute in Stuttgart and had worked for several instrument makers including F.W. Breithaupt & Son of Cassell and Thomas Cooke & Sons of England. In 1866 he made the long voyage to America and worked for Thomas Upham in Boston; four years later he went to work for Edward S. Ritchie, a maker of philosophical (scientific) and nautical instruments. One year after joining the Ritchie firm and with about 14 years of instrument making experience "under his belt," C. L. Berger joined with George L. Buff on October 28, 1871, to form the firm of Buff & Berger, with a factory at 9 Province Court in the heart of downtown Boston. This was just two blocks away from the old State House where less than 40 years earlier Simeon Borden had set up the large Troughton & Simms theodolite in its spire to make triangulation observations for the world famous *Trigonometrical Survey of Massachusetts*. The five-story brick factory building was next door to the Province House, which was the residence of the Royal Governors before the American Revolution.

Christian Louis Berger very quickly gave the firm of Buff & Berger a reputation for high quality work and a willingness to design and build precision instruments to meet the needs of the engineering and surveying professions. The firm soon had dividing engines that would give least counts of 5 seconds of arc. Berger Instrument Company in the Mattapan section of Boston still has the 24" Jesse Ramsden dividing engine that Buff & Berger acquired in the last half of the 19th century. It was also during the Buff & Berger years that the company acquired a dividing engine made by William Wurdemann, the first "Chief Mechanic" of the U.S. Coast Surveys' instrument shop.

C. L. Berger had two sons: William Albert Berger was born December 16, 1876, and Louis Herman Berger was born about 1½ years later on March 21, 1878. During the partnership of Buff & Berger both of C. L. Berger's sons were sent (between 1894 and 1898) to apprentice as instrument makers with L. Tesdorph of Stuttgart, Germany. It is interesting that John L.

Saegmuller, son of the famous instrument maker George N. Saegmuller, apprenticed to Tesdorph at the same time as the Berger boys did.

During the 1890's the relationship between Christian L. Berger and George Buff became strained, and finally reached a point of no return. On October 18, 1898, they dissolved their partnership and each formed his own company; Berger, with his two sons, formed the partnership C. L. Berger & Sons. The Bergers retained the 9 Province Court establishment while George Buff and his sons moved to the Jamaica Plain section of Boston.

At the time of the dissolution of the partnership, the Buff & Berger company had issued approximately 2900 serial numbers; some of the 2900 instruments were not complete and were distributed between the two men. From time to time, instruments appear with either C. L. Berger & Sons or Buff & Buff's name on them, with serial numbers under 2900 — these were obviously started during the Buff & Berger partnership.

It is clear that C. L. Berger worked closely with both the practitioners and teachers of surveying, navigation, and engineering. He developed many geodetic instruments in close cooperation with the United States' premier surveying organization, the U.S. Coast and Geodetic Survey. He gave lectures at such Boston area universities as M.I.T. and Northeastern. Many surveying texts used "cuts" from C. L. Berger's catalogues. The Bergers continued to improve and create new instrument designs; and although the company has changed hands twice since 1947, it has until recently, under its chief engineer David St. John, been doing prototype design and construction.

This reprint allows both aspiring and advanced students and collectors to acquire a wealth of information about great American mathematical instruments from a maker located in a city that has been one of America's centers of science since colonial times. This reprint deserves a place in the library of every student or collector of American scientific instruments.

David C. Garcelon
Millbury, Massachusetts
December 1992

ASTRONOMICAL INSTRUMENTS.

ENGINEERS' TRANSITS.   ENGINEERS. LEVELS.

———

HAND-BOOK AND ILLUSTRATED CATALOGUE

OF THE

Engineers' and Surveyors'

INSTRUMENTS of PRECISION

MADE BY

C. L. BERGER & SONS,

SUCCESSORS TO BUFF & BERGER,

No. 9 Province Court, Boston, Mass., U.S.A.

Written and Edited by C. L. BERGER.

PART I.

A FULL DESCRIPTION OF THE INSTRUMENTS AND CONCISE DIRECTIONS HOW
TO TAKE CARE OF AND ADJUST THEM.

PART II.

ILLUSTRATED CATALOGUE AND PRICE LIST OF ENGINEERING, SURVEYING,
AND ASTRONOMICAL INSTRUMENTS, MANUFACTURED AND
SOLD BY C. L. BERGER & SONS.

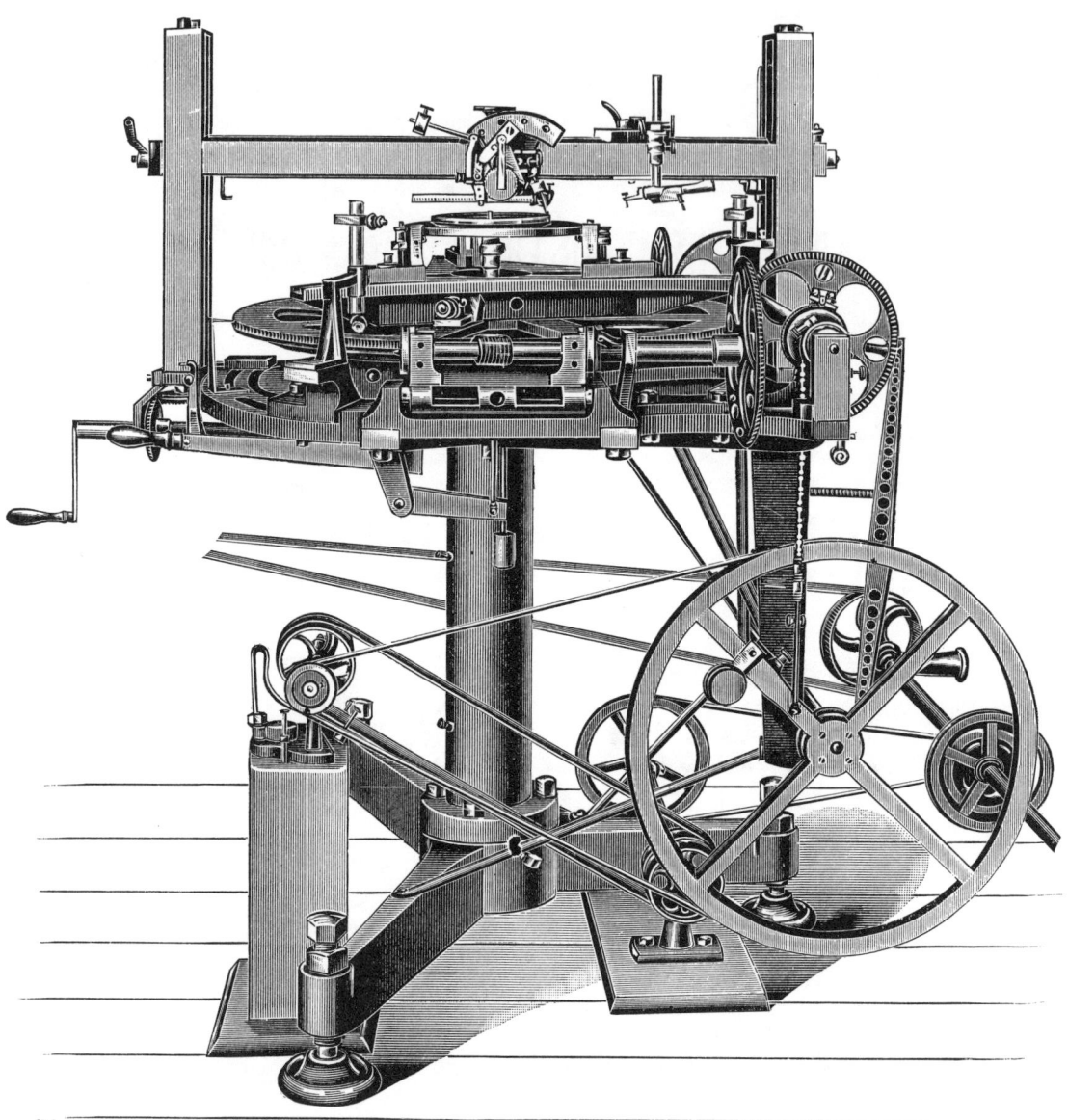

C. L. Berger & Sons' 30″ Automatic Dividing Engine.

Built entirely of Iron and Steel.

This engine is also arranged that circles can be graduated in the centesimal denominations.

Circumference	= 400 centesimal degrees.		
Degree	= 100	"	minutes.
Minute	= 100	"	seconds.

PREFACE.

THE Instruments enumerated in this Catalogue, and described in the Manual, are all of our own design and regular manufacture. Full supplies of Engineers' and Surveyors' Instruments will be kept on hand. The demand, however, is at times so great as to exhaust our supply. To secure an instrument in season, it is best to order it from four to eight weeks in advance of its intended use. Instruments varying from our customary designs, or those of rare inquiry, will be made to order only.

While the illustrations given represent the instruments as they have been made by us, we nevertheless make changes from time to time, as experience and the progress of engineering show them to be desirable.

Our long experience in the manufacture of Engineering and Astronomical Instruments, enables us to unite in our Instruments the high accuracy and finish of the European makers, with the *lightness* combined with *strength, steadiness, practicability* and *portability* required by American engineers.

Having three well-equipped automatic dividing engines, with circles 2, 2½, and 3 feet in diameter (the last-named built by the late J. H. Temple, of Boston, Mass.), we are enabled to make graduations of rare excellence. We also have a machine for engraving the figures, etc., on our circles and plates. Our adjusting apparatus, consisting of nine collimators, for the purpose of testing our lenses, and for correcting and adjusting the telescopes, etc., and instruments of our manufacture, marks a new epoch in the manufacture of accurate Engineering and Surveying Instruments. With the aid of this, we can test our instruments in so thorough a manner as only the most rigid tests in the field could reveal. In the possession of an apparatus of this kind, we believe, we stand entirely alone, here or abroad. It is the invention of Mr. C. L. Berger of this firm.

A careful selection of skilled workmen, increased facilities of steam power, and the application of the most approved tools and machinery, enables us to offer at a moderate cost a very superior article. We make no pretence at manufacturing cheap Instruments, — our prices are as low as is consistent with thoroughness of workmanship and the best material.

IMPORTED INSTRUMENTS. — We have made and do make Scientific Instruments, such as Cathetometers, Spectrometers, etc., but owing to the somewhat limited demand, and the fact that their manufacture entails chiefly the employment of hand labor, for this reason we find that we cannot produce them of equal quality and completeness at prices prevalent with best makers in Europe. In order, therefore, to enable customers and scientific institutions to procure these instruments at the lowest prices, we are giving our special attention to their importation. We refer more particularly to those used in the higher branches of Geodesy and Astronomy. When a greater demand for such instruments exists, justifying the introduction of special tools and machinery for their manufacture in numbers, we will make our own designs and improvements, and incorporate them in our regular line of manufacture. These instruments we can import to the order of schools and colleges free of duty. Approximate cost will be given on application.

C. L. BERGER & SONS.

C. L. Berger & Sons' 36″ Automatic Dividing Engine.

PART I

DESCRIPTION

OF THE

Essential Features of Our Instruments.

Graduation.

This very important part of a good instrument we guarantee *exact* and *accurately centered,* opposite verniers reading the same. The lines are straight, thoroughly black and uniform in width. There are two double verniers in every transit to read angles with great rapidity as well as to make four separate readings at every sight, when extreme accuracy in the repetition of angles is required. The horizontal circle is graduated from 0° to 360° with *two* sets of figures, running in opposite directions (unless ordered differently,) and the verniers are marked **A** and **B.** The figures are large and distinct, and to avoid mistakes in reading, the figures of these two sets of graduations, and those on the verniers, are *inclined* in opposite directions, thus indicating the directions in which the verniers should be read.

Instruments intended for mining and mountain use can have the verniers so placed that they may be read without changing the position of the engineer after sighting through the telescope.

Glass covers protect the arc and verniers from exposure. For ease in reading the verniers, we have added to most of our instruments *two plates* of *ground* glass, which cast a very clear light on the verniers, in any position. We recommend this addition to all of our more complete transits. The cost will be $3 additional.

The graduations on our transits are either on brass and silvered, or else graduated on *solid silver.* The former we can only recommend for the more ordinary instruments, since imperfections in the brass or composition castings frequently impair the graduations, and the silvering is apt to tarnish with time and exposure.

To graduate on *solid silver* adds $10 to the first outlay for the instrument, but its many advantages, great permanency and smoothness of surface render it the only satisfactory surface for fine graduations.

The Telescope.

All of its lenses are ground especially for us, by the best opticians. The telescope is perfectly *achromatic*, and designed to furnish a *large, flat* field of view with *high power* and yet without loss of light. For this purpose the curves of all our lenses are ground by special formulæ. The telescopes show objects right side up, unless ordered otherwise. *

The object-glass has a very large aperture, and is focussed by rack and pinion,† but the eye-piece is focussed by simply turning its head to the right or left in an improved screw-like manner.

By a method of construction peculiar to ourselves, we are enabled to guarantee the line of collimation correct for *all distances* without making use of the very objectionable adjustment for the object-slide by means of inner rings, which time and experience has proven to wear loose too readily, thus rendering this adjustment worse than none at all.

The eye-pieces are thoroughly achromatic, and their lenses are mounted in such a perfect manner (a method also peculiar to us) as to require no further adjustment with regard to the axis of telescope.

*It should be remembered that the focal length of the object glass is limited in engineering instruments and that a high power is obtained only at the sacrifice of light. To obtain the fullest satisfaction, telescopes intended for close work, as in stadia measurement, etc., should invariably be ordered to be inverting. The brilliancy with which objects appear in such a telescope, owing to the amount of light gained by saving two lenses in the eye-piece is very marked as compared with one of the same power and focal length showing objects erect.

†This rack and pinion motion is now so placed upon our telescopes that it is more easy of access by either hand than when placed at the side, as shown in most of our cuts.

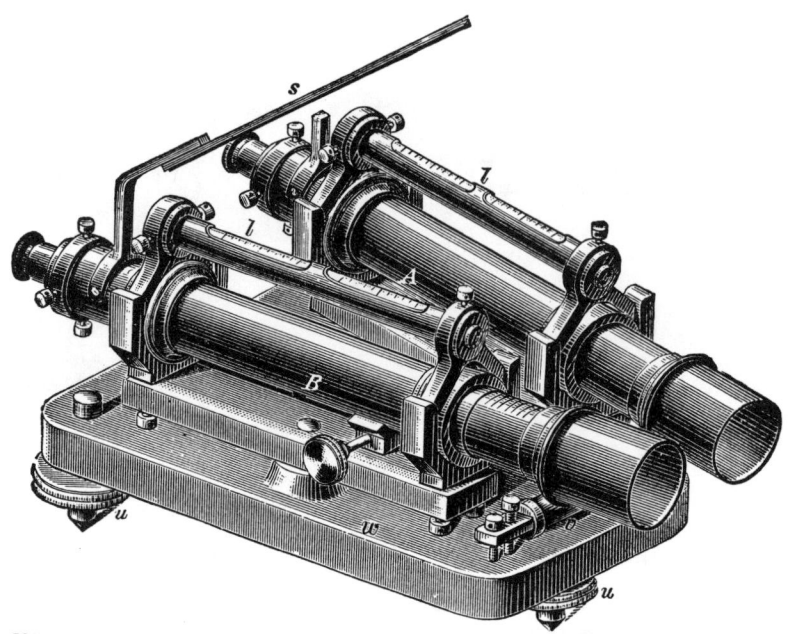

The Collimator Apparatus for testing Objectives and adjusting Telescopes.

The Collimator used for testing Objectives of larger Telescopes.

C. L. Berger & Sons' Auxiliary Apparatus.

Used during the construction of their Instruments of Precision.

The telescope of the transit reverses at both the eye and object ends, and is thoroughly balanced when focussed for a mean distance.

The telescope of the wye and dumpy level is also balanced *each way* from the center of the vertical axis when *focussed for mean distance* and with the *sun-shade attached to it*.

Spirit-Levels.

The Spirit Levels used in our instruments are carefully ground, filled and tested by us in person.

Those for the highest class of engineering work are sometimes provided with an air chamber by which the length of the bubble can be regulated according to temperature. The levels for astronomical instruments have air chambers, and are filled with ether, but in field instruments ether is not admissible, owing to the high degree of expansion and contraction in that fluid with changes of temperature. For these we use a composition fluid that we have found to be more sensitive and quick-acting than that used in instruments we have seen of other makers.

Our astronomical levels are so ground that a depression through one second of arc causes a displacement of the bubble through about $\frac{1}{50}$ of an inch. The curvature or sensitiveness of our levels for field instruments we adapt carefully to the instruments and the kind of work to which they are to be applied. With too sensitive a level the position of the bubble would be too uneasy to work with, while too low a sensitiveness would not reveal the full qualities of an instrument. Persons ordering instruments of us will confer a favor by stating for what purpose they are intended, whether for water works, for railroads, or for general use, so that we can use our judgment for their benefit.

Gradienter Screw.*

[Description to be found elsewhere]

This is attached to the clamp of telescope of all of our transits except the plain transit. This attachment was first introduced by Prof. Stampfer, of the Vienna Polytechnic School. It does not add to the weight of the instrument, and once used we have found it to be universally approved by our customers. By means of it *grades* can be established, and *horizontal distances*, *vertical angles* and *differences of level* can be measured with great rapidity. Indeed this attachment to an engineer's transit is one of the most useful introductions in practical engineering. It is so universal in its application to railroad and general work, that when once used it *will afterwards form an indispensible part of an engineer's outfit*.

Fixed Stadia Wires for Distance Measurements.

We have specially devised an optical and mechanical apparatus for the purpose of placing fixed, or non-adjustable stadia wires so accurately upon the diaphragms of our telescopes that their distance apart will read 1′ : 100′ † on any leveling rod, as with the gradienter screw, thus dispensing with a special rod.

It is well known that adjustable stadia wires are so apt to change their distance apart with every change of temperature, that no reliance can be placed upon them unless previously adjusted. With fixed stadia wires, annoyances of this kind are obviated — *they are reliable at all times*.

As regards the degree of accuracy attainable by the use of fixed stadia wires, experiments with our powerful telescopes, made optically as perfect as the most advanced optical and mechanical skill enables us, warrant to say that with some experience and proper care the results obtained will approximate and even equal those obtained by chain measurements. The price for this accessory in any new instrument is only $3.00, but if inserted into a telescope sent to us for that purpose, we must charge $10.00. We advise to order both the gradienter screw and the fixed stadia wires, as each in itself, separately or jointly, will prove of great value.

* Stampfers Gradienter attachment for leveling instruments, as introduced by him in the year 1838 (see Bauernfeind's Vermessungskunde), is, however, mechanically more complicated as compared with our own. As to simplicity of design and manipulation, we believe, our Gradienter Attachment, as applied to our transits, is unequalled.

† In all stadia work, the constant, which is the distance from the center of the instrument to a point in front of the object glass equal to its focal length, must be added to every measurement. Thus the constant in our transit No. 1, with inverting telescope, measured from centre of the instrument, is 1.3 feet; same instrument, telescope erecting, 1.15 feet. Transit, size as in No. 2, telescope inverting, 1.15 feet; same instrument, telescope erecting, 0.94 feet. In our 18-inch Wye level, telescope erecting, this constant is 1.78 feet.

Tangent Screws.

These are made of Aluminum bronze, or phosphor bronze, and sometimes of german silver, and are provided with strong *spiral springs* of german silver, which take up all the dead motion, no matter how long the screw may be in use, or how worn. They are *less* liable to get out of order, by blows or accidents, than any of the existing tangent screws, and require *little* or *no* attention on the part of the engineer. There is no *strain* on either plate when the instrument is clamped, so that the levels are unaffected. They are set and turned with the *greatest ease*, following the movements of the finger instantaneously with mathematical precision, and do not *scratch* the plate in revolving instrument. We confidently recommend this form of construction to those who have not used our instruments, as the best possible; superseding the usual methods by means of two opposing screws, or ball tangent screw, greatly in point of convenience and accuracy, and equalling them in point of steadiness. By this construction we are also able to fit our upper and lower circle plates so snugly that it is impossible for dust to enter between them. Our leveling instruments have the clamp and tangent screws so placed that they can be reached by either hand with the same readiness.

The Compass.

The Compass circles are graduated to half degrees in quadrants from 0° to 90°. The needles are made of superior steel, and tempered all over. A coil of fine wire attached to the end pointing South balances the needle for our latitude, which must be re-balanced if the instrument is used further north or south of this latitude, and must be entirely reversed if used on the southern hemisphere of our earth. At a cost of $10.00 a variation plate can be placed upon our surveyors' transit to set off the variation of the needle for any particular locality. A stationary pointer just above the graduated ring at the South end, and protected by the glass-cover of the compass, indicates the line joining the vertical plane of the line of collimation of the telescope. By means of a milled-headed nut, also at the South end of compass, serving both as a handle and as a clamp-screw, the graduated ring can be turned past this pointer towards East or West as the case may require.

Tripod.

The form we adopt for our instruments is an improvement over what is commonly termed the "split leg" tripod, used extensively in Europe, which unites the *greatest strength and steadiness* with the least weight. The tripod-head is cast in a single casting, to avoid all small screws, as well as to attain greater stiffness. For the legs we use the best fine grained white ash, taking particular pains that the grain of the wood runs in the direction of the leg. They are still further guarded against all possible accidents by having wooden tongs inserted at their top. When folded, our tripod is better adapted than the ordinary form, for carrying on the shoulder without irritating the place on which it rests. The good qualities of this over the ordinary round leg tripod provided as that is with unyielding brass cheeks to "tighten" the legs, are so great that there is but one opinion regarding its real advantages, and we gladly bear the greater expense incurred in its manufacture. The cast-steel shoes have projections for the foot, to aid in pressing the legs into the ground. Our levels and transits both fit the same tripod, and are of equal length.

Shifting Tripod.

We have also adapted to *all* our engineers' transits the *shifting tripod* or *shifting center*, by which, after an approximate setting of the tripod, the transit can be immediately brought over a point on the ground. This device we also attach to our instruments with three leveling screws in a most perfect and simple manner, and without impairing their steadiness and portability.

Adjustable Plumb-Bob.

We furnish with all our transits a small brass chain and hook, which are connected to the centers of the instruments. The cord of the plumb-bob can be readily attached or detached from this hook, and by means of a neat, small and simple device, (also furnished with every instrument,) the plumb-bob can be adjusted over the ground at any height, with hardly any effort on the part of the engineer.

Illumination of Cross-Wires.

For Mining and Tunnel Transits.

This consists of a small hole drilled through the transverse axis of the telescope, and closed at each end with small glass plates, to prevent dust entering the telescope. In the center of the telescope is placed a small adjustable reflector, by means of which the cross-wires can be very readily illuminated in the mine or tunnel by the reflection of the light of a lamp placed on a small table, which is attached to the standard. This lamp is provided with a ground lens. This method of illuminating wires is the best known to astronomers; it is the easiest to operate without assistance, or a change of lamp or position of the telescope. It can be applied to all our transits.

[*See Wood Cut Astronomical Instruments.*]

Arrangement for Offsetting at Right Angles.

Upon unscrewing the small adjustable reflector in the center of the telescope, which is explained in the foregoing paragraph, a perfect line of sight is had at right angles to the telescope. By simply sighting through the axis, offsets may be conveniently established without disturbing either clamp or telescope when the eye is brought close to the instrument; its application is, however, limited to even ground. To use it on an uneven ground it is necessary to place the eye at a distance of twelve or fifteen inches from the instrument. The head should then be moved until the eye is in line with the openings of the transverse axis. An offset can then be aligned irrespective of the height of the instrument.

Quick Leveling Attachment.

This we can apply to any of our Mining and Mountain Transits and Leveling Instruments. It adds about 1 lb. to the weight and $8.00 to the cost of an instrument.

Protection to the Object-Slide, &c.

A rain and dust guard for the object-slide is now furnished with all of our telescopes, and to insure smooth working of the object slide and telescope tube both are made of a non-friction metal. The graduation of the horizontal circle, the centers and such other important parts that are liable to injury by the action of dust and water in the field-use of an instrument, are entirely protected.

General Construction.

In regard to the general construction of our instruments, the dead weight is removed wherever it is shown to be not essential to the stiffness of the instrument; but we have at the same time strengthened the parts most likely to be injured by an accident or fall. Thus the *base of the standards*, the *vernier plate* and *circle*, the *parallel plates* for leveling screws, the *telescope axis*, the *flanges* of *centers*, *cross-bar of level*, etc., are made especially rigid and provided with ribs. Instead of finishing the smaller pieces of an instrument separately and then joining them with small screws, or solder, each screw or joint being a *weak* place in an instrument, we have adopted the opposite principle, (at an increased expense to us,) and aim to unite as many pieces as possible in a single casting, which casting, by means of *ribs* is made as light as consistent with strength.

We also call attention to the exceptionally *hard bell-metal* and *phosphor bronze* used for our centers and telescope axis, which are *long* and *unyielding*, and the remaining parts are of a *composition metal*, which is itself *harder* than hammered brass, or red composition, used ordinarily for centers, etc. It is more difficult to work, but we avoid the objectionable softer brass in its use. Experience has proven that soft, or *hammered* yellow brass is unfit for a good field or astronomical instrument, since it is more liable to fretting and yielding generally, and in the hammered state its unequal expansion and contraction at different temperatures may be so marked as to impair the reliability of the adjustments.

Aluminum bronze containing 90% copper, is also extensively applied in our instruments on account of its great tensile strength.

Almuminum alloyed with small percentages of silver or copper must be used with caution on account of its softness. (See Aluminum for Instruments of Precision, page 27.)

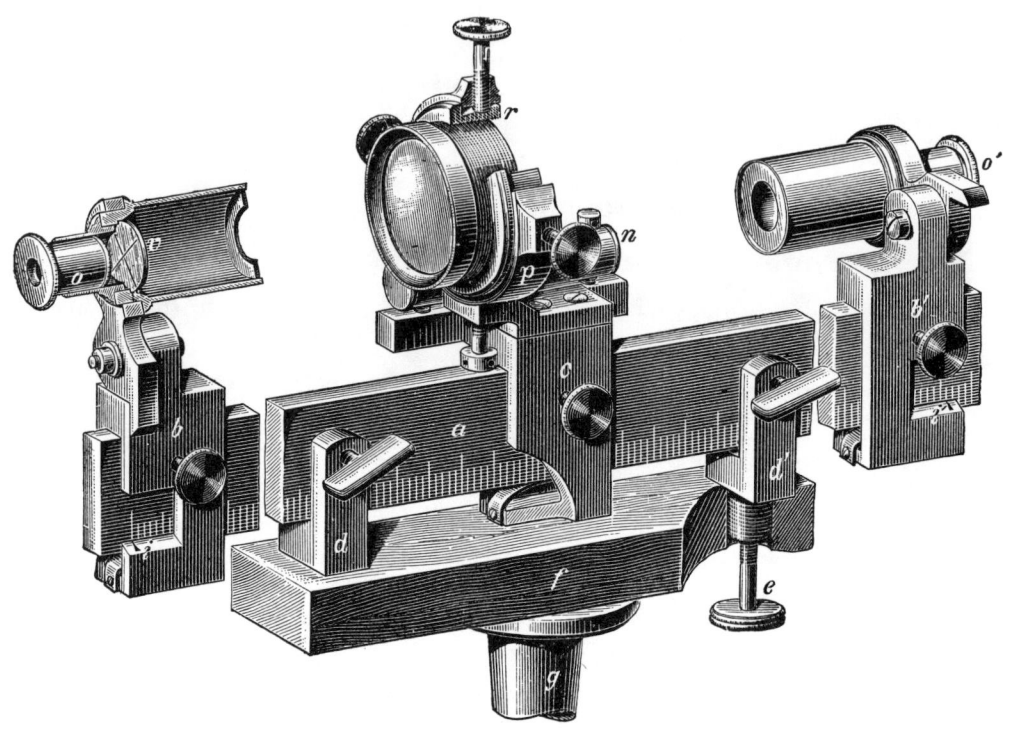

The Focal Length Apparatus.

Used to determine the focal length of an objective from the optical center, as well as from the second principal point. The latter by attaching a micrometer microscope at one end of the graduated bar and measuring with it through the objective the wire interval of the collimator A.

C. L. Berger & Sons' Auxiliary Apparatus.

Used during the construction of their Instruments of Precision.

The Finish.

It is a well-known fact that the black finish has one objection. It absorbs the heat readily, and therefore is apt to expand an instrument unequally, and thereby deranges its adjustments. We therefore consider it necessary to finish certain parts of an instrument in a bright but *not glaring finish* —including the upper plate, the standards and the telescope in the transit; the cross-bar and the telescope in the wye level, etc. All other portions may be finished and bronzed before lacquering. This finish gives a very fine appearance to the whole instrument; it wears better than black, and is in better taste.

Customers desiring to have their instruments finished entirely in bronze, however, can do so by notifying us of their wishes.

Cloth-Finish.

It is so called because the parts of an instrument so finished have the feeling to the hand of being covered with cloth of a very close texture,—there is no further resemblance to cloth however.

The principle is borrowed from astronomical instruments, where it is necessary to cover the surfaces with some non-conducting material in order to avoid disturbances in instrumental adjustments caused by suddenly varying temperatures.

We have adopted this principle with the view of securing the same results for our finer transits, wye and dumpy levels. Some of these levels are sensitive to a depression of a single second of arc.

Instruments finished in this manner heat up or cool down very gradually, causing the minimum derangement of the adjustments, and being of a dark brown color, this finish unites all the advantages of a bright finish with the convenience of having a dark colored instrument to use in the sunshine.

As regards durability, it will not quite equal the bright finish, but is superior to the bronze or black; this fact, coupled with the ease with which it can be restored at any time, leads us to recommend it in all cases where engineers do not care so much for an elegant appearing instrument after a number of years, as for an instrument in which every precaution is taken to avoid the influence of sudden changes of temperature.

In finishing an instrument in this manner, we are not obliged to polish its surfaces so finely, and thus can offer our transits with standards finished in this manner at $5 less than when finished in the other ways.

Packing.

In putting our instruments in their cases, none of them separate above the leveling screws. They stand *erect*, and are *ready for use* upon unlocking the case.

The cases are provided with rubber cushions, to check severe jarring arising from transportation over rough roads.

In conclusion, we wish to say that we aim to secure in our engineers' instruments —

1. *Simplicity in manipulation.*
2. *Lightness, combined with strength.*
3. *Accuracy of division.*
4. *Achromatic telescope, with high power.*
5. *Steadiness of adjustments under varying temperatures.*
6. *Stiffness; to avoid any tremor even in a strong wind.*

And we would add, that since all our leveling, tangent and gradienter screws are cut with precision in our engine lathes, and then run through a size plate to ensure uniformity and perfect smoothness, that we are able to replace any such part of our instruments by mail. The spiral springs, and most other small parts of the instrument, can be supplied in the same manner.

The Longitudinal Dividing Engine.

Apparatus for graduating the grooves for spider threads on the diaphragm of telescopes.

Care of Instruments.*

Do not allow the legs of your tripod to play loose on the tripod head; keep nuts and bolts always well tightened up against the wood. Examine the shoes from time to time, and sharpen them if necessary, also screw the shoes tight, if wear and tear loosen them. Be sure your instrument is well secured to its tripod before using it. Bring all four leveling screws to a seat before shouldering instrument. Let the needle down upon its pivot as gently as possible, and allow it to play only when in use; if too far out from its course check movements of needle carefully by means of lifter. Never permit playing with the needle, especially not with knives, keys, etc. Be sure to arrest the needle after use, and screw it well up against the glass cover before shouldering instrument. Do not clean the glass cover or the lenses with a silk handkerchief; breathe over the compass-glass and reading lens if one is used, after cleaning. Examine the buttons of your coat with regard to iron that may be concealed in them, also beware of nickel-plated watch chains, etc. To clean the object-glass and the lenses use a fine camel hair brush. If dust, or sticky or fatty matter cannot be removed with the brush, take an old clean piece of soft linen, and carefully wipe it off. Do not unscrew the object-glass unnecessarily,—this is apt to disturb the adjustment of line of collimation. The lens nearest the eye of eye-piece, as well as the front side of the object-glass, need careful brushing with *fine brush* from time to time.

If dust settles on cross-hairs and become troublesome, unscrew the eye-piece and object-glass, and gently blow through the telescope tube, cover up both ends and *wait a few minutes* before inserting the eye-piece and object-glass. Be sure to have the object-glass cell *screwed well up against its shoulder*, and then examine the adjustment of line of collimation (see adjustment of line of collimation.) Do not grease the object-slide of telescope, or screws that are exposed to dust; use a stiff tooth brush to clean slides or threads if dusty.

To take out the eye-piece, unscrew the screw at the end of the main tube, take hold of the eye-piece and pull it out.

To focus the cross-hairs, take hold of the *eye-piece cap* and turn it in a screw-like manner until cross-hairs appear distinct, and as if fastened on the object when the head is being moved.

Should there be any fretting in the telescope slide, take it out, and endeavor to smooth the rough part with the back of a pocket knife.

To clean the threads of leveling or tangent screws when *working hard*, use a stiff tooth brush to first clean the threads of all dust, then apply a little oil, and work the screw in and out with alternate brushing to remove dirt and all oil until it moves perfectly free and smooth.

Screws for the adjustment of cross-hairs should not be strained any more than necessary to insure a firm seat; all straining of such screws beyond this simply impairs the accuracy of instrument and reliability of adjustment.

When in the field always carry a Gossamer water-proof for the instrument in your pocket, to put over it in case of a shower or dust cloud. On reaching office, after use of instrument, dust it off generally with another fine brush; examine the centers and all other principal movements to see if they run perfectly free and easy, and oil them if necessary; also examine the adjustments. This will save expense and many hours of vexation in the field.

Care of Centers and Graduation.

As the centers, the telescope axis and the graduations require greater care to preserve their fine qualities, perhaps it is not amiss to say a few words concerning their treatment.

Upon finding that the centers do not revolve as free as usual after exposure of the instrument in an extremely hot or cold weather, they should be cleaned as soon as time permits, and then proceed as follows:

Unscrew the milled-head nut at the extreme end of the cylindrical tube containing a spiral spring, which is opposite the upper tangent screw. Do it somewhat cautiously, or the spring will fly out. Then unscrew a small cylindrical case, which also has a milled edge, and which is at the bottom of the centers. This case contains a small triangular spring to balance the upper weight of the instrument within a

* For additional suggestions see p. 14.

few lbs. Be careful to keep the face of this spring up in its case, which is best indicated by a bright point in its center. After unscrewing the nut attached to the inner center, a gentle pressure upwards will lift the vernier plate out from the lower part of the instrument. Take a fine camel hair brush, and with it clean the graduation, the verniers and the inner part of the instrument,—but do not rub the graduation, *especially not its edge,*—then take a stick of about the same taper as the inner center, wrap some wash-leather slightly soaked in fine oil around it, and clean the insides of the sockets as carefully as possible; then remove this piece of wash-leather and wrap a fresh piece *without* oil around the stick and clean dry. Proceed similar with the centers and their flanges.

Before applying fresh and pure watch oil, however, care should be taken that not a particle of dust or other foreign matter is left in the sockets, on the centers, or on the graduation. This caution having been taken, the fresh oil should be well distributed on all the bearing parts. It will be well to also examine the arm of the clamp screw of the circle and telescope axis, and if necessary clean by removing washer. After the instrument is thoroughly cleaned and oiled, the nuts and springs screwed back to a *firm* seat, the instrument must turn perfectly free and yield at the slightest touch of the hand.

To remove dirt and oxyd that may have accumulated on the surface of a solid silver graduation, apply some fine watch-oil, and allow it to remain for a few hours; take a soft piece of old linen and slightly rub until dry, but without touching the edge of the graduations. If, after cleaning, the solid silver surface should show alternately brighter spots, which would interfere somewhat with the accurate reading of the graduation, barely moisten the finger with vaseline and apply the same to the surface; then wipe the finger dry and lightly rub it once or twice around the graduation. Avoid touching the edges as much as possible. Such cleaning, however, must only be resorted to when absolutely necessary, and then only with the greatest care, as it is too apt to reduce the minuteness of the graduation, and spoil its fine appearance. If, after such cleaning, dirt and grease has accumulated on the inner edge of the graduation and verniers, gently wipe clean before restoring the vernier-plate to its place. Remember, also, that the centering of the graduations of the circle and verniers is a most delicate adjustment to make. These should never be unscrewed from their flanges by anybody except a maker.

Care of Telescope Lenses.

As dust and moisture, as well as perspiration from the hands, will settle on the surface of the lenses of a telescope, it becomes necessary that they should be cleaned at times. A neglect to keep the lenses free from any film, scratches, etc., greatly impairs the clear sight through the telescope. To remove the dimness, produced by such a film, proceed thus:— Brush each lens carefully with a camel's hair brush, wipe gently with a clean piece of chamois leather moistened with alcohol, and wipe dry using a clean part of the chamois skin on every portion of the lens, to avoid grinding and scratching. When perfectly transparent brush again to remove any fiber that may adhere to the lens. The tubes in which the lenses fit should be brushed, and if damp should be dried; this done, restore each lens to its original place as marked. To remove dampness in the main tube of the telescope. take out the eye-piece, *cover the open end with cloth* and leave the instrument in a dry room for some time.

If an instrument has been exposed to a damp atmosphere, or water has penetrated the telescope, moisture may settle between the crown and flint glass of which the object-glass is composed. If such is the case expose the instrument to the sun for a few hours, but if in the winter, leave it in a warm room some distance from the stove, the moisture will then generally evaporate. However, if not successful, unscrew the object-glass from the telescope, and heat it slightly over a stove or open fire. If a film settles between these glasses nothing can be done except sending the instrument to the maker. The two glasses form one lens only and must not be disturbed, as upon their relation to each other the definition and achromaticity of the telescope depends. Much depends also on the stability, with which these lenses are mounted in their cell, as any looseness between them or the cell will affect the adjustment of line of collimation. — Of course, if at any time the object-glass has been unscrewed from the telescope, this latter adjustment must again be verified before the instrument is used.

Additional Instructions concerning the Care of Telescope Lenses, etc.

Ever since the introduction of the high power in the telescopes of geodetic instruments, now used by the best makers, complaints are frequently made of the loss of light in such telescopes and of the hazy appearance of objects viewed through them, the latter in particular when an instrument has seen service in the field for some time. Now, while the loss of light is wholly due to the greater power as compared with the low powers formerly in vogue, and to the use of erecting eye-pieces (see page 33), the "haziness" is produced principally by films of dirt, settled on or between the lenses of a telescope, and becomes even more marked as more lenses are used in a telescope.

Perhaps it is proper to say here, that when comparisons are made between low and high-powered telescopes of geodetic instruments, *other things being equal*, the first named, as a rule, will incite favor, because, as in spy-glasses, the image of an object seen through them has a brilliancy never attained by telescopes of higher power. But, whenever the results of stadia work, or fine levelling, as obtained with the more powerful telescope, are compared with those obtained by a lower power, it will be found that, though less brilliant, the defining power of a high-powered telescope is superior to the other within the customary range of distances had in the ordinary engineer's and surveyor's practice.

On the other hand, owing to the less amount of light with high powers, it is necessary that the fine qualities of the superior lenses required for them should be preserved, and on this account a more frequent inspection and a more careful treatment of them is needed than when lower powers are used, — inasmuch as the least impairment of these lenses by films, or dust, etc., will reduce the defining power accordingly. A little extra care, as consequent upon the use of high-powered lenses, is, therefore, imperative, but in so doing one is more than compensated by the satisfaction of having a finer and more penetrating telescope.

To prevent an untimely settling of a film on the lenses of a telescope, and particularly that apt to form on the inner surfaces of the lenses composing an object-glass that has not been cemented together — such film being so fatal in an object-glass because it cannot ordinarily be reached and without disarranging the cross-wire adjustments — the treatment of an instrument should be strictly in accordance with the instructions given under "Prevention better than Cure," page 21. Unless these conditions are complied with, the greater efficacy of a telescope composed of superior lenses will be entirely lost.

Upon finding that, after carefully cleaning the object-glass and the lenses of the eye-piece, the telescope is not as clear as when first received from the maker, then the cause of it is generally a film between the lenses of the object-glass — we take for granted that the lenses are not scratched or otherwise impaired—but, as a rule, it takes several years (with careful use sometimes many years) before such a film has sufficiently developed to impair the transparency of these lenses. But whenever it is found that a film has settled between them, then it is best, if the distance is not too great, to send *the whole instrument to its maker*, and if this is not feasible, then the telescope, at least, well and soft packed in a box, should be sent.

Cemented Object-glasses. — To prevent the settling of a film between the lenses composing an object-glass, and to avoid disturbing reflections of light from their inner surfaces, such films and reflections imparting to an object viewed through a telescope the *hazy appearance* noticeable in high-powered telescopes, we now, since 1889, cement these lenses together, so as to form one lens only. The lenses so treated are more efficacious in many respects than when separated by three thin pieces of tin foil, as has been the custom of nearly all instrument makers up to date.

The cement, however, needs some five or six months to harden, and until it has hardened sufficiently, an exposure to a cold atmosphere causing a greater contraction of the metal cell than the glass, the lenses are very apt to *warp*, which may lead to a distortion of an object, when viewed through such an objective.

The proper treatment of an object-glass freshly cemented is to keep the instrument, when not in use, in a room having a mean temperature of about 68° F., or slightly above. The same treatment should be followed if it is found that the image formed of an object is slightly distorted; only in this case the temperature in which it is kept over night should be raised to about 75° or 80° F. This treatment applies only to normally mounted objectives. If they are too tightly fitted the lenses cannot be restored to their original efficacy without being attended to by a maker.

Object-glasses that are cemented are very apt to show some specks, or, with ill usage, cracks in the cement, but, unless the specks are very numerous, so as to cover almost the whole area of the object-glass, the opacity caused by them does not sensibly affect the efficacy of the telescope, and therefore need not disturb the mind. Our experience is that the usefulness of an instrument is greatly enhanced when these lenses are cemented together, and that a few specks that may appear after an exposure from a sudden change from hot to a very cold atmosphere, or vice versa, are a lesser evil, as compared with the ill effects produced by a film that in time will settle between these lenses if separated by pieces of tin foil, or even when brought in direct contact with each other, as such a film will have much the same effect as a fog, in preventing vision.

When, after carefully cleaning the lenses of a telescope, the object-glass of which has its lenses separated by pieces of tin foil, it is found that the image is not as clear as originally, it is a sure sign that there is a film between its lenses, and that it has been exposed to a damp or impure atmosphere, either by injudicious use in the field, or by being left too long a time in the packing box, in which it is protected by cushions of paper or shavings, both of which attract moisture, or by storing it away in its box in such an improper place as a basement or cellar. Such film being noticed, it will then be well to send the object-glass, or much better, the telescope, or, best, if the distance is not too great, the whole instrument, to the maker, in order that the lenses may be cleaned by him, and, if deemed advisable, be cemented. The slight expense incurred of a few dollars will be more than justified by the advantage gained.

When the object-glass, or telescope is returned after the cleaning or cementing of its lenses, the cross-wire, spirit level, and vertical arc adjustments of the instrument will require a thorough verification before it should be used. In case the whole instrument has been sent to the maker, these adjustments are attended to by him. If the object-glass has been cemented, the telescope should be watched for a year to see that there is no distortion of the image. If there is a distortion, it will indicate that the object-glass has been too tightly fitted, of which fact we should be informed, as also whether after cementing the object-glass the instrument retains its cross-wire adjustment the same as before the cementing took place. If the cross-wire adjustments have to be more frequently made than before the lenses were cemented, it indicates that the object-glass is not tightly fitted to its cell; and if such is the case it should be sent to us to be more tightly fitted, after a lapse of about ten or twelve months, when the cement will have sufficiently hardened to allow of a tighter fit of the object-glass in its cell.

In telescopes of very high power it is of as great importance to keep the lenses of the eye-piece free from grit and films as of the object-glass. Therefore, whenever the telescope does not appear to be clear, the lenses of the eye-piece need most careful cleaning (if necessary, every four weeks). The cleaning must be done by first wiping gently with a clean piece of old linen barely moistened with alcohol and then wiping dry, using a clean part of the linen on every surface of the lenses. (Please read the various articles on this point on pages 13, 14, and 33, of our handbook and catalogue.) To remove the eye-piece, unscrew the German-silver screw at the eye-end of the telescope. — Of course, after cleaning, every lens must be put back in its tube precisely as marked, and then the outer bearings of the eye-piece in the main tube must be greased with tallow before the German-silver screw is restored to its place.

Additional Suggestions Pertaining to the Care and Protection of Instruments in Field Use.

In field use, an instrument has to be necessarily exposed to the heat of the sun, and to the action of dust and water; all of these, however, singly or combined, have a tendency to affect its accuracy and endurance. While our instruments in particular have been designed to guard against injuries resulting from exposure of this kind, yet glaring abuses, such as to allow it to stand for hours in the hot sun, etc., without a covering or shelter of some sort, may often lead to a permanent injury to its most vital parts. To preserve the finer qualities of an instrument, viz., the telescope slide, the lenses, *the edge of the graduation and verniers*, the centers, etc., any undue unequal expansion of the different parts should be prevented. A bag thrown over the instrument when not in use, or any shelter that can be had, is to be recommended. While in use, an umbrella or screen held over it will insure greater permanency of its adjustments, and the results obtained will be more accurate and uniform than when carelessly exposed.

To protect an instrument from the effects of salt water, when used near the sea coast, a fine film of watch-oil rubbed over the exposed parts will often prevent the appearance of oxyd. To remove such oxyd-spots as well as possible, apply some watch-oil and allow it to remain for a few hours, then rub dry with a soft piece of linen. — To preserve the outer appearance of an instrument, never use anything for dusting except a fine camel's hair brush. To remove water and dust spots, first use the camel's hair brush, and then rub off with fine watch-oil, and wipe dry ; to let the oil remain would tend to accumulate dust on the instrument.

Lubricating, etc. — An instrument used in a tropical or semi-tropical country, or during the warm season in a northern latitude, requires more frequent cleaning and oiling than in the more temperate climes and seasons; but so long as an instrument works well and the centers revolve freely, it is best not to disturb it. However, if necessary, proceed as described under "Care of Centers, etc." A few additional remarks we give here : Should the centers or the object-slide commence to fret, they should be examined as soon as possible. *Once commencing to fret, it grows worse rapidly and oftentimes is then beyond repairing.* Never use emery or emery-paper on them, as this will cause everlasting trouble afterwards. After a thorough cleaning of the slide and tube (taking care not to break the cross-wires), endeavor to smooth carefully the injured parts with the back of a pen-knife, and barely apply enough tallow to grease the surface of the injured part. If this does not remove the trouble, a little scraping of the roughened parts on the slide, and, if accessible, on the inside of the tube, may become necessary, and apply a mere trifle of finely-powdered pumice stone moistened with oil. Replace the slide and grind a little by moving it in and out; *clean thoroughly*, and with a piece of charcoal moistened with oil smooth the parts thus ground on the slide. This process of grinding is a most precarious operation, and generally requires the hand of a skillful workman; it should be resorted to only in case of utmost necessity. Whenever permissible, recourse should be had to a maker. These remarks apply equally to the centers.

The centers of a transit should always be lubricated with *fine watch-oil only*, and after a careful cleaning ; never apply fresh oil before thoroughly wiping off old grit and oil. Rendered marrow is a most excellent lubricant for instruments made of brass and the many kindred alloys of copper and tin. In the varying climes of our northern latitudes this lubricant becomes rigid in cold weather, and an instrument so treated will often become unmanageable in the field. Its application, particularly to the centers of a transit, is therefore restricted to the warmer zones. The use of watch-oil for the finer parts of an instrument, involving freedom of motion, is *imperative in our latitudes.*

Many parts of an instrument, especially those whose metal compositions are closely related to each other, may sometimes cause trouble if simply oiled. If they begin to fret and grind, but are otherwise free from grit, etc., the judicious application of a little marrow may prove very beneficial, but it should be cleaned off again as much as possible. The rack and pinion motion and the telescope clamp should always be greased with marrow, but the clamp, tangent and leveling screws, should receive as little of it as possible in the Northern States.

Vaseline, not having as great a tendency to rigidity under similar circumstances, may prove an excellent substitute for marrow, and may often be applied to level-centers, where watch-oil would not give the necessary rigidity in the use of the more ordinary instruments, but it must be renewed quite often. In the finer class of leveling instruments, the centers should be lubricated with oil only, as in transits.

A great deal of annoyance is caused to the engineer if the eye-piece or the object-slide of the telescope move too freely in their tubes, requiring a re-focussing of the cross-wires and object at every revolution of the telescope in altitude. If the eye-piece can be retained in its socket, with sufficient friction to keep it focussed to the cross-wires, no matter how much it may wabble otherwise, this imperfection (in old instruments) will not lead to any inaccuracy, but if there is not sufficient friction to keep it focussed to the wires, a little rendered tallow or marrow applied to its bearing surfaces in most cases will remedy this evil. Wabbling in the object-slide, however, leading to inaccuracy of collimation, or back-lash in its rack or pinion motion, can be remedied only by a maker; but if the object-slide moves too freely in and out of its tube only, this may be remedied by applying a little tallow to the bearing parts of the rack and pinion, or by tightening the screw in the pinion-head. If not entirely successful, a thin disk made of parchment, or a thin leather-washer, both greased with tallow, and inserted between the flanges of the pinion-head and its socket, will insure the desired result. — These latter remarks apply to transit and level telescopes of the customary design. In telescopes, where the object-glass is mounted permanently to the telescope-tube, the eye-piece tube, *containing the cross-wires*, becomes the slide with which to focus the object. Its motion must be in a line parallel to the optical axis. Any wabbling in this eye-piece slide would lead to inaccuracy in sighting through the telescope, hence it requires the most careful treatment on the part of the engineer.

Care in the Use of Spirit-Levels.

Spirit-levels are very susceptible to the least change in temperature, as will be readily seen by the difference in the length of its bubble in varying temperatures. Hence, to guard against inaccuracies from this source, it is necessary that the bubble should lengthen symmetrically from the center of its graduated scale (supposed to be made by the maker), and that both of its ends should be read. Sufficient time must also be allowed for the bubble to settle before a reading is made.

The fluid ordinarily used for levels is pure alcohol, and requires, according to curvature, diameter and length of tube and length of bubble, from twenty seconds to one minute to attain its equilibrium. The composition fluid used in our levels for field instruments requires only from five to fifteen seconds of time; those filled with pure ether, a few seconds only.

A great source of error in spirit-levels, however, increasing with their greater sensitiveness, is occasioned by an *unequal* heating of the level-tube, *as the bubble will always move towards the warmer spot or end*, thereby imparting to the instrument an inaccurate position. This must be attributed to a changed condition in the adhesiveness of the fluid in the level-tube, and not to a change in the form of the tube itself. Therefore, to guard against inaccuracy resulting from sudden changes of temperature, a spirit-level, while in use, should be protected from the sun, and no part of it or its mounting should ever be touched with bare fingers; neither should it be breathed upon, nor the face of the observer come too close to it. For this reason, in the finer instruments the mountings of our spirit-levels are cloth-finished, and if the levels are detachable they are provided with wooden handles, as the case may require, and glass covers are placed over them whenever deemed necessary.

If at any time during the progress of field-work a spirit-level has been improperly exposed, it is best to cover it with a cloth for from five to fifteen minutes, before proceeding with further work.

Mounting Spirit Levels. — To prevent any undue strain and change of curvature in spirit levels used in astronomical instruments, they are mounted by us in wyes, as shown in the cuts of these instruments, and are protected from injury, or inaccuracy caused by the breath of the observer and other air currents, by a cover of glass placed over them. Such a mounting, while most suitable for such delicate levels, would, however, require constant attention and expose a spirit level to breakage in field instruments. To guard against this danger and to lessen the expense and weight, the spirit levels for field instruments are mounted in a brass tube; but owing to the difference existing in the expansion and contraction of glass and brass at different temperatures, a spirit level so mounted may sometimes become loose, involving inaccuracy and unreliability of adjustment. — Upon finding that the adjustment of a spirit level in an even temperature is not as stable as desirable, the level fastenings, tube, screws, etc. should be examined, to see if any of them are loose. If the trouble is in the screws, tighten them up; but if the spirit level can be shifted in its tube by a touch of the finger, take it apart; soften the plaster of paris in water, and remove it with a sharp pointed stick of wood. Cautiously move the spirit level with your finger, at first only a trifle to and fro, increasing the length of stroke little by little, until it can be safely taken out without breaking; -- clean thoroughly. Cut pieces of white paper, of the width of the radius of the tube, and somewhat shorter than the length of the spirit level, but longer than the opening in the brass tube, and insert these of sufficient quantity at the bottom of the brass tube, to fill up the space intervening between the glass and the brass tube. The uppermost layer of paper should, however, be so wide, as to envelope the spirit level up to the opening in the brass tube. Now insert the spirit level, taking care not to touch the glass ends that are sealed up, and place the division or other marks, indicating where the level has been ground to a true curvature, uppermost in the brass tube. The level must be pushed in with sufficient friction to prevent slipping in the tube, yet not so tight as to cause a crack at a subsequent low temperature, as brass will contract more than glass. No part of the spirit level should touch any part of the metal tube. Now prepare some plaster of paris with water, of the consistency of paste, and pour in at each end enough to fill up the space between the end-pieces and the glass, stirring it sufficiently to make a perfect contact by it and the glass and the brass, but leaving the spirit level ends exposed. Now put the level together, and adjust as described elsewhere.

There are other causes, such as centers and flanges that have been bent by falls, etc., or that have been worn out — unequal expansion or contraction in different temperatures of the metals employed in the construction of an instrument, or a non-symmetrical lenghtening or shortening of the air-bubble at different temperatures — all of which, singly or combined, tend to impair the adjustment of spirit levels on instruments. Of these we will not speak here, as it requires a most thorough mechanician and instrument-maker to trace the cause to its proper source.

Being assured that the level is mounted as explained above, our advice is, not to meddle too frequently with the adjustment of a spirit level. Though it may appear to be out one day, it may be in perfect adjustment other days. It is the function of a spirit level to indicate the changes taking place in an instrument, so that the engineer may make proper allowance and apply his corrections, as the character of his work may require. The finer an instrument, the more sensitive the spirit levels must be, in order to admit of corrections to arrive at closer results. As a rule, a spirit level that does not indicate changes taking place in an instrument, is too insensitive for the character of the instrument, and in many cases entirely unfit for reasonably good work.

Replacing Broken Cross-Wires.

The cross-lines in our telescopes are *bona fide* spider webs (except where platinum wires have been specially ordered). In case they should be broken, they may be restored in the following manner: clean the reticule frame of all foreign matter; put it on a sheet of white paper with the cuts on its surface uppermost. Prepare a little shellac by dissolving it in the best alcohol and waiting until it is of the consistency of oil. From the spider's cocoon, (those from a small black wood-spider preferred), which the engineer has prudently secured at some previous time, select two or three webs, each about two inches long and of the same appearance. Attach each end of these webs to a bit of paper or wood to act as weights, and immerse them in water for five or ten minutes. Remove one web from the water, and very gently pass it between the fore-finger and thumb nails, holding it vertically to remove any particles of moisture or dirt. Stretch the web carefully over two of the opposite cuts in the reticule frame. Fasten one end by a drop of the shellac, — let fall gently from a bit of pointed wood or the blade of a penknife. Wait a moment for this drop of shellac to harden. See that the web is stretched tight across the

The Microscope.

For use in fixing spider threads to the diaphragm of telescopes.

C. L. Berger & Sons' Auxiliary Apparatus.
Used during the construction of their Instruments of Precision.

frame, and apply another drop of the shellac to the opposite cut with its enclosed web. Wait several minutes before cutting off the two ends of the web, and then proceed in the same manner with the web which is to be placed at right angles to this one.

NOTE. — The fine spider-threads used were formerly taken from the cocoons of the small black wood-spider; now, however, we obtain them from the cocoons of a species of spider found in Michigan. These threads are almost opaque, and not apt to relax their tightness if properly placed on the diaphragm, and as they retain their elasticity, they are preferable to platinum wires, which have a tendency to break, owing to their great brittleness. The best spider-threads are those of which the spider makes its nest. These nests are yellowish-brown balls, which may be found hanging on shrubs, etc., in the late fall or early winter. The nest should be torn open and the eggs removed; if this is not done, the young spiders, when hatched, will eat the threads. The fibers next to the eggs are to be preferred on account of their fineness and darker color. As it is important to get the proper kind of spider-web, we subjoin an extract from a letter addressed to us on the subject by Prof. J. B. Davis, University of Michigan, Ann Arbor, Mich., to whom we are indebted for our supply.

" The species of spider of which I send you cocoons is not difficult to find in Ann Arbor — Lat. 42° 26′ N. — as far as my experience goes, and is numerous on Beaver Island, out in Lake Michigan — about 46° N.— at St. James. I have also always succeeded in hunting it in our Michigan woods, in places of concealment, — under bark of dead trees, in cracks and holes, about old stumps, logs, and the like. It is especially partial to painted woodwork. It roosts high, — the higher the gable the more numerous the cocoons; but it is also found on fences quite numerously, as I am led to think it is quiet rather than security this spider seeks. The body of the female is three-fourths of an inch, I guess, long, and nearly half an inch wide across the abdomen. The male is about the same length, but far slimmer. They are both entirely harmless. I never knew any one to get bitten by either, and many persons in my observation have had them freely crawling over their hands, face and body. They may be certainly gently handled without the least harm. They both (male and female) bear a plain escutcheon design on the back of the abdomen; female much the more beautiful, — in browns. Colors all brown and yellowish brown. The cocoon is a snarl of webs, and is attached under ledges of window-sills, cornices, projections of gables, and the like partly sheltered places. The color of the threads you have is of a light corn-color, distinctly separating it from the white cotton-like cocoons so common everywhere. The threads are silky, not like cotton. Of late years I keep one or two nice cocoons where they can be reached. You know one can wrap them in a bit of paper and carry them in the pocket, or any such place, and they are always ready."

Prevention Better than Cure.

IT cannot be denied that instruments frequently meet with serious accidents which, with a little care on the part of the operator, could be prevented. It certainly does not betoken proper care to leave it standing *unguarded in a street, road, or pasture, or in close vicinity to blasting,* or to expose it unnecessarily to the burning rays of the sun, or to dust, dampness, or rain at any time. Such carelessness must inevitably result in deterioration of the accuracy and efficiency, not to speak of the durability, of an instrument.

It should be borne in mind that there are many parts of an instrument which, *if once impaired,* cannot be restored to their original efficiency; and when it is considered that a conscientious maker bestows no little care, time, and expense on his work in order to attain a high degree of perfection, such neglect seems like a wanton waste of human energy and skill.

Legs of tripods, if fitting too loose or too tight, and dull shoes are frequent sources of falls, and loose shoes tend to make an unsteady instrument. The test of the proper degree of the tightness of the legs is this, that if the leg is raised to a horizontal position and left free, it should gradually sink to the ground. If it drops abruptly it is too loose; if it does not sink it is too tight.

When taking an instrument from its box, it is not immaterial where and how to take hold of it. To lift it by the telescope, circles, standards, or wyes is improper, and while it may not be attended at once with any serious consequences, yet it may sometimes lead to some permanent injury, and it certainly is always fraught with danger to the permanency of the adjustments. In handling, it is always best to place the hand beneath the leveling base.

When mounting an instrument on the screw of its tripod, or screwing any of its parts together, it is important to turn the part in the direction of *unscrewing* until it is perceived by a slight jar that the threads have come to the point where they enter; the motion may then be reversed, and the parts screwed together.

To secure an even wear of tangent and micrometer screws, they should be used equally on all portions of their lengths.

Carrying an instrument in cold weather into a warm room, without the protection of its box or bag, will cause a sudden exchange of air within the hollow spaces, and carry with it dust and other substances through the minutest openings. The vapor, also, that will thus condense on the metal surfaces, if it were not protected, will have a tendency to settle a film on exposed graduations, making them indistinct and difficult to read.

Failure to protect the lenses of the eye-piece and object-glass of a telescope, when not in actual use, from the effects of moisture, dust, etc., by the covers provided for them (eyepiece-lid and cap) will result in a more frequent settling of a thin film, which, like the fatty substance left by the touch of the fingers, greatly impairs the clearness of vision. That the too frequent cleaning of the lenses must in the course of time be detrimental to their brilliant polish, and lead to a corresponding loss of transparency so essential to the proper working of a good telescope, is apparent. Too much care cannot be taken to guard the lenses, and particularly the inner surfaces of the lenses comprising the objective, against any film that may settle on them. The ill effects of such a film are especially noticeable in high-powered telescopes of first-class geodetic and astronomical instruments. In short, it should be remembered that the slightest film, scratch, or dirt will, according to their nature and location, impair the sight through a telescope, and often render it unfit for accurate work.

The glass covers protecting the compass, arc, and verniers from exposure need very careful brushing and cleaning, the same as the lenses, as any scratch or film will impair their transparency. If at any time the ground-glass shades should lose their pure whiteness, by either dirt or film, and will not act as *illuminators* of the verniers and graduation, take them out of their frames and simply wash them with soap and water.

To prevent loss of magnetism in the needle of instruments provided with a compass: when storing away, allow the needle to assume magnetic North and South; then, by means of the lifter, raise it from the center-point against the glass cover.

If an instrument has met with a fall, bending centers and plates, etc., it should not be revolved any more, in order to preserve the graduations from still further injury, but recourse should be had at once to the nearest competent maker.

If the box or tripod should have become wet, they should be rubbed dry, and the varnish should be renewed whenever found wanting.

Loose or detached resting-blocks in the instrument-box, or any looseness of the instrument in them, are very detrimental to the instrument and its adjustments. Cracks in the instrument-box, the absence of rubber cushions under it, worn-out straps and defective buckles, hinges, locks, and hooks, should never be tolerated, as the remedy is so easily applied by any mechanic. Such defects and imperfections are known to lead to injury of the instrument.

The place where instruments are kept or stored away should be thoroughly dry and free from gases. The placing of fused chloride of calcium, or caustic lime, in an open vessel in the instrument-box is to be recommended where there is dampness; and if the presence of sulphureted hydrogen is suspected, then, cotton saturated with vinegar of lead, placed in the box, will prove a preventive against the tarnishing of solid silver graduations.

Transportation of Instruments.

DURING the progress of field work the more ordinary and portable transits and levelling instruments, etc., can generally be carried on their tripods for ease and dispatch. Nothing in the way of precise instructions, however, as to the best method of carrying an instrument: whether on the tripod, in the arm without the tripod — placing the hand beneath the leveling base — or in the box, can be suggested here. The nature of the ground, the surroundings, the size and weight, and the distance to be traveled over, and last but not least the fineness of the instrument, will dictate to the engineer the best means of conveying it from point to point in order to protect it from injury, and its adjustments from derangement.

The finer and finest classes of field instruments, such as those provided with micrometer-microscopes, should always be placed in their boxes for safe conveyance —no matter how short the distance—for fear of improper handling, and because of danger of unequal expansion, temporary as it may be, of such parts as would come in contact with the body or fingers.

Carrying an instrument on its tripod **without slightly clamping** its principal motions, will wear out the centers. When carrying on its tripod,

clamp telescope { in TRANSIT, when placed on a line with its centers; { in LEVEL, when hanging down.

When carrying an instrument in the box it is important that it be placed therein exactly in the position and manner designated by the maker. Therefore, upon receiving a new instrument, the first step should be to study its mode of packing, and if necessary a memorandum should be made for future guidance and pasted in the box. This will save time and vexation, as some of the boxes for field instruments must necessarily be crowded to be light and portable.

Before placing an instrument with four leveling screws in its box, the foot-plate should be made parallel to the instrument proper, and then brought to a firm bearing by the leveling screws. The instrument must also be well screwed to the slide-board, if one is provided, as is the case in most of our transits. Having put the instrument in the box in such a position, that no part of it will touch the sides, the principal motions are now to be checked by the clamp screws, to prevent motion and striking against the box. With instruments not standing erect in their boxes, but which are laid on their sides in resting-places, padded with cloth, specially provided for that purpose, their principal motions must not be clamped until the instrument has been secured in a complete state of repose in these receptacles, so as to be entirely free from any strain. Care must be taken, too, that all of the detached parts of an instrument, as well as its accessories, are properly secured to their receptacles before shutting the box.

When shipping an instrument over a long distance it is commendable to fill the hollow space between it and its box with small soft cushions made of paper, or of excelsior or shavings wrapped in soft paper, taking care not to scratch the metal surfaces, nor to bend exposed parts, nor to press against any adjusting screws.

For greater safety in transportation by express, the instrument-box itself should always be packed in a pine-wood box one inch larger all around. For the ordinary size of field instrument the packing-case should be provided with a strong rope handle, which, like the strap of the instrument box, should pass over the top of the case and through holes in the sides, the knots being within the case and strongly secured. In cases where the gross weight of the entire package, as prepared for shipment in the above manner, exceeds 40 or 50 lbs., then two men should handle it, and two strong rope handles, one at each end of the packing-case, should be provided. In order to check jars and vibrations while en route, the loose space between the instrument-box and the packing-case is to be filled with dry and loose shavings.

The cover bearing the directions should always be screwed on and marked thus, in large black letters:

THIS SIDE UP.
HANDLE WITH GREAT CARE.
Scientific Instrument.

Mr. George Brown,
36 West Street,
Cleveland,
Ohio.

Value $

From JOHN SMITH, Chicago, Illinois.

The upper halves of the four sides also should have **'CARE'** and **'KEEP DRY'** marked in large letters on them. These precautions are *indispensable* for safe conveyance while in the hands of inexperienced persons, as without them messengers will often carry them wrong side up.

The tripod needs packing simply in a close-fitting box. If not placed in a box, it often happens that legs or shoes are broken off while en route, or that the tripod head becomes bent.

Many hundreds of instruments, packed as explained above, have been shipped by us, travelling over thousands of miles, over rough roads, on stages and on horseback; and the instances are so rare where one has become injured (and then only through gross carelessness), that this mode of packing must be regarded as the only proper one for conveying instruments of precision by express or other public carriers.

Arriving at its destination, an instrument should not remain packed up with cushions, etc., any longer than absolutely necessary. The atmosphere in such boxes naturally must be close and often moist, and consequently has a tendency to produce the ill effects by moisture mentioned in preceding paragraphs.

Some Remarks Concerning Instrument Adjustments.

The mechanical and optical condition of instruments used in geodesy, and their adjustments, although satisfactory when they leave the maker's hand, are liable to become disturbed by use. It is therefore of vital importance that the person using an instrument should be perfectly familiar with its manipulations and adjustments. He should be able to test and correct the adjustments himself at any time, in order to save trouble and expense, as well as to possess a thorough knowledge of the condition of the instrument. It is evident that if the character of an instrument is not properly understood or if the adjustments are considerably out, the benefit due to superior design and workmanship may be entirely lost. Under these circumstances an instrument may be little better than one of lower grade.

In the best types of modern instruments the principal parts are so arranged that they can be adjusted by the *method of reversion*. This method exhibits an existing error to double its actual amount, and renders its correction easy by taking one-half the apparent error. Thus errors of eccentricity and inaccuracy in the graduations are readily eliminated by reading opposite verniers and *reversing* the vernier plate 180° on the vertical center and taking the mean of the readings, and by repeating the measurement of an angle by changing the position of the limb so that the measurement will come on different parts of the graduation. The striding levels and levels mounted on a metal base are readily tested by reversing their position end for end. In the transit plate-levels the adjustment is assured by turning the vernier plate 180°. Errors of the line of collimation are detected or eliminated by reversing the telescope over the bearings, or through the standards, as the case may be. In short, an instrument, the important parts of which are not capable of reversing in one way or another, cannot be examined quickly and accurately.

The adjustments of an instrument, and particularly those of its cross-wires, should be taken up successively in a systematic manner. The proper way is to select a place from which they can be conducted in succession without moving the instrument, as none of the adjustments should be completed independently of the others. This method is followed by the maker, and will save time and vexation. Any auxiliary apparatus that may be available, such as collimators, etc., will be of great service and expedite the work. One of the most important considerations in making adjustments (when the same are greatly disturbed, as when new wires are to be inserted), is to place all the respective parts in an approximate adjustment without introducing any strain except what properly belongs to the action of the adjusting screws themselves. The more natural the method, and the less internal strain introduced in bringing these adjustable parts into position, the more lasting will be the final adjustments, provided the instrument is otherwise in good condition.

It is important that all adjusting screws and nuts should fit truly on the surfaces against which they operate, with only a mere film of tallow between them, so as to insure a true metallic contact, and that they be brought to a firm bearing, yet without excessive strain. Opposing screws and nuts should always work somewhat freely, so that one can feel when they come to a true bearing. A moderate pressure applied with an adjusting pin about one and one half inches long, and held between the thumb and forefinger, will then make a perfect contact. For instance, after the opposing capstan-headed screws of the cross-wire reticule have come to a bearing, it is only necessary to give them each a slight turn, say from 20° to 30° (with the usual pitch of these screws) in order to insure such a tightness that a moderate pressure of the finger upon these screws, or an accidental gliding of the hand over them, cannot change their relative position. On the other hand, if one pair of these opposing screws be fastened tightly during the tentative process of adjustment, there will be, in all likelihood, at the end, an excessive strain exerted upon the pair of opposing screws at right angles, which will make itself felt at any change of temperature, or whenever any external pressure may be momentarily applied to them. It is but natural that these continual changes in the resultant pressure must affect the adjustments in a like manner. To obviate such changes the procedure should be as follows : —

Having placed approximately in position the principal wire of an instrument : viz , in a transit, the vertical wire in a plane perpendicular to the horizontal axis of revolution, in a level, the horizontal wire in a plane perpendicular to the vertical axis of revolution, the other wire should be approximately adjusted for collimation, with the capstan-headed screws *only moderately tightened.* This accomplished, the capstan-headed screws of each pair in succession should be unscrewed about one-quarter turn, and again screwed tight the same amount. Now if the two pairs of opposing screws have exerted no undue strain upon themselves, the telescope tube, or the wire reticule, the principal wire will still be in the perpendicular plane ; but if the screws have been used too much the wire will have slightly moved out of the perpendicular plane. Therefore all four capstan-headed screws will have to be released again, say about ⅛ turn, so that they may be moved simultaneously until the principal wire is again in a plane perpendicular to the axis of revolution, and then each pair in succession must be again tightened an equal amount. The adjustment of the wires for collimation must now be made in turn — the less important wire should always be taken up first — by slightly releasing the capstan-headed screw *away* [1] from which the wire must be moved, and tightening the opposite screw the same amount, and repeating this process until the adjustment is gradually perfected. If during this operation either or both of these wires have become so much displaced that the capstan-headed screws have to be moved more than a quarter turn, it would be advisable to slightly release all four of them again, in succession, and commence anew.

It should be said here, that the force applied by the capstan-headed screws cannot break or affect the tightness of the wires in any case, since the reticule, as made by us, although very light in weight, is of a very stiff form. Too great pressure exerted by the capstan-headed screws against the outer tube of the telescope may, however, change the form of the main tube, thereby affecting the true fitting of the object-slide, and creating friction of so serious a nature as to lead to the fretting of the object-slide mentioned in other paragraphs.

In following the above-described course, the cross-wire reticule occupies a position in the telescope free from any excessive side strain ; the result of which is found in the greater permanency of these adjustments ; and although it may require a little more time for an inexperienced person to make the adjustments in this manner, the satisfaction derived from their greater permanency will more than recompense for the extra time spent on them. The adjustments should be made at leisure, and should not be meddled with, unless they appear to be permanently deranged ; when, ordinarily, the adjustments will merely require a very slight turn of the capstan-headed screws and opposing nuts *in the proper direction.* [2] Unequal exposure of the instrument to the sun, or exposure to sudden changes of temperature, may for a time expand some parts more than others, so that the instrument may seem to be slightly out of adjustment. In such a case it would be better to stop temporarily and cover the instrument with a bag to allow the temperature to become equalized, instead of attempting adjustments that would need to be repeated when the instrument is again in a normal condition. The use of metals of different co-efficients of expansion in the construction of corresponding parts of an instrument will naturally lead to a

---

[1] We refer here exclusively to the more common instruments of American manufacture, where the shoulders of the capstan-headed screws bear against the outer tube of the telescope, and where the adjusting threads are contained in the wire reticule. In other designs where, as in most instruments of Continental Europe, the capstan-headed screws are made to butt against the wire reticule, the capstan-headed screws *towards* which the wire must be moved, must first be loosened. In the latter case this action is identical with that of opposing nuts used for the adjustment of most telescope levels on American instruments.

[2] See foot-note on page 49.

permanent derangement of adjustment; such also will be the case when the temperature of an instrument is greatly altered after the adjustments have been completed. A similar result is caused if the bubble of a spirit level should not lengthen symmetrically from the center of its graduated scale in varying temperatures. These imperfections, however, seldom occur in instruments of modern make (or if they occur, they are generally caused because the principal constituents, glass and metal, are substances of widely differing co-efficients of expansion), and are generally so slight in well made instruments, as to be of little practical value, and may be overcome by adjusting the instrument while at a mean temperature of an entire season.

If an instrument does not remain in adjustment a reasonable length of time, the cause that leads to the trouble, such as a loose object-glass or cell, loose object-slide, worn out screws or bearings, etc., must be found and remedied. If this is beyond the scope of the operator the corrections should be made by an instrument maker.

Some Facts Worth Knowing.

The Line of Collimation.

The expression "Line of Collimation," usually defined vaguely in treatises on geodetic instruments, generally means any line of sight in a telescope given by the inter-section of the cross-wires, whether they are in perfect adjustment or not. The term "Line of Collimation," should, however, be confined solely to the line of sight defined by the cross-wires when they are in perfect adjustment, with reference to the optical axis of the object glass; and any difference existing between the optical axis of an object glass and the actual line of sight as delineated by the geometrical axis of the instrument is the "Error of Collimation."

The principal optical axis of an object-glass is the line passing through the optical centers formed by the curvatures and the thickness of the two lenses composing it. Thus it will be seen that it is a well defined axis, giving direction to the light passing through an object glass, and that, when the intersection of the cross-wires is placed in its prolongation at the focus of the object glass, it becomes the axial or fundamental line by and from which all measurements by telescopic sighting are made. *It is the line of collimation.*

To make a good instrument, therefore, it is necessary that the outer circumference of the lenses composing an object glass shall be truly concentric with the optical centers. The aim of the maker is to so construct his instruments that this optical axis shall be truly concentric with the geometrical axis of the telescope and that the latter shall also occupy a normal position with regard to the geometrical axis of all other important parts : upon this depends the proper working of an instrument.

In the larger geodetic and stationary astronomical instruments, the telescopes of which are arranged only for distant sighting, this condition is readily obtained ; but it becomes very difficult of attainment in the smaller geodetic instruments, since, owing to the varying position of the focussing slide when set for *different distances*, the optical axis may not always remain truly coincident with the geometrical axis of the telescope. Hence in these instruments, *carefully adjusted for distant sights*, there is frequently an error of collimation when nearer sights are taken. In the latter case the intersection of the cross-wires remains no longer exactly in the optical axis, its displacement being the cause of the error observed — disregarding momentarily the other and more complicated features of different instruments, upon which the line of collimation also depends.

In the Engineer's transit, however, the line of collimation must also lie exactly at right angles to the axis of revolution of the telescope, so that when this axis is placed in a horizontal position, the line of collimation shall describe a truly vertical plane, whether the telescope be mounted in the centre of the instrument or outside of the plates, or whether it be focussed for long or short sights. In the more common instruments of this class, where the telescope is situated in the center of the instrument, the intersection formed by the line of collimation and the horizontal axis of revolution is also required to lie truly in the prolongation of the vertical axis of revolution, so that there be no eccentricity between the vertical axis of revolution and the line of collimation when sights are taken at objects nearer than 200 feet.

In transits of this latter type, and in which the above conditions are fulfilled, the sights taken would at once define the true angle, and no reversing of the telescope would be necessary, were it not for other reasons. On account of the necessity for eliminating the eccentricity and error of graduation and verniers, as well as for eliminating errors arising from an inaccurate adjustment of the line of collimation and of the adjustment of the telescope in the vertical plane, an instrument should be reversed and an angle should be repeated. These remarks apply equally to transits made with the telescopes in an eccentric position. If the line of collimation is truly at right angles to the horizontal axis of revolution, the amount of the offset from the line through the center of the instrument to the line of collimation will equal the eccentricity of the latter, and will remain the same whether the sights be long or short. As a rule, however, the small geodetic instruments of the latter class cannot be constructed with the same degree of perfection as those with the telescope in the center: and in consequence the engineer using such instruments will have to rely upon methods of observing that will eliminate all instrumental errors.

In the engineer's wye level the line of collimation must be truly concentric with the object-slide and outer rings; and it is also necessary that the telescope be well balanced from the center of the instrument, in order to project a truly horizontal line.

Difficult of attainment as the foregoing conditions may seem, it is proper to say that improved tools, and a generally better understanding of the principles governing a telescope and its relation to the instrument, have done so much toward the perfection of geodetic instruments, that while it may not always be possible to make an instrument in which the line of sight for both wires remains true for all distances, that result can generally be secured, for at least the principal wire, without requiring any other but the regular cross-wire adjustment.

By the foregoing explanation it will be readily understood that it is of great importance to have the focussing slide of such a telescope *truly fitted*, in order that the optical axis of the object-glass may coincide with the geometrical axis of the telescope, whether this slide moves in the main tube and carries the object-glass, as is the custom now in the smaller instruments; or whether it moves in special rings provided for it in the main tube at the eye-end, where it will contain the eye-piece and the cross-wires, as is the case in all larger instruments. Any lateral motion in the *focussing slide* that carries the *object-glass* or the *cross-wires*, will, therefore, derange the adjustment of the line of collimation. However, it is equally as clear that a wabbling of a focussing slide carrying an eye-piece which serves only the purpose of a compound microscope for close observations of the wires and the image of an object, is of no account save that such lateral motion may be so great that the obliquity which the optical axis of the eye-piece may at times have with respect to the optical axis of the telescope, may cause some parallax, if the wire and image under observation are not sharply focussed together. In concluding, it may not be considered amiss for a full understanding of this subject, to also mention in this connection, that any transparent substance, such as *prisms,* lenses, or shade-glasses,*introduced between the object sighted at and the object-glass, will deflect the line of sight from its true course, unless such parts can be made optically and mechanically perfect, which is rarely the case without elaborate adjusting apparatus. The introduction of a lens or lenses between the object-glass and wires, or that of a glass micrometer, will also have the tendency to deflect the optical axis and affect the line of collimation. For this reason " Porro's telescope," which requires a lens between the object-glass and the wires, complicates the above conditions of a measuring telescope: and while it may prove of some value in stadia measurements, can never be adapted for the engineer's transit so long as the proper functions of the transit telescope, as explained above, are considered of the greatest importance. The successful performance of an instrument should not be sacrificed for the sake of some doubtful novelty.

The proper way of attaching prisms and colored glasses necessary to make sun and star observations is to put them upon the eye-piece of a telescope. After the rays from an object have passed through the object-glass and the plane containing the wires, the line of sight as fixed by the object, optical axis, and the wires, cannot be changed by additional refraction. The best way, therefore, is to apply prisms and shade-glasses between the eye and the lens nearest the eye.

Aluminum for Instruments of Precision.

In consequence of recent improvements in the production of pure aluminum and a corresponding great reduction in its cost, we frequently receive inquiries as to the adaptability of this metal for the manufacture of engineers' and surveyors' field instruments.

We may be permitted to say, that while we were among the earliest advocates of aluminum and its alloys for mathematical instruments (see *Scientific American,* Feb. 1, 1868), we are not so sanguine concerning its adoption for the finest class of

* The object prism, so called, attachable to the object end of a mining telescope to aid in steep sighting, from its position between the object glass and the object sighted at, must of necessity be of very limited usefulness, since the slightest change of the prism or its mounting or a change of the position of the telescope itself or of its object slide will almost certainly deflect the line of sight from its true course and give no satisfactory results.

geodetic instruments, as these inquiries would warrant us to be. There are certain advantages derived from the use of the lighter aluminum instead of copper and its alloys,—the metals now employed for field-instruments; but the disadvantages are that pure aluminum, although very rigid, is nevertheless a very soft metal like tin, and that, when alloyed with 10 per cent. copper, to make it harder, it becomes very brittle, but when alloyed with 20 per cent. or 30 per cent. of copper, it becomes so brittle as to break like glass. Therefore, we believe, in the present state of its development it is not a suitable material for precision instruments.

An alloy of 95 parts aluminum and 5 parts of silver by weight has been found to give good results, being more rigid and harder than the pure metal, and but little heavier, while it is almost as resistent to corrosion, polishes well, and is said to be better for graduation; but, the fact that it contains silver, will, of necessity, limit its use to the more exceptional class of work.

Very little is gained in the way of reducing the weight of an instrument by employing aluminum bronze (90 per cent. copper and 10 per cent. aluminum). The parts of instruments made of the latter metal might be easily reduced somewhat in thickness on account of its greater rigidity as compared with copper alloys; yet to lessen the tendency to vibration, and also in order to withstand the wear and tear of the field use of an instrument, such parts need a little more mass, or dead weight as it may be called. It is then found that the weight of an instrument remains materially the same as ever. An exception to the rule may exist in the construction of the larger and stationary astronomical instruments, where aluminum bronze may be used to a certain extent to advantage. Its adoption is, however, restricted to non-revolving parts, since, when closely fitted into bearings made of the softer copper and tin alloys, the friction and wear of these parts is so marked that we would never think of substituting it for steel, bell metal or phosphor bronze, or for any work requiring a smooth and accurate motion.

There can be no doubt that aluminum possesses great utility over brass in the construction of instruments of minor importance. Sextants, reflecting circles, and the more ordinary compasses,* parts of plane-tables, etc., can be made of it with propriety. We have used it occasionally for many years, but for reasons already stated above, we are not prepared to advocate its general adoption for instruments requiring greater precision, such as the finer transits, theodolites, etc. It is only in rare cases when a judicious use of this metal may be a necessity for the successful construction of an instrument, as for instance in our new style of mining transit, permitting of vertical sights up and down a shaft without the use of an extra side telescope, where certain detachable parts of the instruments are mounted in an excentric position, and unless such parts are made of aluminum they would require a heavy counterpoise.

It is principally the indiscriminate use of aluminum that we would warn against. We are aware that transits have been made of aluminum, but aside from their novelty as such, little or no merit can be claimed for them. To make this fully understood, it will be necessary to explain that all the finer bearings of an instrument made of aluminum, such as *centers, object slides, leveling* and *micrometer screws*, etc., will have to be *bushed* with a harder and non-friction metal, to guard against friction and wear and to obtain the close fitting of such parts, and permanency of adjustments so necessary in instruments of precision. Now, to make the principal bearings of an instrument of different metals will have the tendency to weaken the parts so treated, to make them less secure, and to render the adjustments more liable to derangement on account of unequal contraction and expansion between the two metals. It simply means, then, that the present high state of perfection in geodetic instruments, which retain their adjustment in the varying temperatures and climes of our zone, shall be abandoned, and we go back many years to when the indiscriminate use of widely different metals often made an instrument entirely unreliable, except when used in the temperature in which it was adjusted.

Modern instrument-making has, however, already achieved great results in reducing the weight of field instruments. By improved designs and by the use of harder metals in place of the soft brass, remarkable changes have been brought about in the weight of instruments. They are no longer the heavy and formless structures of soft or hammered brass as of yore, but are of the type and character of a long-span steel bridge, as compared with an old-fashioned wooden structure. Every important member of an instrument is now calculated with regard to its strength, and the materials are particularly chosen for the part they are to perform.

* Commercial Aluminum, unless obtained from reliable sources, often contains a small amount of iron.

Owing to the many improvements made in the designs, the use of better materials, the application of specially designed tools and machinery, it is no longer necessary to use large and heavy instruments. An instrument of about two-thirds the size and weight of those made ten or fifteen years ago will now do the same class of work. It is by these methods that lightness has been gained, and to them we must look for advances in the future. Unless the size of an instrument is decreased, the resistance of its exposed surfaces to wind pressure, causing sudden vibrations or tremor in the instrument, will of necessity require a certain amount of weight to secure the needed steadiness, and if this weight is not in the instrument proper, it will have to be in its tripod legs. This is especially true in this era of high telescope powers and sensitive spirit-levels. What is needed is that engineers and surveyors should have more confidence in instruments of smaller size as made by the best makers.

Wherever less weight is of great importance our patrons should not hesitate to order our smaller Transits Nos. 2, 3, or 4, weighing 10½ and 5 lbs. respectively, in preference to a larger instrument made of Aluminum and divided to single minutes, but of equal weight. These small instruments are just as durable and capable of doing just as close work as the larger ones. Being made of a like metal throughout, whose coefficient of expansion* is lower, they will retain their adjustments better than larger ones made in whole or in part of Aluminum. — Suppose an instrument is adjusted in-doors and immediately is taken into the cold atmosphere of winter : other things being equal, if the coefficient of expansion of some parts differ the adjustments will very likely be deranged. — Besides, the instrument being smaller, the boxes are likewise smaller, thus reducing the weight and making it more portable at the same time. The same, in a measure, can be said of the tripod, although it is against our convictions to use a lighter tripod with a small transit than is used on the larger ones.

The only exception to the above exists in the Telescope, which, of course, being correspondingly shorter in a small instrument, will have a smaller aperture and less power. However, to secure the same aperture and power for Transits Nos. 2 and 3 (No. 4 being inverting), as for our Transit No. 1, with an erecting eyepiece, it is only necessary to order an inverting telescope to attain these conditions.

There are other reasons why makers should be somewhat conservative in the adoption of aluminum as a material for the finer class of surveying instruments, but as they relate principally to the treatment of aluminum during construction : graduating process, etc., they may be omitted here. In conclusion we wish to say that the future developments in alloying it as a base with other metals or combination of metals, will be watched by us with due care, and that whenever such developments will warrant their adoption in the various parts of instruments, we will only be too glad to avail ourselves of any superiority such alloys may possess.

* *The Ideal metal for a Surveying Instrument is that which has a coefficient of expansion equal to that of its glass parts, so as to retain the adjustments in varying temperatures.*

Coefficient of glass per linear foot, for 1° F. .	. . . . . .	0.000054 inches.
" " steel " " " .	. . . . .	0.000076 "
" " brass " " " .	. . . . .	0.000125 "
" " aluminum per linear foot, for 1° F. .	. . . . .	0.000148 "

Aluminum is farthest removed from the above requirements, steel or cast iron being nearest, and also lighter and harder than brass; and non-friction metals would be more generally adopted were it not for the use of the compass and the liability to rust in the field.

Repair of Instruments.

We are often applied to for correcting new and repairing old instruments made by other makers. We will here remark, that as workmanship, material and construction of different makers' instruments vary from one another, it is oftentimes impossible to repair them in an entirely satisfactory manner without going into an unwarrantably great expense, or without making such alterations as would practically make a new one. We will always guarantee in such cases to put the instrument in as good order and adjustment as the character of its construction, workmanship and material, the extent of damage and the general wear will permit, and that all repairs are promptly and conscientiously made. The charges will be according to time consumed, and as low as is consistent with good work. Parties sending instruments should point out in detail whatever parts they wish to have repaired; but the best course to be pursued is to have the instrument *put in thorough order and adjustment*, implying, as it does, that the firm should make such warrantable repairs as will make it as serviceable as possible. This course is always more expensive, but the most satisfactory to insure good work, and it is also the cheapest in the end. — Our own instruments, whenever practicable, should always be sent to us for repairs to insure fullest satisfaction. Much time and money is frequently saved by so doing, as we are in a position to duplicate parts from stock on hand. In sending an instrument to us from a distance it should be carefully placed *in its box* and then again in *a packing box*, as explained under "Transportation of Instruments," Part I., in order to conform to the rules of most of the large Express Companies, which will admit it to single rates.

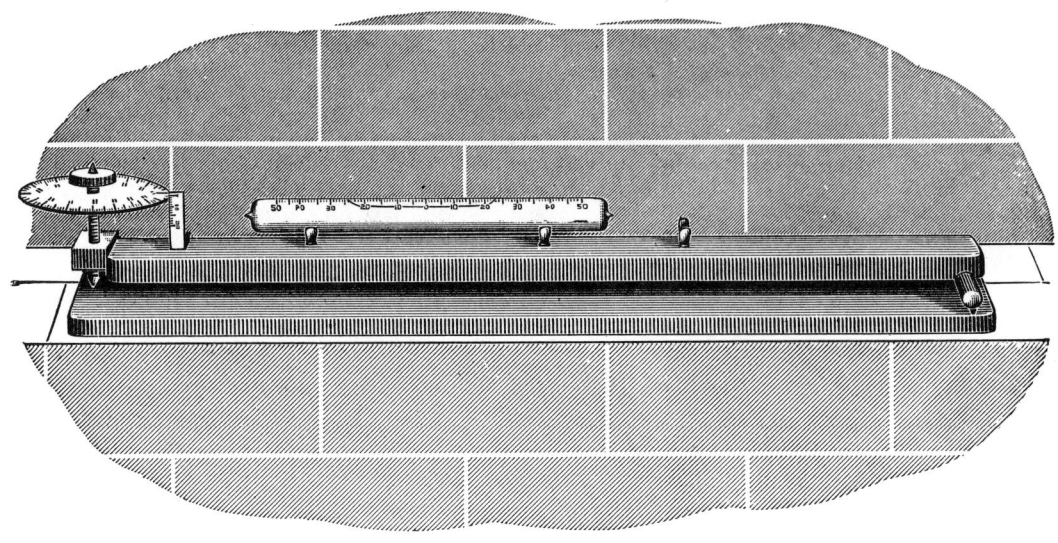

The Level-trier.

Apparatus used in the manufacture of spirit-levels to try the character and approximate sensitiveness of the curvature ground in the glass tubes before they are finally filled and sealed. Also to accurately determine the value of *one division of level* in seconds of arc.

C. L. Berger & Sons' Auxiliary Apparatus.

Used during the construction of their Instruments of Precision.

Engineers' Instruments and Their Adjustments.

Written especially for this Catalogue by Dr. LEONARD WALDO.

General Remarks.

THE OPTICAL PART.

In the construction of telescopes for engineers' instruments, several difficulties present themselves. To be portable, the telescope must be of small aperture, and of short focus. To make it of short focus and yet retain sufficient aperture to give the light necessary with the eye-pieces used, requires especial care on the part of the maker, both in securing the true curves for the crown and flint glass lenses, which make up the achromatic object-glass, and in adapting an eye-piece which will secure a flat field, with the least distortion.

Of the many forms of eye-pieces known, Messrs. C. L. BERGER & SONS, after careful experiments with the formulas suggested by the distinguished astronomer, Sir George B. Airy, and the late Mr. Kellner, of Wetzlar, (the two best formulas known,) have adopted the latter. Mr. Kellner's formula employs four lenses, mounted separately, and so arranged as to secure a flat field of the sharpest definition, to the very edge.

The magnifying power of the telescope depends upon the relation between the focal length of the object-glass and the focal length of the eye-piece, considered as a single lens: Thus—

If

$\mathbf{F}$ = focal length of the object-glass,

$\mathbf{f}$ = " " " eye-piece,

Then

$\dfrac{\mathbf{F}}{\mathbf{f}}$ = magnifying power of telescope.

It is readily seen that the magnifying power may be increased or diminished by altering the focal length $\mathbf{f}$, of the eye-piece; but if the maker increases the power too much, since only a fixed amount of light can enter the object-glass, this fixed amount of light is spread over too much surface in the field of view, and the object seen is therefore too faint. If the maker gets the magnifying power too small, then the engineer has a difficulty in pointing the telescope accurately. Some other points in regard to the magnifying power will be referred to in the description of the transit telescope. Messrs. C. L. BERGER & SONS, have found about twenty-four diameters to be the most satisfactory power for their Engineers' Transit Telescope; and for levels the powers increase in proportion to the size of the instruments.

Very much depends upon the optical part of any instrument, and very little has been put into the hands of the practical engineer by which he may rigidly test it. The following suggestions may be found convenient.

The telescope should come sharply into focus, and a very little movement of the focussing screw, either way, should cause the image to blur. When it is sharply focussed, covering any part of the object-glass without altering the focus, should not alter the sharpness of definition but merely cut off light. The pencil of light which enters the object-glass, should come out at the eye end. To ascertain this, see whether a pointer which you place just in contact with the edge of the object-glass, can be wholly seen in the small disc of light which you will notice at the small opening of the eye end when you draw your head back some inches from the telescope, and point the telescope towards the sky. If the pointer cannot be seen up to the very edge, then the maker has inserted a diaphragm which cuts off light from the object-glass, and, very probably, to conceal the faults in making. In this case the real aperture of the telescope is found by moving the pointer over the object-glass until its point is just visible, and measuring from the inner edge of the brass cell holding the object-glass to the pointer. Twice this distance subtracted from the distance between the two edges of the brass cell, will give the real or clear aperture of the telescope. The clear aperture, divided by the diameter of the small circle of light at the eye end, when the telescope is focussed on a distant object, will give the magnifying power of the telescope. Thus the clear aperture of a telescope, measured by means of a pair of dividers and a scale, was $1^{in.}$ 35, while the diameter of the circle of light at the eye end, was, $0^{in.}$ 06. In this case, the magnifying power of the telescope was $\dfrac{135}{6} = 22.5$ diameters.

Another way to determine the magnifying power, is to measure the angular distance between two points with a transit, and then measure the same distance with the telescope of which the power is to be ascertained, placed so that the transit must point into its object-glass and see the same angular distance through the second telescope *inverted*. Then calling the first angle **A,** and the angle as seen diminished through the introduction of the second telescope inverted **a,** we have the magnifying power of the second telescope $= \dfrac{tan. \frac{1}{2}\ \mathbf{A}}{tan. \frac{1}{2}\ \mathbf{a}}$. Thus the angle subtended by a window sash, several hundred feet away, was measured by a transit instrument direct, and found to be, $1°58'50''$. When a Y level, previously focussed on a distant object, was set before the transit, with its object-glass towards the transit, the same sash was measured and the angle was found to be but $3'30''$. In this case, therefore,

$$\text{the magnifying power of Y level} = \frac{tan.\left(\dfrac{1°58'50''}{2}\right)}{tan.\left(\dfrac{3'30''}{2}\right)} = \frac{tan.\ 0°59'25''}{tan.\ 0°\ 1'45''} = 34.0 \text{ diameters.}—$$

Or, for an approximation, a card cut one inch wide may be set up across a room by the side of a measure graduated to inches. Then, the number of inches on the measure seen by one eye, covered by the image of the white card seen through the telescope by the other eye, will give, roughly, the magnifying power.

It is difficult, without months of use, to fully test an instrument in all its parts; but in choosing an instrument the engineer should bear in mind that the making of the transit and the level are considered to be feats of mechanical skill. It should be remembered that there is no machine so delicate that it can *finish* the essential parts of an instrument. The last stages in its making must depend upon the personal skill of some mechanic, who has a reputation for that particular work; and we are sorry to add, that so difficult is it to secure the mechanical skill and patience required in the finishing of the interior parts, the only essential ones, and so easy is it to add the lacquer and polish of the outside, that the market is full of instruments sold at a price enough lower than the best makers can work, to seem to effect a large saving of the first cost; but such a saving is money borrowed at the highest rate of interest, when the cost of annual repairs is considered. It is better at the outset to buy of a maker who is noted for the conscientious accuracy of his work. An imperfect rack motion; a screw turned home on the wrong thread; a wabbling of the object-slide or eye-piece; a slight space between the edge of the vernier and the limb of the circle; in fact, any mechanical defect, no matter how slight it may seem, may be taken as a pretty sure indication that the work has been slighted in other parts as well, and should have a strong influence in guiding the selection of an instrument, in the absence of a test by work in the field.

The Engineer's Transit.

In the first part of this catalogue, Messrs. C. L. BERGER & SONS, have pointed out the peculiarities and improvements in this instrument, as constructed by them. In speaking of the adjustments of these instruments it is well for the engineer to remember that the construction is aimed to be such that if the telescope and levels are carefully adjusted they may remain so for even a number of years to come, if the instrument suffers no rough usage.

Description of the Telescope.

THE object-glass is achromatic, being made of two lenses, one of crown and one of flint glass. Both these lenses are made of the celebrated " Jena " glass (introduced about 1885), which has a greater index of refraction and power of dispersion than known before this time. For the most part, that is, whenever the diameter of these lenses is not too large, we — since 1889 — cement them together so as to make one lens only. In so doing the disturbing reflections from their inner surfaces, and the settling of a film between them is prevented, besides securing to the telescope an additional amount of light equal to about 8 per cent. The curvatures are computed from special formulæ, so that the telescope may have the largest aperture possible with a short focal length.

The engineer will appreciate the slightest gain in the diameter of the object-glass, since the amount of light received from any object varies as the square of that diameter. Thus an object-glass $1\frac{1}{4}$ inches in diameter will admit half as much light again as an object-glass one inch in diameter.

The eye-piece, or ocular, as it is sometimes called, is the combination of lenses used in the telescope with which the image formed at the focus of the object-glass is viewed.

The simplest and most commonly used eye-piece in the telescopes of instruments of precision, where spider-threads and micrometers are used in making measurements, is the Ramsden astronomical or positive eye-piece. It consists of two plano-convex lenses, commonly of the same focus, placed apart at a distance of two-thirds the focal length of either, the convex sides facing each other. It has the advantage of being placed behind the focus of the object-glass. It is almost free from spherical aberration, and gives a perfectly flat field of view, so that the spider-threads can be seen distinctly throughout their entire length. Unfortunately it is not entirely free from chromatic aberration, that is, not strictly achromatic, and therefore the Kellner and Steinheil eye-pieces are frequently preferred, as in them the chromatic aberration is sensibly eliminated, so that a bright object viewed with a normal eye will appear achromatic, a condition as important in the eye-piece as in the object-glass.

The Kellner eye-piece, also, consists of two lenses. The one nearest the eye, or eye-lens, is a compound lens composed of crown and flint-glass, as in the objective. Both are cemented together so as to make one, to prevent loss of light consequent upon a ray passing from one substance into another. In its common form the eye-lens is plano-convex, with the plane side nearest the eye, while the second or field-lens is double-convex.

In the Steinheil eye-piece both lenses are compound, as in the eye-lens of the Kellner. The parts of each lens being cemented together, they form two double-convex lenses, and therefore it may be designated as an achromatic double eye-piece. There are some deviations in the construction of the three eye-pieces mentioned above, but mainly as to the proper curvature of the lenses and their proper distances apart, depending as they do on the index of refraction and power of dispersion of the glass used in the construction of the object-glass and eye-piece, but the principle as above explained, by which an achromatic image is obtained, underlies all of them.

The Ramsden eye-piece is generally preferred on account of its greater simplicity and its flat field of view, which latter condition is more difficult to be obtained with the Kellner and Steinheil eye-pieces in powerful telescopes of limited length, on account of the somewhat larger field of view possessed by these eye-pieces. Moreover, the compound lenses are liable to be affected after a while by opacities caused by a crystallization, as it were, of the cement uniting the parts composing them.

Objects seen through the above-mentioned eye-pieces are, however, inverted, and telescopes so constructed are often objected to on this account. It nevertheless is the most proper telescope to use where fine telescopic measurements must be made, as the image is more brilliant than when the objects are shown upright, and it requires but little practice to get accustomed to its use. The inverting telescope has some other advantages that should be mentioned here. The eye-piece being shorter, an object-glass of greater focal length is obtained in the same length of telescope, thereby favoring the conditions imposed to secure the best definition where the telescope must be short and powerful. Any increase in the focal length of an object-glass adds to the magnifying power in the direct way, without entailing the loss of light consequent upon the use of an eye-piece made unduly powerful. On the other hand, an increase in the magnifying power of the eye-piece magnifies the least imperfection that may exist in the object-glass, and makes the cross-wires appear too coarse.

In practice, however, many engineers prefer the erecting or terrestrial telescope. Such telescopes must be made with an eye-piece consisting of four lenses, as by adding two more lenses, objects are shown right side up, as viewed with the naked eye. In the construction of an erecting eye-piece the chromatic aberration can be corrected by the two additional lenses required to secure an upright image; but in the case of short and powerful telescopes the difficulties presenting themselves to secure a perfectly flat field of view are very great, and recourse must often be had to a compound lens. In the Kellner terrestrial eye-piece the third lens, reckoning from the eye, is therefore compound, and both parts are cemented together.

The Huyghenian eye-piece is used to a very limited extent in the more modern telescopes of instruments of precision. It is most frequently met with in the large telescopes used in physical astronomy, where objects are merely viewed, but no measurements made. The field of view is large, but not quite flat. The amount of light is greater than in the other eye-pieces. The eye-piece consists of two plano-convex lenses with their convex sides facing the object-glass. The main features are, that in this eye-piece the second lens is placed between the object-glass and its focus, and that it brings the image to a focus at a point half-way between the two lenses of the eye-piece. The focal length of the second lens is three times larger than that of the eye-lens, and they are placed apart at a distance equal to one-half their combined focal length. The image is viewed by the eye-lens. It is called a negative eye-piece, because the image is formed at a point between the lenses.

The magnifying power of a telescope must be proportional to the aperture. If the magnifying power is too high for the aperture, ordinary objects will appear too faint; and if the magnifying power is too low, the objects will appear so small that the engineer cannot point upon them with sufficient accuracy.

The magnifying power should be such that the least perceptible motion of the bubble of a level, or change in the reading of the verniers, should cause sufficient movement of the cross-wires over the object in the field of view to be readily noticeable. A higher power than this is worse than useless, since objects are less brilliant. A lower power would not develop the full capacities of the instrument. Messrs. C. L. Berger & Sons adapt, therefore, the magnifying power and aperture of their instruments to the sensitiveness of the levels, and the fineness of the graduation.

In the telescopes of the instruments manufactured by C. L. Berger & Sons, the main tube has a much smaller diameter than is usual in proportion to the size of aperture. This is accomplished without cutting off any light derived from the object-glass, since the pencil of light within the telescope is continually diminishing in diameter until it comes to a focus at the plane of the spider-lines. The danger of an increase of reflections caused by bringing the interior surface of the telescope-tube nearer to this pencil of rays, is neutralized by the introduction of several more diaphragms properly placed, and by the use of a specially dead black coating for the interior. By this method of construction the weight of the telescope is greatly reduced compared with the large apertures used by them, and therefore there is less wear on the horizontal axis of revolution, and less friction of the object-slide. There is, also, on this account, less surface exposed to the wind, and the instrument is consequently more steady. C. L. B.

The Graduations.

Engineers' transits have various graduations on their circles, according to the requirements of the different branches of civil engineering. These various graduations are read by opposite verniers, which may be either single or double. American instruments have usually double opposite verniers, commonly reading the circle to single minutes or to thirty seconds. For a higher grade of work, required in the larger cities and on extended land surveys, they should, however, read to twenty or ten seconds.

The customary graduations of C. L. Berger & Sons' instruments are, First, — the circle divided to half degrees, the verniers reading to single minutes. Second, — the circle divided to twenty minutes, the verniers reading to thirty seconds. Third, — the circle divided to fifteen minutes, the verniers reading to twenty seconds. Fourth, — the circle divided to twenty minutes, the verniers reading to twenty seconds. Fifth, — the circle divided to ten minutes, the verniers reading to ten seconds.

To express the relation between the vernier and circle divisions, let $d=$ the value of one division of the circle; $d'=$ the value of one division of the vernier; $d-d'=$ the least count of the vernier, or, in other words, the smallest reading of the circle.

$n=$ the number of spaces of the vernier which correspond to $(n-1)$ spaces of the circle.

We then have the three formulas;

$$(1.) \qquad n = \frac{d}{d-d'}$$

$$(2.) \qquad d' = \frac{n-1}{n}d$$

$$(3.) \qquad d-d' = \frac{1}{n}d$$

Thus, for example, suppose the circle was divided to 15′, and it was desired to read to 20″. Here, $d=15'$

$$d-d', \text{ or, the least count} = 20''$$

Then, by formula (1)

$$n = \frac{15'}{20''} = \frac{15 \times 60''}{20''} = 45$$

Therefore, 45 spaces of the vernier must correspond to 44 or $(n-1)$ spaces of the circle.

Suppose again the arc to be divided to 20′, and to be read to 30″. In this case we have

$$n = \frac{20 \times 60}{30} = 40$$

Therefore, 40 spaces of the vernier must correspond to 39, or $(n-1)$ spaces of the circle. These are the graduations which Messrs. C. L. Berger & Sons usually adopt for engineers' transits.

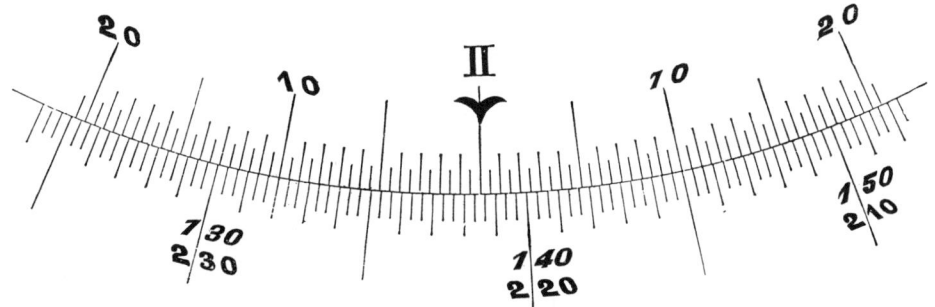

The cut shows a portion of the circle and vernier, to illustrate the method of reading to thirty seconds.

The lines marked 130, 140, and 150 denote 10° each. The shorter lines half way between them denote 135° and 145°. The next shorter lines denote whole degrees, while the shortest lines are one-third of a degree, or 20′ apart.

The vernier comprises the upper series of lines. Of this series only that half lying to the right of the vertical arrow, or zero, and having the figures 10 and 20 inclined in the same direction as the 130, 140, and 150 of the arc, is to be used in connection with these figures. The vernier is double,—one half to be used with one set of graduations of the arc, the other half to be used when angles are laid off in the opposite direction, and then the lower set of figures, 210, 220, and 230 are used.

It is to be especially remembered that the figures on the vernier are inclined in the same direction as the figures on the arc to which they belong.

To read the vernier, first note the whole degrees, and 20′ spaces lying between the last 10 degree division and the zero division of the vernier.

Thus in the cut, using the upper line of figures, the zero of the vernier has passed the 130° division, and moved on until it is between the 20′ and 40′ space beyond the 138° mark. The first part of our reading will therefore be 138° 20′.

Second, look along the vernier, beginning from the zero point, and in the direction in which the graduation of the arc runs, until one line of the vernier is found which seems to be a prolongation of an opposite line on the arc.

Consider each of the vernier spaces between the vernier zero and such a line, as equal to 30″ of arc.

Add the number of minutes and seconds thus obtained to the first reading. The result will be the reading of the circle.

Thus we notice that the vernier zero is a trifle over half-way of the distance between the 20′ and 40′ marks of the arc.

And looking along the vernier to the right, we notice that the lines of the vernier gradually approach the lines on the arc until the twentieth line of the vernier is precisely opposite a line on the arc. Of course, since each vernier space denotes 30″, the alternate ones made a little longer in the cut will denote single minutes, and on the vernier therefore the twentieth line would correspond to 10′ 00″, and since our first reading was between 20′ and 40′, this vernier reading is to be added to that first reading.

Thus,

$$138° \quad 20′$$
$$10′ \quad 00″$$
$$\overline{138° \quad 30′ \quad 00″}$$ will be the reading of the vernier, using the *upper* graduation.

In the same manner we proceed to the *left* in reading the lower graduation, in which the figures are inclined to the left. Thus in the cut, we should find the zero point of the vernier is beyond the 221° 20′ mark, and the line of the vernier, which is seemingly a prolongation of a line of the arc, corresponds to 10′ 00″. Then we have

$$221° \quad 20′$$
$$10′ \quad 00″$$
$$\overline{221° \quad 30′ \quad 00″}$$ for the reading of the vernier, using the *lower* graduation.

Practically, in reading the vernier, the engineer decides which line is in coincidence by the position of the lines on both sides.

He first notices, roughly, what fractional part of a space on the limb lies between the vernier zero and the last graduation mark it has passed. This enables him to look immediately to that part of the vernier in which the coincidence occurs.

Thus in the figure the vernier zero is about half way between 221° 20′ and 221° 40′, the engineer therefore immediately looks about half way along the vernier and finds the 10′ 00″ division to be the one sought.

When the graduation is to thirty seconds, the engineer will find that if he only chooses, he can work to minutes with this graduation quite as rapidly as with a transit graduated to minutes, by simply disregarding the shortest lines of the vernier.

The second vernier, which is distant 180°, or exactly opposite the one read first, may also be read. Not so much to eliminate any eccentricity of the circle and verniers as to afford a valuable check upon the angle measured.

Greater accuracy in the measurement of any angle may be obtained by the principle of repetition. In this case, before and after an angle has been repeated a number of times, all four of the verniers should be read, and if, for example, the graduations proceed from right to left, the left hand side of each double vernier should be read as usual; but in the right hand side the line now marked 20 on the vernier should be considered 0, and the arrow on the vernier 20. Then, with this convention, only the minutes and seconds of the second vernier should be used.

But it should be here remarked that the repetition of angles is not now held in such repute by our best engineers, as it was before the present perfection of the art of graduating and centering the circles and verniers of engineering instruments.

The engineer who has not used them will find the ground-glass shades a great convenience in reading the vernier. They are so placed as not to be readily broken, and they shed a clear, white light upon the graduations.

Graduations on solid silver are much to be preferred to graduations on any known brass alloy. The surface of the silver can be worked very plane, since it is of uniform texture. The graduations can be cut with the utmost uniformity in width of line and spacing.

The Centers.

Quite as important as the graduation, is the exact fitting of what the makers call the *centers* of the instrument; *i.e.*, the two vertical metal axis, about which the circle and the vernier plate turn.

Both axes must be exactly concentric with the center of the graduated circle, and the center of the horizontal axis of the telescope in any position of the instrument. The most sensitive level about the instrument should not show any displacement when the circle-plate is held, and the lower plate moved by the hand.

In the construction of the inner center, the hardest bell-metal should be used, and for the outer center a red composition metal of the best quality. To insure a true concentricity of the axis, and consequently of the limb and vernier, it is necessary that they should each be turned in a dead center lathe, each about its own axis. In fitting the centers, they should turn without the slightest play, and yet with very little friction.

Messrs. C. L. Berger & Sons take the precaution of casting the outer center, circle and vernier plate in the same mould, to avoid any difference in the composition of the metal.

The upper plate should not be hammered, since this would also effect an unequal expansion of the metals in extreme temperatures, causing the vernier to read too long or too short.

After the plates are put together, the vernier and limb should revolve in the same plane, to avoid parallax. The space between the limb and vernier should have the appearance of a uniform, fine, black line.

The Compass.

In running old lines, and as a check in running new ones, the compass is frequently a very important part of the transit. Its needle should be tempered throughout, and of hard steel, to retain its magnetism. It should be thin, and yet at the same time have enough surface to be strongly magnetic. It should be swung upon a jewelled center, and so nicely fitted that when at rest, with the instrument levelled, the two extreme points should just clear the graduation of the compass box, and read precisely 180° different in any part of the graduated arc. The pivot on which it swings should be conical, and hardened so that it may swing upon a sharp point, without having this point weak.

The needle should also be so sensitive, that when drawn from its pointing by the outside attraction of a piece of iron held in the hand a foot or so away, it will settle to the same reading several times in succession.

This sensitiveness depends upon the form and sharpness of the pivot, the strength of its magnetism, and its bearing on the jewelled center.

If it should be found that a needle has lost its sensitiveness, it is probably not so much owing to its loss of magnetism, as to a dulling of the pivot. Since this may happen when the engineer is without access to the maker, and an instrument otherwise be in good condition, it should be remarked that the pivot can be sharpened after removing the needle, by taking a fine oil-stone, and while turning the instrument with one hand, grinding the pivot, with the oil-stone in the other; being careful to incline the grinding surface about 25° to the pivot. The pivot is originally turned and sharpened in a lathe, and in grinding by hand, great care should be taken to preserve its conical form.

The two extreme points which lie next the graduation, together with the point of suspension, should lie in one straight line.

The center of gravity of the needle should be as far below this line as possible.

The quivering of a needle so constructed is not annoying, since the center of its quivering motion is in the line through its two extreme points, which are, therefore, stationary.

To determine whether the transit itself has any iron in it to disturb the needle, it is a good plan, after setting the instrument so that both compass-needle and vernier reads 0°, to go round the circle, setting the vernier ten degrees ahead each time, and noting whether the compass-needle also describes an arc of precisely ten degrees. If it does not, there is some local attraction.

The graduations on the compass box should begin at the North point, and run 90° in both directions; then decrease to 0° again at the South point. In order that the needle reading may indicate the direction of the telescope, the line joining the zeros of the ordinary compass ring must be in the same vertical plane, with the line of collimation of the telescope; and the letters denoting the cardinal points, East and West, must be transposed; *i. e.*, when the letter N is towards the North, the letter W should be towards the East. Of course the needle indicates magnetic north, and in the case of instruments unprovided with means of setting off the local variation of the needle, all the readings of the needle must be corrected for this local deviation.

Spirit-Levels.

The spirit levels, as regards their sensitiveness, should be in strict keeping with the optical power, and the graduations of the instrument, but the quality should be of the best. A level-bubble should move uniformly over the same distance, when the telescope is made to point on two objects alternately, differing slightly in altitude, by the leveling screws alone. In change of temperature the bubble should lengthen symmetrically from the center; and no matter what its length, it should move quickly, without any of the hitching, which is caused usually by a little dirt introduced when it is filled.

Of the three levels attached to the complete transit, the telescope level is the most sensitive. It should be sensitive enough for ordinary leveling, such as good railroad work. The level in front, or at right angles to the standards, should be sensitive enough to make a line plumb by it to any height; while the third level on the standard is used in leveling up the instrument, and to establish the zero point for the vernier correctly when vertical angles must be measured.

The test of the fitness of the various levels for the capacity of the instrument should lie in this: that after carefully bi-secting an object in the field of view, in such a position of the instrument that all the levels can be read, and then slightly deranging them all with the leveling screws, the bi-section will be accurately made after restoring the levels to the exact position they before occupied, by the leveling screws alone.

Leveling Screws.

Messrs. C. L. Berger & Sons usually cut their leveling screws with 32 threads to an inch provide the usual four screws in opposing pairs. The plates once set firmly apart by tightening two of these screws on the *same side*, the leveling of the instrument is easily accomplished by turning the two screws of an *opposing* pair so that both thumbs shall move toward each other (when the bubble will go toward the right), or both thumbs away from each other, when the bubble will move toward the left. Instruments intended for triangulation, *i. e.*, reading to 10″ or less, should however be supported on three, instead of upon four screws. In this case the instrument is rapidly leveled by bringing one level parallel to two of the screws, the other level will now be at right angles to it. Level both levels at the same time by turning one of the screws to which the first level is parallel and the screw which is at right angles to this level. Of course the instrument may now be reversed to guard against non-adjustment of the levels.

Three Leveling Screws versus Four.

To the student of the progress in Engineers' field instruments, the question often presents itself as to the comparative merits of an instrument provided with three, over one having four leveling screws. It should be here remarked that the greater portability existing in instruments provided with four leveling screws still commends itself to all using the more customary class of instruments. However, the finest class of field instruments, requiring spirit-levels corresponding to the fineness of graduation, cannot be advantageously manipulated with four leveling screws. The results thus obtained would be little better than those obtained with a more ordinary instrument. To insure the full benefit of a finer instrument, such as used in triangulation, the maker will prudently apply three leveling screws, mounted on *a basis larger* then is usual in instruments with four screws. So, while four leveling screws have the advantage of greater compactness and less weight three screws have the advantage for closer setting, giving better results. The maker will therefore adapt either the one or the other kind to his instruments as the case may require.

Quick Leveling Attachment.

[For illustration see elsewhere.]

As all devices of this kind detract more or less from the stability of an instrument, it seems they never have been regarded with much favor by the engineering profession at large. There are cases, however, where the use of such a device, in a mountainous country, or in underground work of a close character, becomes very desirable. Messrs. C. L. Berger & Sons' device, unlike devices of a similar kind forming a part of the instrument proper, consists of a coupling with a ball and socket joint which can be screwed between the instrument and tripod. As this intermediate piece forms no part of the instrument itself it can be readily attached or detached at will, thus adapting the instrument to the circumstances and to the class of work in hand. For this purpose the threads of this coupling or quick-leveling attachment, and those of the instrument and tripod are identical; and as all their transits and levels with four leveling screws are interchangeable on any of their tripods, one such coupling is sufficient for an engineer's outfit. In fact one extra tripod permanently provided with this quick-leveling attachment may be kept ready for occasional use in an office where there are a number of their instruments.

To use this quick-leveling attachment proceed as follows:—Screw it to the instrument, and then screw both to the tripod in the usual manner, taking care that the coupling becomes firmly fastened thereto. Now to operate it, slightly unscrew the instrument from its hold upon the flange of the coupling by means of the milled edges provided for this purpose, and move it approximately into a level plane, then again screw the instrument firmly to the coupling same as before. This being accomplished, move the instrument over the given point on the ground by means of the centering arrangement described later on, and level up carefully by the leveling screws alone. It will be seen that this quick-leveling attachment is operated entirely independent of the leveling screws or centering arrangement. Of course, when this device is to be used for several days in succession, it is not necessary to detach it from the tripod every time the instrument is to be removed. In such cases the instrument only should be detached from the coupling. Whenever it becomes desirable to detach the coupling from the tripod, it can best be performed by allowing the instrument to remain fastened to the coupling, then by taking hold of the milled edge of the coupling unscrew in the usual manner. In cases where the coupling has been permanently attached to a tripod, the small screws connecting it to the tripod head must first be removed.

To secure the greatest possible stability to the instrument, the outside diameter of the hollow hemisphere is equal to the distance between the leveling screws of the instrument; and to secure a smooth and ready action, leather washers are provided in the socket which act against the hemisphere. However, when the instrument is clamped to the flange of the coupling these washers recede, and the metal surfaces are brought into direct contact with each other.

The Gradienter Screw.

This very convenient attachment consists simply in a screw working against the clamping arm suspended from the horizontal axis, and on the opposite side from the vertical arc. A strong spiral spring is set directly opposite the screw, and presses the clamp arm against the end of it. This screw is cut with great care in a lathe. It has a large silvered head graduated into fifty equal parts. As the screw is turned, the head passes over a small silvered scale, so graduated that one revolution of the screw corresponds to one space of the scale.

Obviously then, the number of whole revolutions made by the screw, in turning the telescope through a vertical arc, can be ascertained from this scale. The clamp arm of the telescope has its clamping screw just above the horizontal axis, in the usual manner. When this screw is free, the telescope may be revolved; but when it is clamped, the telescope can only be moved by the gradienter screw, which thus takes the place of the ordinary vertical tangent screw. The screw is cut with such a value of a single revolution, as to cause the horizontal cross-line of the telescope to move over a space of $\frac{1}{100}$ of a foot, placed at a distance of 100 feet, when the screw is turned through one of the smallest spaces on its graduated head; and since there are fifty such spaces on the head, it follows that one revolution of the screw is equivalent to $\frac{50}{100}$ of a foot, at a distance of 100 feet. The numbered graduations on the screw head are then each equivalent to $\frac{1}{10}$ of a foot in 100 feet; and two entire revolutions of the screw would be twice $\frac{5}{10}$, or 1 foot to the 100. It is readily seen that grades can be established with great rapidity with this screw. It is only necessary after setting the gradienter screw to zero, and leveling and clamping the telescope, to move it up or down as many spaces of the head of the gradienter screw as there are hundredths of feet to the hundred, in the grade to be established. Thus, to establish a grade of 1.^{ft} 85, the screw head is turned through three whole spaces of the scale, which corresponds to 1.^{ft} 50, and through three of the numbered divisions, and five of the shortest ones to make up the entire reading of 1.^{ft} 85.

For measuring distances this screw takes the place of stadia lines, and is more convenient; since for any approximately horizontal distance, the space on an ordinary leveling rod expressed in hundredths of feet, included in two revolutions of the screw, will be the number of feet the level rod is distant from the center of the instrument. Thus the difference between two readings of the level rod was 2^{ft}.965 when the telescope was moved in altitude through two revolutions of the screw. The rod therefore was distant 296.5 feet.

It is unnecessary even that a leveling rod be used. A ranging pole or walking stick, or any arbitrary length which can afterwards be measured, will suffice. Thus a stick, which was afterwards measured and found to be 3^{ft}.38 long, was found to be subtended by $3\frac{8}{50}$ revolutions of the screw at an unknown distance.

In this case the distance was —

$$\frac{3\,.\,38}{1\,.\,58} \times 100 = 213\,.\,9 \text{ feet.}$$

In case, however, the distance to be measured is not approximately in the same level plane with the transit telescope, it is necessary to compute the distance, from the readings of the rod. In taking such readings at an altitude, it is customary to incline the rod towards the telescope, and by trial find the least space subtended by two stadia lines. A skilful rod-man will *plumb* a rod more readily than he can *incline* it at the proper angle, and a reading of the plumb rod can be taken with greater accuracy, and in less time than with the inclined rod; but it ordinarily involves some additional computing to reduce such vertical readings to horizontal distances. With the view of reducing the computation to a simple multiplication, the following table is appended with the trignometrical argument on which it depends. The engineer will notice the solution is not rigorously exact, but is sufficiently so for all cases in practice.

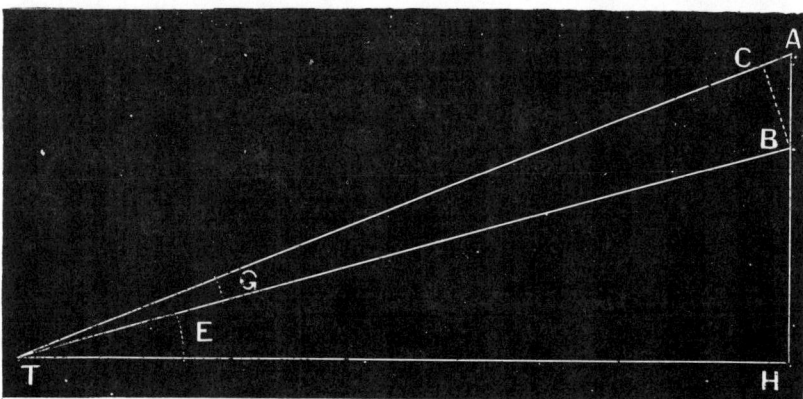

In the above figure,

TH = the transit horizontal sight line.

The angle HTB = the angle of elevation of the telescope to the foot of the rod
= E.

" " BTA = the angle subtended by any number of revolutions of the gradienter screw = G.

AB = the length of the rod included by the angle G, when the rod is vertical = R.

CB is drawn perpendicular to TB.

Then, $\qquad$ CBA = BTH = E $\qquad$ TAH = 90° — (E + G)

$$\frac{BC}{AB} = \sin\frac{\left(90° - (E + G)\right)}{\sin(90° + G)} = \frac{\cos E \cos G - \sin E \sin G}{\cos G}.$$

$$\therefore BC = R\ (\cos E - \tan G \sin E.)$$

$\tan G = \dfrac{nh}{a}$ where h is the height above a horizontal line, subtended by *one* revolution of the gradienter screw at a distance a.

n is the number of revolutions made in any given case.

$$BT = \frac{n}{nh} BC = R\frac{a}{nh}\ (\cos E - \frac{nh}{a} \sin E)$$

$$\therefore BT = R\left(\frac{a}{nh}\cos E - \sin E\right) \quad \cdot\ \cdot\ \cdot\ \cdot\ \cdot\ \cdot\ \cdot\ \cdot \quad I.$$

and

$$HT = BT \cos E$$

$$\therefore HT = R\left(\frac{a}{nh}\cos{}^2E - \tfrac{1}{2} \sin 2\,E\right) \quad \cdot\ \cdot\ \cdot\ \cdot\ \cdot\ \cdot\ \cdot \quad II.$$

Formulas I and II are general formulas for *any* gradienter screw. In C. L. Berger & Sons' transits the screw is cut and placed so that when $a = 100$, for $n = 2$ and $h = \frac{1}{2}$, by substitution these formulas become,

$$BT = R\ (100 \cos E - \sin E.)$$

$$HT = R\ (100 \cos{}^2E - \tfrac{1}{2} \sin 2\,E.)$$

Where BT = the direct distance from the center of the horizontal axis of the transit to the foot of the vertical rod.

HT = the horizontal distance from the center of the horizontal axis of the transit to the plumb line dropped from the foot of the vertical rod.

R = the space included on the vertical rod by two revolutions of the gradienter screw.

E = the elevation of the foot of the rod above the horizontal sight line of the telescope.

When the angle **E** becomes an angle of *depression* instead of elevation, then the point **B** is the *upper* end of the part of the rod used, **A B.** The distance **B T** in this case is the direct distance between the center of the horizontal axis of the telescope and the upper reading of the vertical rod in the valley.

The distance **H T** is, as before, the horizontal distance between the center of the horizontal axis of the telescope, and the plumb line prolonged in this case upwards from the upper end of the vertical rod. The plumb line in all cases coincides with the direction of the rod.

By means of the following table, it is only necessary to multiply the factor opposite the angle of elevation, by the space included upon a vertical rod by two gradienter screw revolutions, to obtain either the direct or horizontal distance of the center of the horizontal axis of the telescope from the foot of the rod; or the same distance from the upper reading of the vertical rod in the case of an angle of depression.

Gradienter Screw Table I.

Factors to be multiplied by the space on the vertical rod expressed in feet and decimals, included in two revolutions of the gradienter screw, to find the distance of the foot of the rod from the center of the horizontal axis of the transit telescope.

Angle of Elevation E.		Factor for the Direct Distance (100 cos E − sin E)	Factor for the Horizontal Dist. (100 cos ^{2} E − ½ sin 2 E)	Angle of Elevation E.		Factor for the Direct Distance. (100 cos E − sin E)	Factor for the Horizontal Dist. (100 cos ^{2} F − ½ sin 2 E)
°	′			°			
0		100.00	100.00	15		96.33	93.05
1		99.96	99.94	15	30	96.09	92.59
2		99.90	99.84	16		95.85	92.14
3		99.81	99.67	16	30	95.60	91.66
4		99.69	99.45	17		95.34	91.17
5		99.53	99.15	17	30	95.07	90.66
6		99.34	98.80	18		94.80	90.17
7		99.13	98.39	18	30	94.51	89.63
8		98.89	97.93	19		94.22	89.09
9		98.61	97.41	19	30	93.93	88.54
10		98.31	96.81	20		93.63	87.98
10	30	98.15	96.51	20	30	93.32	87.41
11		97.97	96.17	21		93.00	86.83
11	30	97.79	95.82	21	30	92.67	86.22
12		97.60	95.47	22		92.34	85.62
12	30	97.41	95.11	22	30	92.01	85.01
13		97.21	94.73	23		91.66	84.37
13	30	97.01	94.33	23	30	91.31	83.75
14		96.79	93.92	24		90.94	83.08
14	30	96.56	93.48	24	30	90.59	82.43
15		96.33	93.05	25	00	90.21	81.76

In practically applying this table, it is preferable to take the mean of several readings of the rod in each position of the gradienter screw.*

Thus, with the target near the foot of the rod, and then moved to correspond to two revolutions of the gradienter screw, three readings in each position were as follows:

	I.	II.
Altitude 18° 20′		
	ft.	ft.
	0.625	3.380
	0.627	3.376
	0.625	3.378
Means, . . .	0.626	3.378
		0.626
Difference, ... 2.752		

Factor for direct distance for 18° $= 94.80$ For Horizontal Distance $= 90.17$
"   "   " 18°30′ $= 94.51$ "   "   " $= 89.63$

Differences, . . . $= 0.29$ 0.54

Therefore, the factor for 18° 20′ will be for the *direct* distance 94.80,—⅔ of 0.29 $= 94.61$, and for the horizontal distance, 90.17 — ⅔ of 0.54 $= 89.81$.

Then we have, 2.752 × 94.61 $=$ 260.37 $=$ the *direct* distance.
2.752 × 89.81 $=$ 247.15 $=$ the *horizontal* distance.

This direct distance being the distance from the position of the foot of the rod or the lower target to the center of the horizontal axis of the telescope,† and the horizontal distance, the one usually desired, that distance reduced to a level line.

The mean value of two revolutions of the Gradienter Screw in arc, is 34′ 23″. Hence the value in arc of one of its smallest divisions on the head is 20″ nearly. Vertical angles therefore may be laid off with facility when they are confined to the range of the screw.

---

*To insure at all times accurate results, the telescope axis should revolve free, but without any looseness in the bearings. The engineer should examine these bearings from time to time, and, if necessary, fresh and pure watch oil must be applied.

To make a measurement with a micrometer screw, its graduated head should be set back slightly, then bring it up to the readings in the same direction in which the measurement must be effected.

† Should the engineer desire the direct distance between the foot of the rod, and *the point over which the plumb-bob is suspended*, it may be found by the following formula.

$$x = \sqrt{d^2 + p^2 + 2\,pd \sin E}.$$

or putting it in a shape adapted for logarithmic computation,

$$x = \frac{(d - p)}{\cos q}. \quad \text{Where } \tan q = \frac{2 \sin \frac{1}{2}(90 + E)}{(d - p)}\sqrt{dp}.$$

Where $x =$ the distance from the point under the plumb-bob to the foot of the vertical rod.

$d =$ the direct distance obtained as above.

$p =$ the distance from the center of the horizontal axis is to the point under the plumb-bob.

$E =$ the angle of elevation of the foot of the rod, as above.

The subjoined table affords a ready means of expressing any number of revolutions, and parts of a revolution, in arc; and the converse, of degrees, minutes and seconds, in revolutions of the screw:

Gradienter Screw Table II.

To convert a reading of the Screw into Arc.				To convert Arc into a reading of the Screw.			
Gradienter Screw.	Arc.	Gradienter Screw.	Arc.	Arc.	Gradienter Screw.	Arc.	Gradienter Screw.
Rev. Div.	° ' "	Rev. Div.	° ' "	° ' "	Rev. Div.	° ' "	Rev. Div.
0 0	0 0 0	2 0	0 34 25	0 0 0	0 0.0	0 8 00	0 23.5
0 1	0 20	3 0	0 51 35	0 0 10	0 0.5	0 8 30	0 25.0
0 2	0 40	4 0	1 8 45	0 0 20	0 1.0	0 9 00	0 26.0
0 3	1 0	5 0	1 25 55	0 0 30	0 1.5	0 9 30	0 27.5
0 4	1 25	6 0	1 43 10	0 0 40	0 2.0	0 10 00	0 29.0
0 5	1 45	7 0	2 0 20	0 0 50	0 2.5	0 20 00	1 8.0
0 6	2 5	8 0	2 17 35	0 1 00	0 3.0	0 30 00	1 37.0
0 7	2 25	9 0	2 34 45	0 1 30	0 4.5	0 40 00	2 19.0
0 8	2 45	10 0	2 52 0	0 2 00	0 6.0	0 50 00	2 55.5
0 9	3 5	11 0	3 9 10	0 2 30	0 7.5	1 00 00	3 24.5
0 10	3 25	12 0	3 26 20	0 3 00	0 9.0	2 00 00	6 49.0
0 20	6 50	13 0	3 43 30	0 3 30	0 10.5	3 00 00	10 23.5
0 30	10 20	14 0	4 0 45	0 4 00	0 12.0	4 00 00	13 48.0
0 40	13 45	15 0	4 17 55	0 4 30	0 13.5	5 00 00	17 22.5
1 0	17 10			0 5 00	0 15.0		
1 10	20 40			0 5 30	0 16.0		
1 20	24 05			0 6 00	0 17.5		
1 30	27 30			0 6 30	0 19.0		
1 40	30 55			0 7 00	0 20.5		
2 0	34 25			0 7 30	0 22.0		

Thus, the telescope being leveled, the gradienter screw was turned through a space of 11$^{rev.}$ 23$^{div.}$ required the arc:

$$
\begin{array}{lrrr}
\text{11 revolutions,} & =3° & 9' & 10'' \\
\text{20 divisions,} & =0 & 6 & 50 \\
3 \quad `` & =0 & 1 & 0 \\
\hline
\text{The whole arc, . . .} & =3° & 17' & 00''
\end{array}
$$

Conversely, it was desired to turn off a vertical angle of 4° 35′ 40″.

Then we have —

$$
\begin{array}{rrrr}
4° & 0' & 0'' & =13^{rev.} \quad 48^{div.}.0 \\
30 & 0 & =1 & 37 \quad .0 \\
5 & 0 & = & 15 \quad .0 \\
40 & = & & 2 \quad .0 \\
\hline
\end{array}
$$

The space on the head of the screw $=16^{rev.}$ $2^{div.}.0$

The engineer will bear in mind that the examples given are purposely given in detail: that in practice the operations may be mental ones.

It will be seen that the vertical gradienter can be used for a variety of purposes; measuring distances, grades, differences of levels, vertical angles, and is a useful check against errors of rod or chain measurement.

Messrs. C. L. Berger & Sons have also applied the same principle to their horizontal tangent screws. By graduating a silver head attached to these screws subdivisions of one minute of arc are readily made.

For constant use with these screws it is better to have a rod with two movable targets, or a rod painted with white and black squares as used in the coast survey.

Stadia Lines.

The gradienter screw is so universal in its application and can be so readily used for angular, distance or grade measures, that it will generally be found best to have it upon transits designed for current work. There are some cases however where stadia lines are more expeditious in use than the gradienter screw, and give quite as exact results.

Stadia lines, for instance, where an instrument is to be used for distance measures alone, commend themselves for their greater simplicity. For such work, non-adjustable lines, in connection with an inverting eye-piece, give the best results. If the lines are adjustable, in the field usage of an instrument they may alter their distance apart; and there is a rapidity of work with fixed lines, and a rod graduated for telemetrical work, which is not reached in any other way.

These lines may be webs, or platinum, or they may be ruled on glass. The latter are extremely accurate, but the use of them is necessarily limited in the telescopes of field instruments for the following reasons: thin as the glass may be on which the lines are ruled, and intercepting only a small amount of light, yet the film of dampness and dirt soon collecting on it will intercept a great amount of light which in time may become a very serious impediment in the use of the telescope. Another objection to their general adoption consists in the fact that as the image of an object is focussed in the plane of these glass-lines, a portion of the light of the image will become reflected from the polished surfaces of this glass, causing at times a disturbance in the clearness of vision. Besides, this glass-"micrometer," as placed in most telescopes, is very difficult of access and must needs be removed for cleaning, thereby increasing the liability of becoming broken, or detached from its mounting.

Cross
of Telescope
Wire
in

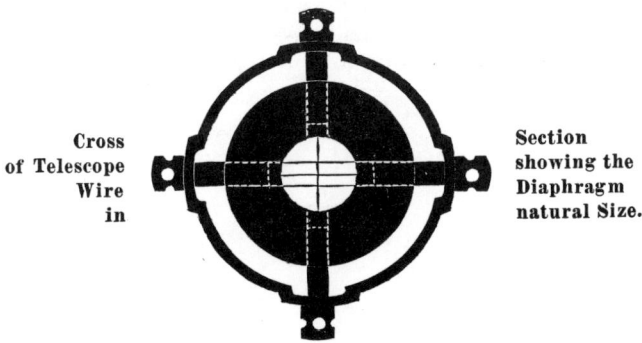

Section
showing the
Diaphragm
natural Size.

Plumbing and Centering Arrangements.

It now remains to speak of several conveniences of the instrument under consideration. By a simple mechanical contrivance the plumb-bob when suspended from the instrument can be set immediately at any desired height. It is suspended directly from the *center* of the instrument, and not from the tripod head. This precaution should be taken with every instrument, since otherwise, when there is difficulty in setting up an instrument, and the legs are unsymmetrically placed, the plumb-line will not pass through the center of the instrument.

The instrument is provided with the shifting tripod, better known as the shifting center, by means of which, when the plumb-bob of the instrument is within a fraction of an inch over a point on the ground, it may be brought immediately over it, by moving the body of the instrument on its lower level plate. This is probably the greatest time-saving arrangement which modern makers have introduced in engineers' transits.

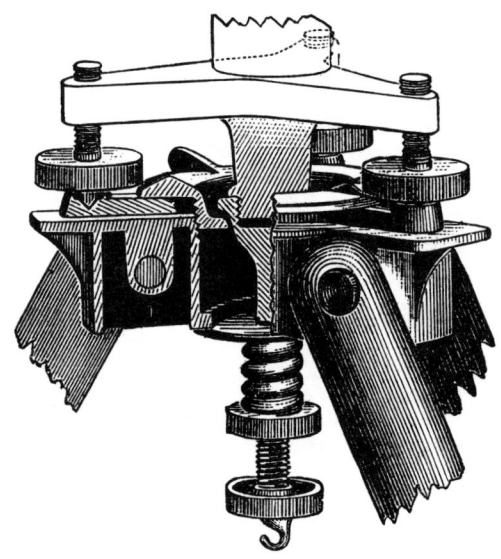

Shifting Center for a Transit with Three Leveling Screws.

As made by C. L. Berger & Sons.

In this device, as shown in the cut, the leveling screws rest in receptacles on the piece made to slide on the tripod head. A clamp-nut, provided with a large flange and handles, serves to secure this sliding piece to the tripod in any position in the range of its lateral motion. The center-piece, bearing the hook for the plumb-bob, is like that shown in the p. 179 cut; the milled-head, at its lower end, serving to fasten it to the instrument; while its milled-headed nut, acting against the spiral spring, fastens the instrument to the tripod. In use, the pressure of this spring needs to be sufficient to take up the back lash or any looseness that may exist in the leveling screws; but to secure the necessary stability of the instrument to the tripod, the clamp-nut should be well fastened to the sliding piece. To operate the shifting center, both the spiral-spring and the clamp-nut must be released slightly from their hold upon the tripod and the sliding piece, when the instrument can be moved over the given point on the ground. This device adds about 2 lbs. to the weight of the tripod.

C. L. B.

Arrangement for Offsetting at Right Angles.

The most common off-set with the transit is one at 90° to the line of sight. Several methods have been proposed for doing this without disturbing the telescope.

Messrs. C. L. Berger & Sons have a very neat one; it consists in simply perforating the horizontal axis, so that by drawing the head back fifteen or twenty inches from one end of the axis, the eye may be placed so that the eye, the horizontal axis of the telescope, and a rod set beyond, may be readily placed in the same straight line, at right angles to the line of sight of the telescope, no matter at what altitude the telescope may be pointing.

In off-setting by the arrangement proposed above, the rod is made plumb by lining it with the plumb-line of the instrument itself. The advantage of this method is, that it holds equally well for any inclination of the telescope. The disadvantage is, that the engineer is obliged to leave the eye-end of the telescope at each off-set made. Where the engineer is willing to bring his telescope nearly level before each off-set is made, Messrs. C. L. Berger & Sons will adapt a simple combination of two prisms to the telescopes, by which the rod may be made plumb, and set at an angle of 90° to the line of sight.

Setting Up.

In setting up a transit, push the iron shoe of one leg firmly into the ground, by pressing on the other two legs near the tripod head. Having secured a firm foundation for this leg, separate the other two legs, at the same time drawing the tripod head toward you. Then set the two remaining legs in the same manner as the first one. If the ground is pretty level, merely noticing that the tripod feet are equidistant, will insure that no unsightly appearance will be given to the leveling screws. If the ground is uneven, however, nothing but practice can produce a graceful position of the instrument. The plumb-bob attached to the instrument should swing within say half an inch of the point on the ground, and the plate on which the leveling screws rest, if possible, should be approximately horizontal, when this stage is completed.

Now with the level screws not tightened up, after leveling approximately, bring the plumb-bob exactly over the point on the ground, by moving the body of the instrument on its shifting head. Then complete the leveling of the instrument, and it is ready for work.

The Adjustments of the Transit.

In a theoretically perfect transit instrument, the following points are established:

1. The object and eye-glasses are perpendicular to the optical axis of the telescope at all distances apart.

2. The line of collimation coincides with the optical axis.

3. The line of collimation is parallel to the telescope level.

4. The line of collimation passes through, and is perpendicular to the horizontal axis of revolution.

5. The vertical circle is perpendicular to the horizontal axis.

6. The center of its graduated arc lies in the horizontal axis.

7. The arc reads zero when the line of collimation is perpendicular to the vertical axis of the upper plate.

8. The pivots of the horizontal axis of the telescope are circles.

9. The bearings for these pivots are of the same diameter or otherwise exactly similar.

10. The line of collimation moves in a plane perpendicularly above the center of the horizontal graduated circle.

11. The horizontal axis is perpendicular to the axis of the upper plate.

12. The upper plate is perpendicular to its axis.

13. The radial lines which form the graduations of the circle and verniers are equi-distant at the same distance from the axis of the upper plate, and pass through this axis.

14. The levels of the upper plate are perpendicular to its axis.
15. The vertical axis of the upper plate coincides with the axis of the lower plate.
16. The lower plate is perpendicular to its axis.
17. The center of the vernier plate lies in the axis of the lower plate.
18. The axis of the plumb-bob coincides with the vertical axis of the instrument.

Of the above points the maker attends to numbers 1, 2, 4, 5, 6, 8, 9, 11, 12, 13, 15, 16, 17, 18, as a part of the skillful manufacture of the instrument; and the engineer has no facilities for adjusting them, away from the shop. Points numbered 3, 7, 10 and 14 are attended to by the maker when the instrument leaves the shop; but owing to their liability to derangement, from accidental rough usage, the maker leaves it in the hands of the engineer to restore them at any time. It is to these adjustments only that the following remarks are confined.

Adjusting.

If the instrument is out of adjustment generally, the engineer will find it profitable to follow the makers in not completing each single adjustment at once, but rather bring the whole instrument to a nice adjustment by repeating the whole series.

After setting up, bring the two small levels each parallel to a line joining two of the opposing leveling screws. Bring both bubbles to the center of the level tubes, by means of the leveling screws. In doing this, place the two thumbs on the inner edges of the two leveling screws, parallel to the bubbles, and the fore fingers of each hand on the outer edge. Turn the leveling screws so that both thumbs move inwards or both outwards. In the former case the bubble will move toward the right, in the latter case toward the left.

Now turn the instrument 180° in azimuth. If the small levels still have their bubbles in the center of their tubes, these levels are adjusted, and the circles are respectively as nearly horizontal and vertical as the maker intended them to be.

If the bubbles, however, are not in the center of their tubes, then bring them half way back by means of the leveling screws, and the remaining half by means of the adjusting screw at the end of each of the level tubes.

It may be necessary to repeat this adjustment several times, but when made, the instrument once leveled will have its small levels in the center of their tubes through an entire rotation of the circle.

There is one adjustment common to all telescopes used in surveying instruments, that of bringing the cross hairs to a sharp focus, at the same time with the object under examination, the adjustment *for Parallax.*

Point the telescope to the sky, and turn the eye-piece until the cross hairs are sharp and distinct. Since the eye itself may have slightly accommodated itself to to the eye-piece, test the adjustment by looking with the unaided eye at some distant point, and while still looking, bring the eye-piece of the telescope before the eye. If the cross hairs are sharp at the *first glance*, the adjustment is made. Now focus in the usual manner upon any object, bringing the cross hairs and image to a sharp focus by the rack-work alone. A point should remain bi-sected when the eye is moved from one side of the eye-piece to the other.

To make the vertical cross-line perpendicular to the plane of the horizontal axis, simply bi-sect some point in the center of the field of view of the telescope, and note whether it continues bi-sected by this cross-line throughout its entire length when the telescope is moved in altitude. If it does not, and the point is to the right of the line in the upper part of the field, the adjustment is made by loosening the four capstan-headed screws, and rotating the reticule in the direction of a left-handed screw, until the cross-line is moved over half the distance between the point and the line. Again, bi-sect the point by means of one of the tangent screws. It should now remain bisected throughout the length of the cross-line.

The following method of adjusting the horizontal line in a dumpy level has its advantages, and it is given in the words of its propounder, Mr. Gravatt.

"On a tolerably level piece of ground, drive in three stakes at intervals of about four or five chains, calling the first stake *a*, the second *b* and the third *c*. Place the instrument half way between the stakes *a* and *b* and read the staff (leveling rod) A, placed on the stake *a*, and also the staff B, placed on the stake *b*; call the two readings A' and B'; then, although the instrument be out of adjustment, yet the points read off will be equidistant from the earth's center, and consequently level. (Supposing the instrument to have its vertical axis vertical.) Now remove the instrument to a point half way between *b* and *c*. Again read off the staff B, and read also a staff placed on the stake C, which call staff C (the one before called A, being removed to that situation.) Now by adding the difference of the readings on B (with its proper sign) to the reading on C, we get three points say A′, B′, C′, equidistant from the earth's center, or truly level. Place the instrument at any short distance, say half a chain beyond A, and using the bubble to merely to see that you do not disturb the instrument, get a reading from each of the stakes a, b, c, call these three readings A″, B″, C″. Now, if the stake b be half way between a and c, then ought

$$C'' - C' - (A'' - A') = 2 \left[B'' - B' - (A'' - A') \right]$$

but if not, alter the screws which adjust the diaphragm, and consequently the horizontal line, until such be the case, then the instrument will be adjusted for collimation."[*]

"To adjust the spirit-bubble, without removing the instrument, read the staff, A, say it reads A‴, then adding (A‴ — A′) with its proper sign to B′ we get a value, say B‴.

"Adjust the instrument by means of the parallel plate-screws (in a transit use the telescope's tangent-screw), to read B‴ on the staff B.

"Now, by the screws attached to the bubble-tube, bring the bubble into the centre of its run.

"The instrument will now be in complete practical adjustment, for level, curvature, and horizontal refraction, for any distance not exceeding ten chains, the maximum error being only $\frac{1}{1000}$th of a foot."

EXAMPLE. — The instrument being placed half way between two stakes a and b (at one chain from each), the staff on a or A′ read 6.53, and staff on b or B′ read 3.34, placing the instrument half way between the stakes b and c, (three chains from each) the staff on b read 4.01, and the staff on c read 5.31.

Hence, taking stake a as the datum, we have

Stake.	Above Datum.	Stake.	Above Datum.	Stake.	Above Datum.
a or A′	= 0.00	b or B′	= 3.19	c or C′	= 1.89

This instrument being now placed at d (say five feet from a, but the closer the better,) the staff on a or A″ read 4.01, on b or B″, 1.03, and on c or C″, 3.07. Now had the instrument been in complete adjustment (under which term curvature and refraction are included), when the reading on staff a was 4.01, the readings on b and c should have been respectively 0.82 and 2.12.

The instrument therefore points upwards, the error at b being 0.21, and the error at c, 0.95: now were the bubble only in error, (as is supposed in all other methods of adjustment) the error at c ought to be four times as great as at b, but $4 \times 0.21 = 0.84$ only, there is an error therefore of $0.95 - 0.84 = 0.11$ not due to the bubble.

For the purpose of correcting this error (and be it remembered contrary to former practice, for this purpose only), we must use the capstan-headed screws at the eye-end of the telescope, and neglecting the actual error of level we are only to make the error at b one-fourth that of c.

After a few trials, whilst the reading at a continued 4.01, the reading on b became 0.75, and that on c, 1.84. Now $0.82 - 0.75 = 0.07$, and $2.12 - 1.84 = 0.28$.

And as $4 \times 0.07 = 0.28$, the telescope is now adjusted for collimation.

All that remains to be done, is to raise the object-end of the telescope by means of the parallel plate-screws (or the telescope's tangent-screw), until the staff at c reads 2.12, and then by means of the nuts which adjust the bubble-tube to bring the bubble into the centre of its run.

It is the vertical wire, however, which in the transit is the most important. When that is to be alone adjusted in the field, it is usually done according to the following simple directions : Select two distant points in opposite directions from the instrument, such that the vertical cross-line will bisect them both when the telescope is pointed upon one, and then the telescope is reversed around its horizontal axis After bisecting the second point selected, revolve the instrument 180° in azimuth, and bisect the first point again by means of the tangent screw. Reverse the telescope around its horizontal axis again, and if the second point is now bisected the adjustment for collimation of the vertical wire is correct. If it is not bisected, move the vertical wire one-fourth of the distance between its present position and the point previously bisected. Again bisect the first point selected, reverse the telescope and find a new point precisely in the new line of sight of the telescope ; these two points will now remain bisected when the instrument is pointed upon them in the manner described above, if the adjustment is correctly made. If the two points are not now both bisected, the adjustment must be repeated until this be the case.

Perhaps the most elegant method of adjusting for collimation, and one which recommends itself because it is best performed by lamp light, is the following : set up the transit and level with their object glasses toward each other, and they need not be but a few inches apart. Cover the eye-piece of the level with a piece of white paper, and illuminate this paper with a lamp. By a slight motion of the two telescopes, and use of the transit focussing screw, the cross-lines of the level will be seen sharply defined against the white background of the illuminated paper over the eye-piece. Bring the transit cross-lines so that their intersection is precisely over the intersection of the cross-lines in the level, which had better be turned in its wyes so that its cross-lines make an $\times$ with the horizon. If now either instrument is not collimated when the focus of the level is altered by its focussing screw, half an inch or more, and the transit is again sharply focussed on its cross-lines, the intersection of the transit cross-lines will no longer exactly cover the intersection of the level cross-lines. If the level has been carefully adjusted by the methods given further on, then the displacement is wholly due to error of collimation in the transit. This must be corrected by the adjusting screws. When the

---

[*] To adjust the line of collimation in a telescope *showing objects erect*, the diaphragm bearing the wires must be moved in the direction in which the error is observed (as if to increase the error); in telescopes *showing objects inverted*, the wires must be moved in the direction lessening the error observed. — To move the vertcal wire, slightly loosen the respective capstan-headed screws at the side of the telescope, and draw to a corresponding degree the one on the opposite side. To move the horizontal wire, make use only of the vertical capstan-headed screws on the telescope in the manner described. — C. L. B.

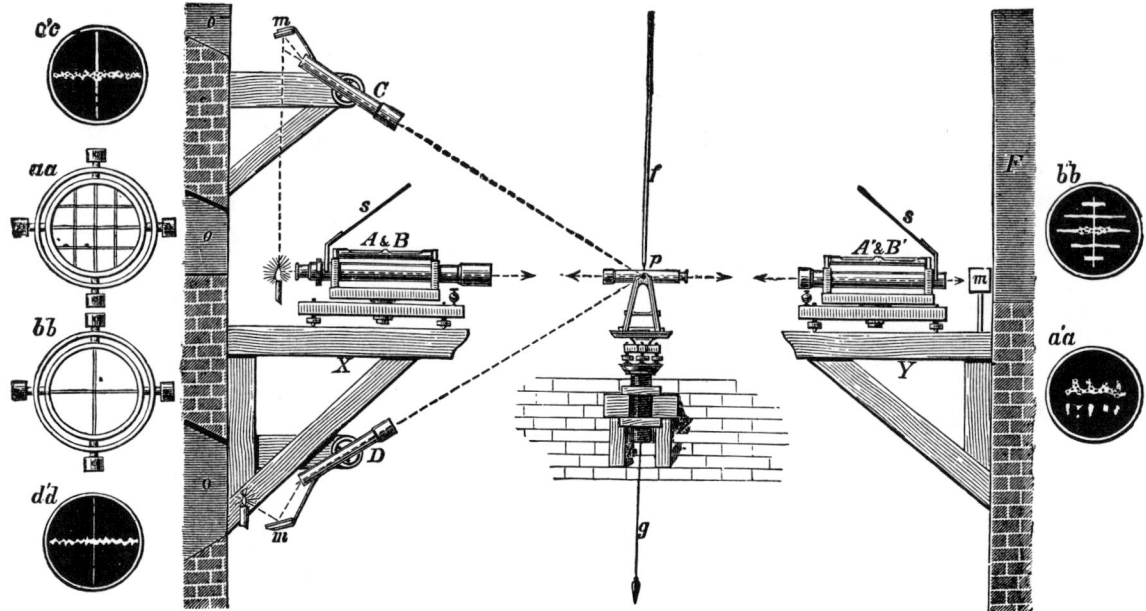

The Collimator Apparatus.*

For adjusting the Line of Collimation, etc., in the Telescopes of Geodetic Instruments.
In use since 1872.

NOTE. — A collimator is an optical instrument whose function is to furnish a pencil of rays of light the axis of which is in a given direction. The collimator is commonly arranged so that the rays shall be parallel, as if emanating from an object at an infinite distance; or it may be arranged so that the rays shall be divergent, as if emanating from an object at a given finite distance. These changes are made by moving the object-slide of the collimator to previously determined positions, as in focusing a telescope.

The pencil of rays emanating from the collimator is received by the telescope of the instrument to be adjusted, precisely as if it had emanated from a fixed star, or from an object at a given finite distance, and the telescope must be manipulated precisely as if it were pointing at natural objects. This renders the adjustment of the telescope, its wires, level, etc., very simple. Even where the necessary long sights are convenient and the state of the atmosphere is favorable, the use of a collimator is to be recommended on account of the greater ease, sharpness and accuracy with which the collimator objects can be seen as compared with ordinary objects.

C. L. Berger & Sons' Auxiliary Apparatus.
Used during the construction of their Instruments of Precision.

* For a more detailed description, see article in *Zeitschrift für Instrumentenkunde*.

adjustment is made, the two sets of cross-lines should coincide throughout the entire focussing motion of both telescopes. It must be borne in mind, however, that this method is only applicable to instruments like Buff and Berger's, in which the method of construction precludes the idea of sensible change in the telescope axis, when the focussing screw is turned from one end of its rack-work to the other.

It should be remarked here that whenever the engineer has a level and transit with him at the same time, j is a great convenience to use either one for adjusting the other. It requires a little patience at the outset to point one into the other, so that the cross-lines in both instruments may be black and sharp against a dead-white background ; but once accustomed to using these methods, the cross-lines are so sharp, and their motions so easily controlled by the screws of the instruments, the engineer usually prefers them.

To determine whether the standards are of the same height, suspend a plumb-bob by means of a long cord from a height say of from thirty to forty feet. The plumb-bob may swing in a bucket of water to keep it steady. Level the instrument carefully, and point upon the plumb-line at its base. If the plumb-line remains bi-sected throughout its entire length when the telescope is moved in altitude, and then the telescope reversed and again made to bi-sect the line throughout its length from its base upward, the adjustment is correct. Otherwise make the adjustment by means of the capstan-headed screw directly under one of the telescope wyes, loosening or tightening the small screws in the pivot-cap at the same time.

To adjust the telescope level in the field, set up the transit in the middle of a tolerably level piece of ground, and carefully level it. At equal distances, in opposite directions from the transit, drive two stakes, so that the readings of a level rod held successively on each of them will be the same when the telescope level bubble is brought to the center of its tube by the vertical tangent screw in each case, and the instrument is turned in azimuth. Take up the instrument and re-set it over one of the stakes; measure the vertical distance from the center of the horizontal axis of the telescope to the top of the stake over which the instrument is set. Set the target of the rod to read this distance, Hold the rod on the distant stake, and bi-sect the target with the horizontal cross-line. With the target thus bi-sected, turn the cylindrical nuts at the object-glass end of the level, till the bubble plays in the middle of its tube. Test the adjustment by re-setting half way between the two stakes, and noting that the bubble remains in the same position, and the rod gives the same reading when the instrument is turned in azimuth alone upon the two stakes. Sometimes it is convenient to use a sheet of water for the same purpose. Two stakes are driven into the water bed at different distances from the transit, until their tops are even with the surface of the water. The transit is leveled up near one of them, and its telescope altered in altitude until a rod held on each successively gives the same reading. Then with the telescope clamped in this position, the adjusting nuts are altered as before until the bubble plays in the middle of its tube. The methods of this paragraph assume the horizontal wire to be adjusted for collimation.

This adjustment also permits of being made by an auxiliary level in the office. Set up the transit and a level, as described in the adjustment for collimation, and after both instruments are in collimation : take the precaution to set up the instruments so that when the telescopes are approximately level they will point into each other, and the cross-lines may be made to coincide by means of the leveling screws of the transit instrument, after the level has been carefully leveled. Now make the transit bisect the intersection of the level cross-lines, and bring the bubble of the telescope level into the middle of its tube by means of the capstan-headed screws, it is obvious the telescope and level axis are both truly level.

It now remains to adjust the vernier of the vertical arc to read zero when the telescope is level, to complete the adjustments of the transit. Bring the telescope level bubble in the middle of its tube, and with the bubble in this position, set the zero of the vernier to coincide with the zero of the vertical arc; loosening the capstan-head screws, which secure the vernier to the standard, in so doing.

NOTE.—If the vernier for the vertical arc is single, made to read both ways, in reading it proceed to the right or left on the upper line of figures in the direction of the graduation used, and if the coincident line of the vernier is beyond the 15′ line, continue on the lower line of figures on the other half of the vernier, so that the whole graduation from 0′ to 30′ lies in the same direction. Messrs. C. L. Berger & Sons now make double verniers for their vertical arcs, similar to those described on pp. 34 and 35.

The Wye Level.

The description of the telescope of the engineer's transit applies with the following modifications to the telescope of this level.

It has a clear aperture of 1⅜ inches focus, and is 17 or 18 inches long over all, the sun-shade excluded.

The bell-metal collars which rest in the wyes are about 10½ inches apart and 1¾ inches in diameter.

On account of the extreme length of the telescope tube. four capstan-headed screws are provided for centering the eye-piece.

The object-glass focussing screw is in the middle of the tube. The eye-piece is focussed by turning a milled ring at the eye-end. The level attached to the telescope is about 8 inches long, with about 5½ inches exposed, over which is placed the metal scale for reading the position of its bubble. The level-tube is suspended from the telescope-tube in such a manner that at the object-glass end it can be moved in azimuth, with reference to the telescope axis, and at the eye-piece end it can be moved in altitude with reference to the same axis.

Its graduated scale has its graduations set carefully opposite each other on its two sides, and they are numbered from 5 to 0 to 5 at each end of the bubble.

Since it is not necessary to construct a level which shall have absolutely the same value in arc for the same motion of its bubble throughout its length for engineering purposes, the graduated scale is so set that the slight deviations from the arc of a circle may be equally distributed on each side of the zero of the scale. The bubble tube is ground cylindrical.

The level-bar is about 12 inches long over all, and at its two extremities supports the two wyes which rise about $3\frac{1}{2}$ inches from its upper surface. One of these wyes is adjustable in altitude. The level-bar is attached to a long conical center of the hardest bell-metal, which may be clamped to the upper level plate, and then a slow motion in azimuth may be given to the telescope, by a slow motion screw which presses the clamping bar against a stiff spiral spring. With the sunshade on the telescope, the weight is equally distributed from the center, each way. This is necessary, since a sensitive level, in the nicest work, is affected by any unequal strain, though it may seem to be, practically, imperceptible.

The base, on which the leveling screws rest, has as great a diameter as portability will permit; and the leveling screws are cut with a fine thread. These two points add to the ease with which the instrument may be accurately leveled.

A stop is so arranged that the telescope may be readily set with its horizontal cross-line level, when the instrument is in adjustment.

The instrument complete is not separable when put into its box. Messrs. C. L. Berger & Sons, believe this condition to be necessary to protect one of the essential adjustments of the level—the adjustment of the wyes—from needless derangement.

This instrument is sometimes made by Messrs. C. L. Berger & Sons in a different form. One of the wyes is movable in a vertical line by a milled-head screw. This enables pointing to be made with greater accuracy and facility.

The Adjustments.

In a theoretically perfect level the following points are established:
1. The object and eye-glasses are perpendicular to the optical axis at all distances apart.
2. The optical axis coincides with the axis of rotation in the wyes.
3. The axis of collimation coincides with the optical axis.
4. The axis of collimation is parallel to the telescope level.
5. The collars resting in the wyes are circles of the same diameter and concentric with the line of collimation of the telescope.
6. The wyes are exactly similar, and similarly placed with reference to the line of collimation of the telescope.
7. The level bubble moves over equal spaces for equal displacements of the telescope in altitude.
8. The level bubble expands or contracts equally from the center in both directions, during changes of temperature.
9. The vertical axis of revolution is perpendicular to the line of collimation of the telescope.

Of the above, the maker establishes points numbered 1, 2, 5, 7 and 8. The remaining points, 3, 4 and 9, are established when the instrument leaves the shop, but being liable to derangement from rough usage, they are made adjustable in the field.

Adjusting.

After the engineer has set up the instrument and adjusted the eye-piece for parallax, as described under the engineer's transit, the horizontal cross-line had better be made to lie in the plane of the azimuthal rotation of the instrument. This may be accomplished by rotating the reticule, after loosening the capstan-headed screws, until a point remains bi-sected throughout the length of the line when the telescope is moved in azimuth. In making this adjustment, the level tube is to be kept directly beneath the telescope-tube. When made, the small set screw attached to one of the wyes may be set so that by simply bringing the projecting pin from the telescope against it, the cross-lines will be respectively parallel and perpendicular to the motion of the telescope in azimuth.

The first collimating of the instrument may be made using an edge of some building, or any profile which is vertical. Make the vertical cross-line tangent to any such profile, and then turn the telescope half-way round in its wyes. If the vertical cross-line is still tangent to the edge selected, the vertical cross-line is collimated.

Select some horizontal line, and cause the horizontal cross-line to be brought tangent to it. Again rotate the telescope half-way round in its wyes, and if the horizontal cross-line is still tangent to the edge selected, the horizontal cross-line is collimated.

*Having adjusted the two wires separately in this manner, select some well defined point which the cross-lines are made to bi-sect. Now rotate the telescope half way round in its wyes. If the point is still bi-sected, the telescope is collimated. A very excellent mark to use is the intersection of the cross-lines of a transit instrument.

Center the eye-piece by the four capstan-headed screws nearest the eye end. This is done by moving the opposite screws in the same direction until a distant object under observation is without the appearance of a rise or fall throughout an entire rotation of the telescope in its wyes. The telescope is now adjusted.

Next, bring the level bar over two of the leveling screws, focus the telescope upon some object about 300 feet distant, and put on the sun-shade. These precautions are necessary to a nice atjustment of the level tube. Throw open the two arms which hold the telescope down in its wyes, and carefully level the instrument over the two level screws parallel to the telescope. Lift the telescope out of its wyes, turn it end for end and carefully replace it. If the level tube is adjusted, the level will indicate the same reading as before. If it does not, correct half the deviation by the two leveling screws and the remainder by moving the level tube vertically by means of the two cylinder nuts which secure the level tube to the telescope tube at its eye-piece end. Loosen the upper nut with an adjusting pin, and then raise or lower the lower nut as the case requires, and finally clamp that end of the level tube by bringing home the upper nut. This adjustment may require several repetitions before it is perfect.

The level is now to be adjusted so that its axis may be parallel to the axis of the telescope. Rotate the telescope about 20° in its wyes, and note whether the level bubble has the same reading as when the bubble was *under* the telescope. If it has, this adjustment is made. If it has not the same reading, move the end of the level tube nearest the object-glass in a horizontal direction, when the telescope is in its proper position, by means of the two small capstan-headed screws which secure that end of the level to the telescope tube. If the level bubble goes to the object-glass end when that end is to the engineer's right hand, upon rotating the teleseope level toward him, then these screws are to be turned in the direction of a left-handed screw, as the engineer sees them, and *vice versa*. Having completed this adjustment, the level bar itself must now be made parallel to the axis of the level.

To do this, level the instrument carefully over two of its leveling screws, the other two being set as nearly level as may be; turn the instrument 180° in azimuth, and if the level indicates the same inclination, the level bar is adjusted. If the level bubble indicates a change of inclination of the telescope in turning 180°, correct half the amount of the change by the two level screws, and the remainder by the two capstan-headed nuts at the end of the level bar, which is to the engineer's left hand when he can read the firm's name. Turn both nuts in the same direction, an equal part of a revolution, starting that nut first which is in the direction of the desired movement of the level bar. Many engineers consider this adjustment of little importance, preferring to bring the level bubble in the middle of its tube at each sight by means of the levelling screws alone, rather than to give any consideration to this adjustment, should it require to be made.

* See Note p. 49.

The Dumpy Level.

The dumpy level differs from the wye level in being attached to the level bar by immovable upright pieces; in having the level tube firmly secured to the uprights of the level bar, in being provided with an inverting eye-piece (unless ordered otherwise), and in the absence of the tangent and slow-motion screws. In regard to the level itself, and the optical power of its telescope, it is fully the equal of the more elaborate wye level.

Compactness is the object aimed at with the dumpy level, and this must be secured at the sacrifice of the parts of the wye level which may be considered more in the light of conveniences than necessaries.

Adjusting.

A theoretically perfect dumpy level has the same points established that are mentioned under the head of wye level; but since its construction differs from the wye level, the methods of adjustment are not so convenient, resembling closely the adjustment of the transit telescope and its attached level. After adding the sunshade and setting up as nearly as level as may be, and setting the eye-piece so to be rid of parallax, the two cross-lines should be set one at right angles to line the telescope axis describes in its horizontal revolution, and the other cross-line parallel to such a line. This is accomplished by loosening the four capstan-headed screws near the eye-piece, and rotating the reticule until a point remains bi-sected when the telescope is moved in azimuth.

To adjust the level, bring the level over two of its foot screws, and bring the bubble to the middle of its tube by means of the foot screws alone. Revolve the instrument 180° in azimuth, and if the bubble remains in the middle it is adjusted, if it does not, then correct half its deviation by the capstan-headed adjusting screw at the eye end, and the remaining half by the two foot screws. Repeat the operation over the other two screws, until the instrument may be revolved in any position, and the level bubble will remain in the middle of its tube.

To adjust the telescope for collimation, any of the methods given for the horizontal cross-line of the transit telescope (see page 48) will apply to the dumpy level. The usual method is to use a sheet of water, or where that is not available, two stakes which are driven with their surfaces in the same level plane.

To make the adjustment with the stakes, set up the level half way between two points lying very nearly in a horizontal line, and say 300 feet apart. Point upon a rod held at one of them, and bring the level to the middle of its tube. Drive a stake at this point, and take the reading of the rod upon it. Point the telescope in the opposite direction, again bring the level to the middle of its tube, and drive a second stake at the second point selected until the rod held upon the second stake gives the same reading as when held upon the first stake. The tops of these two stakes now lie in the same level line.

Take up the level and set it within a few feet of the first stake. Read the rod upon the first stake, and then upon the second. If the two readings agree, and the level is in the middle of its tube, the collimation is correct. If the two readings do not agree, correct nearly the whole of the disagreement shown when the rod is held on the distant point, by means of the upper and lower capstan-headed screws near the eye end of the telescope, and repeat the operation until both rods read the same with the level in the middle of its tube.

The telescope and uprights are in a single casting, which is finished and fitted to the level bar, so that the line of collimation may be permanently parallel to it.

The dumpy level will then be in adjustment, since the adjustment of its vertical cross-line is of no importance. ———

Adjustment of the Dumpy Level[1] — and attached level of Transit Telescope.

Two-Peg Method.

The following method is simple, direct, and geometrically accurate, requiring no approximate measurement from a peg to the centre of lens, no trial setting of the telescope, no trials to drive a peg just enough and not too far, and no auxiliaries except level-rod and tape or chain.

[1] Contributed by Prof. R. Fletcher, Thayer School, Dartmouth College.

On slightly rising ground locate four points, a, b, c and d, on the same line, nearly, making $bc = ca$, and ad any convenient distance, preferably not much less than ca, and in some simple ratio with it, for ease of calculation afterwards. Set the instrument at c; take readings A′ and B′ on a and b respectively, carefully leveling before each sight. Then, unless the instrument is otherwise much out of adjustment, (B′—A′) is the true difference of level of a and b.

Next set up at d, level carefully, and take readings A″ and B‴ on a and b respectively. [In strictness the centre of the instrument should not be set over d, but beyond, by an additional distance = principal focal length of the object-lens + the distance from that lens to the centre of the tripod. (See the Manual, page 87, Fig. 2.)] Then (B″ — A″) — (B′ — A′) = B‴ = error of collimation in the distance ba, that is the error due to the vertical angle between the line of sight and axis of spirit-level. Now, by similar triangles, we have

$$B''' : ba = B^{iv} : bd \therefore B^{iv} = \frac{B''' \times bd}{ba},$$

which is the error in the distance bd, and is to be applied to the reading B″. Set the rod to read (B″ — B^{iv}). Then:

For Adjustment of a Dumpy Level.

Having first adjusted the spirit-level so that it remains true in all positions about the vertical axis, point the telescope on the rod, properly held at b, with target set to read (B″ — B^{iv}). By means of the capstan-headed screws, raise or lower the horizontal line until it bisects the target. To test the adjustment, set the rod over a, with index reading (B″ — B^{iv}) — (B′ — A′), and see if the target is still bisected.

Adjustment of Attached Level of Transit Telescope.

The rod being held plumb at b, with target set to read (B″ — B^{iv}), move the telescope by *vertical tangent-screw* until the line of sight bisects the target; clamp securely. Then bring the bubble to the middle of the tube by means of the *level-adjusting nuts alone*. Test as in the other case.

REMARKS. — The diagram shows a special case, viz., when (B″ — A″) > (B′ — A′), or the angle subtended by B^{vi} is one of elevation. If (B″ — A″) = (B′ — A′) the line of sight is already level and no adjustment is needed. If (B″ — A″) < (B′ — A′), B^{iv} subtends an angle of depression, and is to be added to B″. In the latter case, if the slope of the ground is slight, the difference (B″ — A″) may be either zero or a very small quantity, positive or negative; but in all cases it is added algebraically to (B′— A′) to obtain B‴.

As in all other methods of adjustment, we assume that the maker has done his part so well that the line of collimation will not be disturbed in any movement of the objective for focusing. Let us suppose that the line of collimation is made truly horizontal, and that in its prolongation we have set the centres of two targets, one over a and one over b, the instrument being at d. If now we focus upon the farther target, the image will be bisected by the horizontal spider-line. Then change the focus so as to view the nearer target. If the centre of the objective has not moved truly in the line of collimation, the new image will not be bisected at the focus, and the nearer target will appear to be out of level, when in fact it is not. Hence, since this adjustment requires change of focus, it cannot be made if the above defect, in the movement of the objective, exists. If, however, the distance ad be not too small and the defect alluded to be only slight, the error in changing focus for b and a may be hardly appreciable. The *adjustment* once made approximately, we need not remark that, in the field work, any further error of objective is avoided when taking equidistant sights.

The Plane Table.

A description of this instrument, as modified in plan by H. L. Whiting, Esq., assistant U. S. Coast Survey, and constructed by C. L. Berger & Sons, may be found in the Coast Survey Report for 1865.

The following description of its adjustments, by A. M. Harrison, Esq., assistant U. S. C. S., is taken entire from the same paper:

"Topography is that branch of surveying by which any portion of the land surface of the earth is mapped in plan on a specified scale or proportion of nature.

With the plane-table such a map is constructed on the ground by at once drawing upon the paper, which is spread upon the table, the angles subtended by different objects, and determining by intersections their relative positions, instead of reading off the angles on graduated instruments and afterwards plotting the lines by means of a protractor, as is done in other methods of surveying. The practice with the plane-table has in this respect a great advantage in directness and precision. The measurement of distances and of vertical angles are used, in conjunction with the method of intersections, to obtain all the data for representing the horizontal and vertical features on the map, which is drawn in the field with pencil, the details being filled in according to established conventional signs.

"**Adjustments.**—From the nature of the service in some sections of the country the plane-table is often necessarily subjected to rough usage, and there is a constant liability to a disturbance of the adjustments; still, in careful hands, a well made instrument may be used under very unfavorable conditions for a long time without being perceptibly affected. One should not fail, however, to make occasional examinations, and while at work, if any difficulty be encountered which cannot otherwise be accounted for, it should lead directly to a scrutiny of the adjustments.

"1. *The fiducial edge of the rule.*—This should be a true, straight edge. Place the rule upon a smooth surface and draw a line along the edge, marking also the lines at the ends of the rule. Reverse the rule, and place the opposite ends upon the marked points, and again draw the line. If the two lines coincide, no adjustment is necessary; if not, the edge must be made true.

"There is one deviation from a straight line, which, by a very rare possibility, the edge of the ruler might assume, and yet not be shown by the above test; it is when a part is convex, and a part similarly situated at the other end concave, in exactly the same degree and proportion. In this case, on reversal, a line drawn along the edge of the rule would be coincident with the other, though not a true right line; this can be tested by an exact straight edge.

"2. *The level attached to the rule.*—Place the instrument in the middle of the table and bring the bubble to the center by means of the leveling screws of the table; draw lines along the edge and ends of the rule upon the board to show its exact position, then reverse 180°. If the bubble remain central, it is in adjustment; if not correct it one-half by means of the leveling screws of the table, and the other half by the adjusting screws attached to the level. This should be repeated until the bubble keeps its central position, whichever way the rule may be placed upon the table. This presupposes the plane of the board to be true. If two levels are on the rule, they are examined and adjusted in a like manner.

"Great care should be exercised in manipulation, lest the table be disturbed.

"3.—*Parallax.*—Move the eye-glass until the cross-hairs are perfectly distinct, and then direct the telescope to some distant well defined object. If the contact remain perfect when the position of the eye is changed in any way, there is no parallax; but if it does not, then the focus of the object-glass must be changed until there is no displacement of the contact. When this is the case, the cross-hairs are in the common focus of the object and eye-glasses. It may occur that the true focus of the cross-hairs is not obtained at first, in which case a readjustment is necessary, in order to see both them and the object with equal distinctness and without parallax.

The Striding Level, its Use and Adjustments.

In transits reading to minutes and half-minutes, the plate-level in front of the telescope is generally sufficiently sensitive to insure good work. However, an instrument of the class as shown and described under No. 1 d, should always be provided with a striding level, to insure a degree of accuracy in keeping with its greater capability. The sensitiveness of this striding level is equal to that of the long level on the telescope. Thus it will be seen that in a transit of this description the plate-levels serve only the purpose of leveling up generally, and that in all cases where the objects vary considerably in height, the striding level only should be depended on at every sight. The striding level of this instrument rests on two cylinders of equal diameters, at points between the standards on the cross-axis of the telescope; and, unlike the method described on page 30, serves also the purpose of adjusting the telescope to revolve in a vertical plane. As shown in the cut, the striding level can be left on the cross-axis when the telescope is revolved in altitude. — To verify the adjustment of the striding level (in other words, to make its axis parallel to the cross-axis) level up the transit and bring the bubble to the middle of its tube, reverse the striding level on the cylinders and see whether it reads the same; if not, remove half the error by the leveling screws, the other half by the capstan-headed screws at the end, and repeat until corrected. To verify the side adjustment of the level, revolve the telescope 20 or 30°, and note whether the reading of the bubble remains the same, if not, correct the error by the capstan-headed screws at the side. To verify the adjustment of the cross-axis of the telescope for right angles to the vertical axis of the transit, revolve the instrument 180° in azimuth, and assuming that both cylinders, on which the striding level rests, are equal in diameter, a change in the reading of the bubble will indicate double the amount of error To correct it, remove half the error by the leveling screws, the other half by the Wye adjustment of the standard. — Remember, that the pressing of the capstan-headed screws against the level tube acts similar to that of the Wye adjustment of the standard. — C. L. B.

For more information on this subject, see pages 98 and following, of this handbook.

C. L. Berger & Sons' Solar Attachment.

Written for this catalogue by H. C. Pearsons, C. E., Ferrysburg, Mich.

The "Solar Attachment," of which the following is a description, is a modification of *Pearson's Solar Transit*.

With the view of reducing the weight and cost of this attachment, the declination arc is dispensed with, using, in its stead, the latitude arc for setting off the declination.

And to attain a greater degree of precision, a small telescope with cross-hairs, and a diagonal eye-piece, have been introduced in place of the lens-bar and focal-plate.

This attachment is an appliance to the surveyor's transit, for the purpose of finding the astronomical meridian. Combined with that instrument, it becomes purely astronomical in its character — indeed, a portable Equatorial, and an Alt.-Azimuth instrument combined, — hence a few astronomical definitions seem to be requisite.

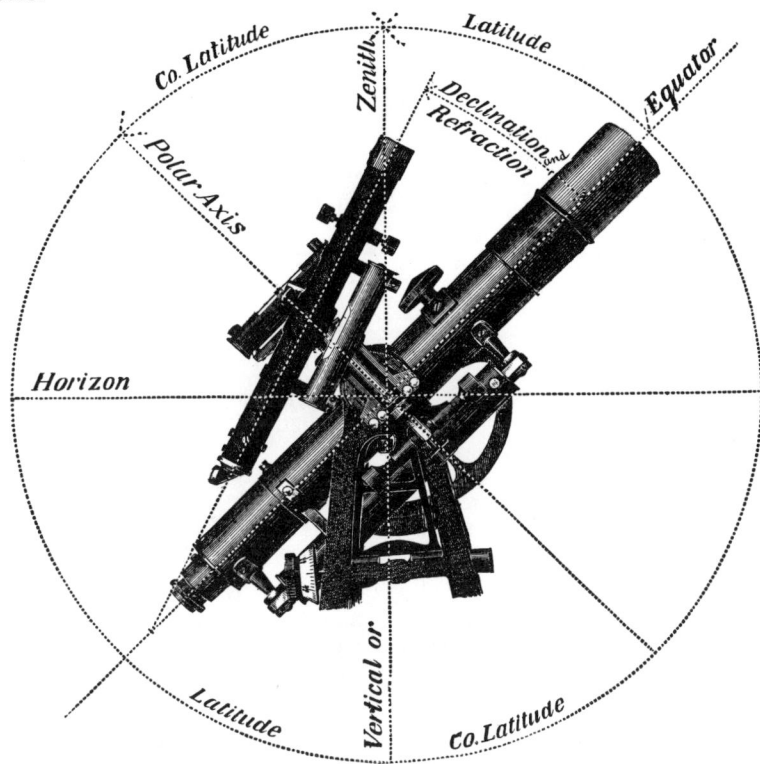

In the accompanying cut, the instrument is represented in position for an observation; and in north latitude (as in these instructions we will suppose the observer to be) the view is as from the west.

(1.) The line through the vertical axis of the transit represents the pole of horizon, and is called the *Vertical*.

The line perpendicular to this represents the *Horizon*.

(2.) The transit telescope, having its optical axis in the meridian, and having its south end (whether object-end or eye-end) elevated so that the vertical-arc reads the co-latitude, will have its optical axis in the plane of the equator also; viz. the optical axis of the telescope will then represent the intersection of the plane of the meridian, with that of the equator. This line is called the *Equator*.

(3.) The line perpendicular to the equator, — that around which the solar telescope revolves, in following the sun in his diurnal course, is the pole of the equator. — It is parallel with the earth's axis, and is called the *Polar Axis*.

(4.) The arc distance from the equator to the vertical is the *Latitude* of the observer, — whence the distance from the vertical to the polar axis, is the *Co-Latitude*.

It will be observed that these arcs occur alternately around the entire circle; so that the student should make himself familiar with their relative position with regard to the horizon, and the vertical, in order to avoid mistakes, when setting the polar axis of the instrument up to the pole of the equator.

(5.) *Astronomical Triangle.* The height of the sun is measured in a plane passing through the "Vertical" and the sun, and is called his *Altitude*, whence his distance from the "Vertical" is his *Co-Altitude*.

In the same manner, the distance from the sun to the "Pole," is his co-declination; and the distance from the "Vertical" to the pole, is the observer's *Co-Latitude*. These three compliments form what is called the *Astronomical Triangle*.

Thus we have the three sides of a spherical triangle, from which to find the several angles.

(6.) The angle at the Pole, contained between the meridian of the observer and that passing through the sun, is called the *Hour Angle*, as it gives the distance from the sun to the observer's meridian, in time or arc, and is usually represented by the letter H.

(7.) The angle at the "Vertical," or at the observer's zenith, contained between the meridian and a vertical plane passing through the sun, is called the *Azimuth Angle*, and is usually represented by the letter Z.

This angle is the one particularly important to surveyors, as from it the place of the meridian is readily determined.

Navigator's look for this angle every day, when an observation can be had, and solve the triangle for Z, by one or both of the following equations.

$$\cos \tfrac{1}{2} Z = \left(\frac{\cos S \;\; \cos (S - p)}{\cos L \;\; \cos h} \right)^{\frac{1}{2}} \quad . \quad . \quad . \quad . \quad . \quad . \quad \text{(a.)}$$

$$\sin \tfrac{1}{2} Z = \left(\frac{\sin (S - L) \;\; \sin (S - h)}{\cos L \; \cos h} \right)^{\frac{1}{2}} \quad . \quad . \quad . \quad . \quad . \quad \text{(b.)}$$

in which

L = Latitude.   Z = the required Azimuth
d = Declination.   p = Polar Distance = 90° — d.
h = Height of the sun's center, corrected for refraction and parallax.
S = $\tfrac{1}{2}$ (L + h + p).

NOTE. — The correction for parallax, which is usually about 6″, and never exceeds 9″, may be neglected except in work of great precision.

To solve these equations numerically requires much computation, but the *Solar Transit* solves them for Z, *mechanically*, with no more computation than that required to deduce the declination for the longitude and local time of the observer, from that given in the Nautical Almanac for the day.

From the above definitions, it is readily seen that the following conditions, or relation between the parts of the instrument, must be established.

(*A.*) The polar axis must be *Vertical*, when the vertical arc (latitude arc) reads zero, and, consequently, perpendicular to the cross axis of the transit telescope.

(*B.*) The horizontal cross-wire of the solar telescope must be parallel with the plane of its rotation around the polar axis; *i. e.* it must be parallel with the plane of the equator.

(*C.*) The plane passing through the vertical wire and the optical axis of the solar telescope must be at right angles to the cross axis of the solar telescope.

(*D.*) The bubble of the level-tube on the solar telescope must be in the middle of its tube, when the optical axis of that telescope is in the plane of the horizon.

These conditions are obtained by the following

Adjustments.

(*Aa.*) Having attached the solar apparatus to the cross-axis of the telescope, as per instructions under the head of "*Remarks,*" and having leveled up the transit (supposed to be in perfect adjustment) carefully, set the vertical or latitude arc to zero, observing that, upon rotating the whole instrument 180° in azimuth, the bubble of the level of the transit telescope is in the middle of its tube. Clamp the polar axis firmly to its collar by means of the clamp screw at the lower end of the polar axis in a position approximately vertical. Then, by means of the opposing tangent screws, also at the lower end of the polar axis, bring the bubble of its level to the middle of its tube, and repeat if necessary.

Revolve the polar axis 180° to see if its level on top is in proper adjustment; if not, remove half the error by means of the opposing tangent screws at the lower end, and the other half by the capstan-headed screw of the level tube. (This condition must be attained, before the polar axis can be set to the latitude of the observer; and being attained, it needs no further attention than to be examined at times for verification.) Now turn the solar telescope 90°, so as to be parallel with the cross-axis of the main telescope, and observe if the bubble remains in the middle of its tube. If not, make the requisite correction by means of the capstan-headed screws near the lower end of the polar axis, some of which *draw*, while others *pull*, taking care not to strain them.

This last adjustment brings the polar axis into a vertical plane perpendicular to that of the first adjustment, and parallel to the vertical plane containing the optical axis of the main telescope. It is made by the manufacturer, and, thereafter, needs only to be examined at times. And if the above adjustments are properly made, *the bubble of the level on the polar axis, that of the level on the transit telescope, and those of the plate levels on the transit, will all be in the centre of their tubes, and the vertical arc will read zero.*

(*Bb.*) Bi-sect some distant object, and turn the solar telescope sufficiently to the right and left, around the polar axis, to make the image of the object traverse the field from one side of the tube to the other. The image should remain bisected by the wire. If not, loosen the four slot-headed screws of the diaphram, carrying the cross-hairs, and turn the diaphram, till the above condition is attained, and fasten the screws securely.

(*Cc.*) Direct the solar telescope to bi-sect a distant object, revolve the transit around its vertical axis exactly 180° in azimuth, and without disturbing the polar axis reverse the solar telescope on its cross-axis. If the vertical wire again bi-sects the same object, the adjustment is made; if not, move the vertical wire one-fourth of the error by the slot-headed screws at the side of the telescope, and by means of the transit's lower tangent screw, again bi-sect the object first selected, and repeat as described above until entirely corrected. The solar telescope showing objects inverted requires the cross-wire diaphram to be moved as described in the foot-note on page 49 of manual. If for the above operation the polar axis revolves too freely in its socket, the operator may insert a thin piece of paper between it and its socket, which afterwards must be removed.

(*Dd.*) Bi-sect any distant object by the main telescope. Then, by means of the clamp and opposing tangent screws on the solar telescope, bring its cross-wires to bi-sect the same object. Now level the main telescope by means of its clamp and tangent screw, then, by means of the capstan-headed screw, bring the bubble of the level on the solar telescope to the middle of its tube. This being done, the optical axes of the two telescopes will lie in parallel planes, and the instrument *is ready for use.*

All these adjustments are made by the manufacturer and need to be verified only occasionally.

Before the solar attachment is available for finding meridian, the observer must know his *Latitude*, and the sun's *Declination* for the day and hour of observation, corrected for refraction, whence the

Reduction of Declination and Refraction.

The sun's *Declination* is given for noon of every day in the year, in the Washington and Greenwich Ephemeris of the sun, for those meridians. The maps and charts in use will give the difference of *Longitude* to all the precision required, and tables in this manual give the required *Refraction*.

An example will best illustrate:

Required a declination table for the different hours of the day for April 25, 1885. Lat. 44° N., and Longitude 97° W. At 15° to the hour, 97° of longitude is about 6½ hours of time, and as this longitude is W., 12 o'clock, or noon, at Greenwich will correspond to 5½ A. M. at the place of the observer.

The declination, as given for that day, in the Greenwich Ephemeris, is *13° 20′ 04″ N.*, and is shown to be *gaining* at the rate of 49″ per hour (see column headed Difference for one Hour, with the signs + for sun going North, and — for sun going South).

If now, to the declination for 5½ A.M., we add the hourly rate of change successively, we shall have the declination for the several hours of the day, observing that the first increment is for only half an hour, thus : —

Form of Daily Declination Table.

APRIL 25, 1885.

Hourly difference Dec. + 49″				Hourly difference + 49″	
Dec. 5½ A. M. **N.** . . .	13° 20′ 04″ +		1 P. M. **N.** . . .	Dec. = 13° 26′ 11″ +
" 6 " . . .	13 20 28 +		2 " . . .	" = 13 27 00 +
" 7 " . . .	13 21 17 +		3 " . . .	" = 13 27 49 +
" 8 " . . .	13 22 06 +		4 " . . .	" = 13 28 38 +
" 9 " . . .	13 22 55 +		5 " . . .	" = 13 29 27 +
" 10 " . . .	13 23 44 +		6 " . . .	" = 13 30 16 +
" 11 " . . .	13 24 33 +		7 " . . .	" = 13 31 05 +
" M. " . . .	13 25 22 +			

The above table must be corrected for the effects of refraction, before it is set off on the vertical arc of the transit. Refraction increases the apparent altitude of an object, and thereby affects the declination of the object — $\left.\begin{array}{l}\text{increasing}\\\text{diminishing}\end{array}\right\}$ the declination when of the $\left.\begin{array}{l}\text{same}\\\text{different}\end{array}\right\}$ name with the latitude.

From the + sign of the "difference" of declination, we see that the declination is of the same name as the latitude, whence the correction is an *increment*, and accordingly the + sign as suffixed. This sign belongs to the refraction.

When the object is in the meridian, refraction affects declination by its full amount; but, if both the observer and the object were in the plane of the equator, refraction would have no effect on the object with regard to refraction; whence, between these limits, only a part of refraction is effective in changing the declination.

Just what portion is effective, is shown by table II. of this paper.

Thus, in the given Lat. 44°, and for, say 4 hours from noon, the position of refraction to be applied is .74 of that corresponding to the altitude of the object at the time of observation. The sign ∓ to be used must be determined, as above, by considering whether the sun is going north or south at the time.

This part of the reduction of declination cannot, of course, be made till the altitude is found at the time of observation.

To Find the Latitude.

Having prepared the declination for the day, as above, level up the transit carefully. Level the main telescope, observing that the vertical arc reads zero, and set the polar axis to a vertical position by means of its level.

These points being attained, set the main telescope, pointing south. Then for a $\left.\begin{array}{l}\text{north}\\\text{south}\end{array}\right\}$ declination, $\left.\begin{array}{l}\text{dip}\\\text{elevate}\end{array}\right\}$ the south end of the telescope, till the vertical arc indicates the declination thus found.

Then, having turned the solar telescope into a vertical plane parallel with that containing the optical axis of the main telescope, level it carefully and clamp it, and see that the friction spring, at the lower end of the polar axis, is sufficiently taut to hold it in position.

A few minutes before the time of the sun's culmination, bring the telescope into the vertical plane passing through the observer and the sun, and "*find the sun*" with the solar telescope. This is readily done by varying the altitude when the sun's image will appear on the diagonal eye-piece.

Having "found the sun," bisect his image with the vertical wire, by varying the azimuth with the tangent screw of the transit plate, or with that of the outer center; and, simultaneously, follow him in altitude — the horizontal wire bisecting the image — till it ceases to rise, then clamp and read the vertical arc. *This reading should be the sum of the co-latitude and refraction,* the refraction being that due to the meridian altitude of the sun, which is the algebraic sum of declination and co-latitude. From this reading the latitude is readily deduced. With the latitude und declination known, we are prepared

To Find the Meridian.

(a.) As for finding latitude, level up the instrument carefully, the vernier of plate clamped, reading zero.

(b.) Point the telescope to the sun to find his altitude for the refraction. This can be found with sufficient accuracy by turning the telescope, till the shadow of a pencil held across the end, or till the shadow of the screws on the side, are parallel with the tube.

(c.) The refraction corresponding to this altitude must be multiplied by the corresponding coefficient, for the time from noon and the latitude, and applied to the *declination*, as per instructions above, for the *corrected declination*.

(d.) Point the telescope to the south, $\begin{matrix} \text{dipping,} \\ \text{elevating} \end{matrix}$ the south end for $\begin{matrix} \text{north} \\ \text{south} \end{matrix}$ declination, till the vertical arc reads the corrected declination, and clamp the vertical arc.

(e.) The main telescope being dipped to the corrected declination, level the solar telescope by means of its level, being careful to do so when it is in a vertical plane parallel with that containing the optical axis of the main telescope, for only when it is in this plane can the declination be properly set off.

(f.) Elevate the south end of the main telescope to the *co-latitude*, by means of the vertical arc, and turn the telescope approximately into the meridian, by means of the magnetic needle.

(g.) "Find the sun" with the solar telescope. This is done by turning the whole instrument in azimuth, on its outer center, simultaneously with a motion of the solar telescope in right ascension, till the sun's image is seen in the eye-end of solar telescope. Bi-sect the image, as nearly as may be, by the two motions above named — clamp and complete the bi-section, by both wires, by means of the transit's lower tangent screw, and by tapping the solar telescope gently with a pencil or other light stick. If the image of the sun should be so large that it cannot all be seen from one position of the eye, look around it by moving the eye around it in such a manner as to see the entire circumference, and bring the cross-wires on the four sides of the image, *normal* to their respective sides, by means of the motion in azimuth, and the motion of the solar telescope, as above described. This being attained, *the optical axis of the main telescope should be in the astronomical meridian*. Refer to an azimuth mark, and repeat the operation. The above is called a *direct* observation.

(h.) To make a *reverse* observation. Having made the direct observation, turn the whole instrument 180° in azimuth, and set the co-declination off, on the opposite side of the vertical arc. Also turn the solar telescope 180°, and proceed as before. The object of repetition is to eliminate personal non-precision, and possible errors in manipulation, while the object of reversing is to eliminate any possible remaining errors of adjustment of the instrument. The prudent surveyor will not trust his work without such verification, and he will take the mean of both observations.

Remarks.

(1.) *To unscrew* the solar apparatus from the packing-piece, at the bottom of the box. First, release both, the clamp and the tangent screws, and then turn the milled nut, at the bottom of the box, a few times to the *left*. To *attach* the solar attachment to the instrument, turn this milled-headed nut, from left to right, around the screw on the instrument (the end of the cross-axis, marked No. 2, on right side of instrument) but *without revolving* the solar apparatus. To insure a perfect contact of the flange of the collar against the shoulder of cross-axis — involving, as it does, the adjustment of the polar axis in the direction of the cross-axis of the main telescope — it is necessary that the parts be kept clean of dust, grit, and dirt of any kind.

(2.) The auxiliary, or latitude level, attaches to the other end of the cross-axis (marked No. 1) on the side of the vertical arc, in the same manner.

(3.) The latitude level is used to facilitate the resetting of the instrument to the co-latitude — the telescope being adjusted to its position, with more facility and precision with the level, than by reading the vertical arc.

NOTE.—The great utility of this auxiliary, or level attachment, is seen in the *setting of grades*. Two of these levels being applied to the telescope of a pivot-leveling instrument — one on each side — or one on each end of the cross-axis of a transit telescope, and one of them being adjusted to the *up*, the other to the *down grade*, the engineer may work in either direction on his grade, with the same facility that he would on a level line.

(4.) In an emergency, the level on the solar telescope could also be dispensed with, by utilizing the level of the main telescope, in the same manner as the latitude arc is utilized for the declination arc. By bi-secting some distant object with both telescopes, their optical axis will be parallel. Then, when brought to a level, their level bubbles should both read zero. Then, the level on the solar telescope being kept level, and the main telescope dipped or elevated to the required declination, the difference of their vertical angles will be the required declination.

(5.) It will be observed, that in this "Solar" the polar axis is *perpendicular* to the optical axis of the main telescope, while in the Pearson's "Solar" these axes are *parallel*, whence the different arcs found by an observation for latitude, and the different angles to be set off on the latitude arc in finding the meridian.

(6.) The latitude having been found, for the initial point of a survey, it may be found for other points, within moderate limits, by allowing 92 chains of northing or southing for 1' of latitude.

(7.) The object of bringing the telescope into the meridian by means of the motion on the spindle, is to have the zero line of the horizontal plate in the meridian, so that the azimuth or bearing of lines can be referred directly to that line.

(8.) If, for any cause, we are obliged to work with an uncertain latitude, it is better to do so with the sun as far from the meridian as practicable, for the following reasons:

It is only when the sun is in the pole of the meridian, that it has its maximum efficiency in pointing out the direction of the meridian. Hence, a large hour-angle, and a small declination, are conducive to the elimination of errors resulting from an incorrect latitude.

Indeed, with the sun precisely in the pole of the meridian, meridian is determinate independantly of latitude.

(9.) In making the several adjustments, or rather in verifying them, the student should have a true meridian established by some other means than by the "solar transit,"—as from the North Star, by some of the methods given in works on surveying. He should compare the results of his observations with this meridian at different times in the day, and under different states of the atmosphere, till he has learned any peculiarity of the instrument and the utmost precision obtainable with it, as well as the ordinary limit of non-precision.

Degree of Precision Required.

(10.) This, of course, depends on the character of the work to be done. In the U. S. Public Land Surveys,—which are, without question, conducted on the best plan the world can afford,—*only compass lines are required.* As a consequence, a wide margin for non-precision is given.

In sub-dividing a block of townships, the surveyor in coursing a random of 6 miles, is required to make his objective point within 3 chains. Charging the half of this error to lineal measurement, we find the error of coursing *must be within 10' of the true course.*

(11.) *In Manitoba*, the authorities, having fallen in love with our system of Public Land Surveys, have adopted it; but they require greater precision. They require clear transit lines, projected with the best six-inch silver lined instruments, graduated to 10".

In coursing a 6 mile random in the sub-division of a township, the surveyor must make his objective point within *one chain*, in order to save reviewing his work, charging, as before, one half of this error to the lineal measurements, we find the maximum error allowed in coursing *to be between 3' and 4'.*

(12) *With the "New Solar,"* as manufactured by Messrs. C. L. Berger & Sons, the surveyor will be surprised and delighted to see the *facility* and *certainty* with which *he can bring his work far within the above limit.*

Inclination of the Meridian.

(13.) In projecting arcs of a great circle with the "solar transit," it is of the utmost importance that the surveyor be able to tell the inclination of the meridians for any latitude, and for any distance of eastings or westings.

As this problem is not treated in elementary works on surveying, perhaps the few following hints may be of use to the young student.

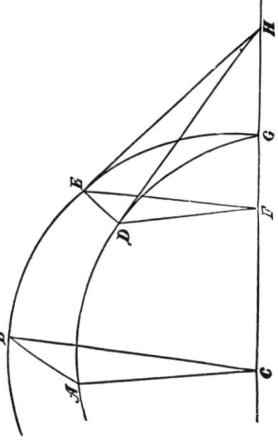

In the following figure, let the two arcs A G, and B G be two arcs of a quadrant of the meridian, 1° of longitude apart. Let A B = the arc of one degree of longitude on the equator = 69.16 miles.

Let D E be an arc of longitude on any parallel of latitude. Also, let E H and D H be the tangents of those meridians meeting in the earth's axis produced, and corresponding to the parallel of latitude D E.

Then the line E F = D F = cos L = cos A D or B E. Also, the angle D F E = 1°, and the angle D H E = the inclination of the meridians, which is the angle we wish to find, and which we will represent by X°. And because the two triangles F D E and D H E are on the same base E D, and isosceles, their vertical angles vary inversely as their sides; and we have the equation,

$$1° \times EF = X° \times EH, \qquad \text{But}$$
$$EF = \cos L, \text{ and } EH = \cot L, \text{ hence}$$
$$X° \cot L = 1° \cos L, \text{ or}$$
$$X° = \cos L \div \cot L = \sin L, \qquad \cdots \qquad \text{(a)}$$

That is to say,

The inclination of the meridians for any difference of longitude, varies as the sine of the latitude.

(14.) Since the sine of the latitude is the inclination in decimals of a degree, for one degree of longitude, if we multiply by 3600″ we shall have the inclination in seconds of arc. Then, if we divide this by the number of miles in one degree of longitude on that latitude, we shall have the inclination due to one mile on that parallel. Thus, for

Latitude 43° . . . . . . log. sine =	9.833783
Multiply by 3600″ . . . . . " =	3.556303
	3.390086
Divide by 50ᵐ66, = 1° long. on that L, log. =	1.704682
48″.46 = inclination for one mile of long.	1.685404

(15.) *The use of the Inclination*, as found by the preceding article, is to show the surveyor how much he must deflect a line of survey from the due east or west, to have it meet the parallel at a given distance from the initial point of the survey,— for it will be remembered that a parallel of latitude is a *curve*, having the cotangent of the latitude for its radius. And the line due east or west is the tangent of the curve.

Thus, on latitude 43°, I wish to project a six-mile line west, for the southerly line of a township.

Remembering that in an isosceles triangle, the angle at the base is less than a right angle by *half the angle at the vertex*, I deflect my line *towards the pole* by the inclination due to *three miles*,— or in this case 48″.46 × 3 = 2′.25″, i. e., *Deflection* = ½ *Inclination*.

(16.) *Table No. III*, which was computed from the formula (a) Art. 37, gives the *Inclination* for one mile, and for six miles on any parallel, from 10° to 60° of latitude; also the *Convergency* for six miles, on any latitude.

(17.) *The Convergency of the meridian* is readily found for any given distance from the corresponding inclination, by multiplying the *Sine* of the inclination by the given *distance*.

Thus, for latitude 43°, the inclination for one mile is 48".46; the sine of which is .000235. This, multiplied by the number of links in a mile, which = 8000, we have the convergency for one mile,= 1.88 links.

Multiplying this by the number of miles in a township,= 36, and we have the convergency for a township = 67.68 links. In this manner were the convergencies of table III computed.

(18.) *Deflection of Range-Lines from meridian.* The second column of table III shows the surveyor how much he must deflect the range lines between the several sections of a township from the meridian, in order to make the consecutive ranges of sections in a township of uniform width, for the purpose of throwing the effects of "convergency" into the most westerly range of quarter sections agreeably to law.

Thus, say between 45° and 55° of latitude, the inclination is practically 1' for every mile of easting or westing. Then, bearing in mind that in the U.S., the surveys are regarded as projected from the East and South to the West and North; the surveyor must project the *first range-line* between the sections of a township in those latitudes, 1' *to the left of the meridian.*

The second, 2'; the third, 3'; and so on to the fifth, which must be 5' to the left of the meridian on the east side of the township.

By this means all the convergency of the township is thrown into the *sixth*, or westerly range of sections, as the law directs.

The fourth column of the above table shows the amount of this convergency. This column is also useful in sub-dividing a block of territory embraced by two "standard parallels" and two "guide meridians" into townships. Thus, starting a meridian from a standard parallel on latitude 43° N, for the western boundary of a range of township,—say the first one west from the guide meridian,—and running North, say 4 townships, the surveyor must make a point that is *East* of the six-mile point on the northern "standard parallel" 4 × 67.7 links = 270.8 links, The second meridian should fall 8 × 67.7 links to the *right* of the twelve-mile point, etc.

(19.) *The Variation of the Needle.* This is easily determined by noting the reading of the needle when the solar transit telescope has been brought into the meridian.

C. L. Berger & Sons' Solar Attachment for Mining Transits.

This Solar Attachment, designed and patented by the firm, illustrated on page 168, consists of an equatorial adapter, an auxiliary telescope, and a striding level.

The equatorial adapter made to fit to the upright post, on **mining transits with style I** interchangeable auxiliary telescope, where the auxiliary telescope ordinarily goes, receives the auxiliary telescope and converts it into a solar telescope, permitting it to move in the equatorial circle about a polar axis, and in the declination circle of the sun. The adapter consists of two plates provided with leveling screws working against opposing springs which permit the polar axis to be adjusted to the zenith when the main telescope is level, and consequently to point to the pole when the main telescope is elevated in the plane of the meridian to intersect the equatorial circle. The lower plate of the adapter screws upon the central vertical post of the mining transit which was originally designed to carry the auxiliary telescope when used as a top telescope. The upper plate of the adapter carries a small level and the socket which moves around the polar axis. This socket carries an arm in which the declination axis can be made to revolve. The declination axis has at one end a threaded stud similar to those provided for the auxiliary telescope when used as a side or top telescope and to which the auxiliary telescope can be screwed. To do this it is only necessary to slightly release the innermost nut from its fastening against the upright and then by turning the declination axis by means of the outer milled-headed screw, the auxiliary, now the solar, telescope can be securely fastened thereto.

To use the solar attachment, screw the equatorial adapter upon the central post, level up the transit by means of the plate levels, and assuming that all the adjustments of the transit and those of its motion in vertical plane have previously been verified, attach the counterpoise and the auxiliary telescope to their screw-studs as already explained, bring the main telescope level-bubble to the middle of its tube, when the zero of the vertical circle should coincide with that of its vernier. Level up the equatorial adapter by revolving it, by means of its level and the two milled-headed screws acting against the opposing springs in the lower plate provided for that purpose. This is necessary in order to make the polar axis truly at right angles to the line of collimation of the main telescope. This adjustment once properly made need only be repeated for verification of adjustment from time to time as deemed necessary. The two telescopes should be placed in the same vertical plane by bisecting with both telescopes some distant object. When the polar axis is clamped, set off on the vertical circle the declination and refraction of the day and hour of observation; bring the auxiliary telescope into a nearly horizontal position and clamp the declination axis tightly to the upright by means of the inner milled-headed nut, previously taking care that the stud between the two opposing screws shall be nearly in the center. Place the striding level upon the auxiliary telescope and by means of the two opposing screws place the bubble in the center of its tube and then remove it. The two telescopes now occupy a position with each other equal to the declination and refraction of the day and hour of observation. Set the vertical circle to correspond to the co-latitude of the place of observation and the solar attachment is ready for work.

The wire arrangement in the auxiliary telescope is a square somewhat smaller than the disk of the sun which is illustrated in diaphragm D on page 168. The outer wires forming the square are equi-distant from the horizontal and vertical central wires and parallel thereto. They are very coarse to distinguish them from the wires marking the line of collimation of the telescope, to avoid mistakes on the part of the observer when the auxiliary telescope is used in the mine.

The striding level will prove a valuable adjunct for the setting of the auxiliary telescope when used as a side telescope to read the same level line as the main telescope.

The equatorial adapter being in part made of aluminum and of brass, weighs only nine ounces, the same counterpoise used for the auxiliary telescope may also be used for the equatorial adapter by the exercise of proper care.

The observations with this solar attachment are exceedingly simple to make. The equatorial adapter raising the auxiliary telescope considerably above the vertical circle, observations can frequently be made, if desired, without the use of a prism, by simply screwing the colored glass furnished with the instrument upon the eye-piece.

The observer should set the tripod firmly, giving the legs an unusually wide spread.

Observation for Meridian with the Berger Solar Attachment.

Written for this catalogue by GEO. L. HOSMER, Massachusetts Institute of Technology.

CALCULATION.

Before beginning the observations the following computations must be made. 1. Take from the Nautical Almanac (table II, for the month) the sun's "apparent declination," for Greenwich Mean Noon of the date of the observation. If it is north prefix a + sign, if south, a — sign. 2. On the same line, in the next column to the right is the "difference for one hour," with the proper algebraic sign before it. 3. The local time corresponding to Greenwich Mean Noon may be found by subtracting the west longitude of the place "from 12h," e.g. at the 75th meridian. This would give 7h A.M.; at the 90th, 6h A.M.: etc. 4. Next compute the declination for each hour by adding algebraically the "difference for 1 h" to the declination for the preceding hour. 5. Next correct each of these declinations for refraction, using the tables given in this catalogue, or such as are given in Prof. J. B. Johnson's work on surveying. Careful attention should be paid to signs.

We will assume for the present that the latitude is known, and proceed to the description of the

FIELD OPERATIONS.

1. Lay off on the vertical arc the declination setting for the time of observation, tipping the telescope in such a direction that the small telescope will point above or below the equator according as the declination is N. or S. 2. Level the small telescope by means of the striding level, and then clamp it. 3. Next change the setting of the vertical circle so that it reads the *co-latitude* of the place. 5. Using both the horizontal and the equatorial motions, point the small telescope at the sun, making the four segments cut off by the cross hairs equal. The main telescope is now in the meridian. To be certain that the settings are correct wait a few moments and see if the disc follows the equatorial wires perfectly. Both plates should be clamped while the image is in the center of the field. The line may then be brought down to the ground and marked.

EXAMPLE OF COMPUTATION.

Long. 5h. West., Lat. + 40°. Jan. 10, 1900.
Decl. for Gr. Mean Noon = — 21° 59′ 04″.
Diff. for 1h. = + 22″.25.

TIME.		DECLINATION.	REFRACTION	SETTING.
7h.	A.M.	21° 59′ 04″		
8	"	58 42	5′ 40″	21° 53′ 02″
9	"	58 20	2′ 51′	21 55 29
10	"	57 57	2′ 07″	21 55 50
11	"	57 35	1′ 51″	21 55 44
12	M.	57 13	(1′ 47″)	(21 55 26)
1	P.M.	56 51	1′ 51″	21 55 00
2	"	56 28	2′ 07″	21 54 21
3	"	56 06	2′ 51″	21 53 15
4	"	55 44	5′ 40″	21 50 04

The co-latitude may be found by measuring the altitude of the sun's lower limb at noon, i.e. by measuring the maximum altitude. This angle must be corrected for refraction, semi-diameter and declination. The result is the co-latitude. The co-latitude may also be found, very nearly, as follows :— Make the angle between the telescopes equal to the declination setting *at noon* in the same way as for any other hour. Bring the telescopes into the same vertical plane, and point the small telescope at the sun. By varying the elevation angle of the main telescope keep the small telescope pointing at the sun until a maximum elevation is reached. This angle is the co-latitude, already corrected for refraction, semi-diameter and declination. This method is not quite as accurate as the former.

A TEST.

The following observations were made by the writer with the Berger Solar Attachment. The plates were clamped at zero degrees and the meridian found by solar observation. An angle was then turned to a mark ¼ mile away. The results are as follows :—

TIME.	AZ. ANGLE
A.M.	
8 :30	240° 07′
8 :40	05½
8 :50	05½
9 :00	06
P.M.	
3 :23	240° 05′
3 :30	03

Clouds prevented further observations.

The true azimuth as found *afterward* by an observation on Polaris was 240° 05′ 30″.

Table I.

Mean Refraction of Celestial Objects for Temperature 50°, and Pressure 29·6 inches.

Alt. ° '	Refr. ' "	Alt. ° '	Refr. ' "	Alt. ° '	Refr. ' "	Alt. ° '	Refr. ' "	Alt. ° '	Refr. ' "
0 0	33 0	5 30	9 8	12 0	4 23	23 0	2 14	46 0	0 55
10	31 22	40	8 54	20	4 16	20	2 12	47 0	0 53
20	29 50	50	8 41	40	4 9	40	2 10	48 0	0 51
30	28 23	6 0	8 28	13 0	4 3	24 0	2 8	49 0	0 49
40	27 0	10	8 15	20	3 57	20	2 6	50 0	0 48
50	25 42	20	8 3	40	3 51	40	2 4	51 0	0 46
1 0	24 29	30	7 51	14 0	3 45	25 0	2 2	52 0	0 44
10	23 20	40	7 40	20	3 40	20	2 0	53 0	0 43
20	22 15	50	7 30	40	3 35	40	1 58	54 0	0 41
30	21 15	7 0	7 20	15 0	3 30	26 0	1 56	56 0	0 38
40	20 18	10	7 11	20	3 26	20	1 55	58 0	0 35
50	19 25	20	7 2	40	3 21	40	1 53	60 0	0 33
2 0	18 35	30	6 53	16 0	3 17	27 0	1 51	62 0	0 30
10	17 48	40	6 45	20	3 12	30	1 49	64 0	0 28
20	17 4	50	6 37	40	3 8	28 0	1 47	66 0	0 25
30	16 24	8 0	6 29	17 0	3 4	30	1 45	68 0	0 23
40	15 45	10	6 22	20	3 1	29 0	1 42	70 0	0 21
50	15 9	20	6 15	40	2 57	30 0	1 38	72 0	0 18
3 0	14 36	30	6 8	18 0	2 54	31 0	1 35	74 0	0 16
10	14 4	40	6 1	20	2 51	32 0	1 31	76 0	0 14
20	13 34	50	5 55	40	2 47	33 0	1 28	78 0	0 12
30	13 6	9 0	5 48	19 0	2 44	34 0	1 24	80 0	0 10
40	12 40	10	5 42	20	2 41	35 0	1 21	82 0	0 8
50	12 15	20	5 36	40	2 38	36 0	1 18	84 0	0 0
4 0	11 51	30	5 31	20 0	2 35	37 0	1 16	86 0	0 6
10	11 29	40	5 25	20	2 32	38 0	1 13	88 0	0 2
20	11 8	50	5 20	40	2 29	39 0	1 10	90 0	0 0
30	10 48	10 0	5 15	21 0	2 27	40 0	1 8		
40	10 29	20	5 5	20	2 25	41 0	1 6		
50	10 11	40	4 56	40	2 23	42 0	1 3		
5 0	9 54	11 0	4 47	22 0	2 20	43 0	1 1		
10	9 38	20	4 39	20	2 18	44 0	0 59		
20	9 23	40	4 31	40	2 16	45 0	0 57		

Correction to the Mean Refraction given in the preceding Table.

Ap. Alt. °	′	20° +″	24° +″	28° +″	32° +″	36° +″	40° +″	44° +″	48° +″	52° −″	56° −″	60° −″	64° −″	68° −″	72° −″	76° −″	80° −″
0	0	2 40	2 18	1 55	1 33	1 11	51	31	10	10	29	48	1 7	1 25	1 43	2 1	2 19
0	20	2 25	2 5	1 44	1 24	1 4	46	28	9	9	26	44	1 1	1 17	1 33	1 49	2 05
0	40	2 11	1 53	1 34	1 16	58	42	25	8	8	24	39	55	1 10	1 24	1 38	1 53
1	0	1 59	1 43	1 25	1 9	53	38	23	8	7	21	36	50	1 3	1 17	1 30	1 43
1	20	1 48	1 33	1 17	1 3	48	34	21	7	6	19	32	45	57	1 9	1 21	1 33
1	40	1 39	1 25	1 11	57	44	31	18	6	6	18	30	41	52	1 4	1 15	1 25
2	0	1 31	1 18	1 5	53	39	29	17	6	5	16	27	37	48	58	1 8	1 18
3	0	1 11	1 1	51	41	32	22	13	4	4	13	21	30	38	46	54	1 1
4	0	58	49	41	33	26	18	11	4	4	10	17	24	31	37	44	50
5	0	48	41	35	28	22	16	9	3	3	9	14	20	26	31	36	40
6	0	41	35	30	24	19	13	8	3	2	7	12	17	22	26	31	35
7	0	36	31	26	21	16	12	7	2	2	6	10	14	19	23	27	31
8	0	32	27	23	19	15	10	6	2	2	5	9	13	16	20	24	27
9	0	28	24	20	16	13	9	5	2	2	5	8	11	14	18	21	24
10	0	26	22	18	15	12	8	5	2	1	4	7	10	13	16	19	22
12		21	18	15	13	10	7	4	1	1	4	6	9	11	13	16	18
14		18	16	13	11	8	6	4	1	1	3	5	7	9	11	14	16
16		16	14	12	9	7	5	3	1	1	3	5	6	8	10	12	14
18		14	12	10	8	6	5	3	1	1	2	4	6	7	9	10	12
20		13	11	9	7	6	4	2	1	1	2	4	5	6	8	9	11
25		10	8	7	6	5	3	2	1	1	2	3	4	5	6	7	8
30		8	7	6	5	4	3	2	1	0	1	2	3	4	5	6	7
35		7	6	5	4	3	2	1	0	0	1	2	3	3	4	5	6
40		6	5	4	3	3	2	1	0	0	1	2	2	3	3	4	5
45		5	4	3	3	2	2	1	0	0	1	1	2	2	3	3	4
50		4	3	3	2	2	1	1	0	0	1	1	2	2	2	3	3
55		3	3	2	2	2	1	1	0	0	1	1	1	2	2	2	3
60		3	2	2	2	1	1	1	0	0	0	1	1	1	2	2	2
65		2	2	2	1	1	1	0	0	0	0	1	1	1	1	2	2
70		2	1	1	1	1	1	0	0	0	0	0	1	1	1	1	1
80		1	1	1	0	0	0	0	0	0	0	0	0	0	1	1	1
90		0	0	0	0	0	0	0	0	0	0	0	0	0	0	0	0
Height of Barometer.					−28·26 inches.	−28 56 inches.	−28·85 inches.	−29·15 inches.	−29·45 inches.	+29·75 inches.	+30.05 inches.	+30.35 inches.	+30·64 inches.	+30·93 inches.			

Example I.

What is the correction for refraction for an altitude of 8° 5′, the thermometer standing at 50·0° and the barometer at 29·6° inches?

Answer (by inspection) ..   ..   6′ 25″:

and therefore,

Apparent altitude   ..   ..   =   8° 5′

Refraction   ..   ..   ..   =   — 6 25″

True altitude 7 58 35

Example II.

What is the correction for refraction for the same altitude, the thermometer standing at 44° and the barometer at 29·45 inches?

		′	″
Thermometer correction for altitude 8° 5′ =		+ 0	6
Barometer ditto =		— 0	2
Correction for both is .. .. .. =		+ 0	4
Mean Refraction .. .. .. =		— 6	25
∴ True refraction .. .. .. =		— 6	21

	°	′	″
Apparent Altitude .. .. .. =	8	5	0
True refraction .. .. .. =		— 6	21
True altitude	7	58	39

Table II.

Coefficients showing the per cent. of Refraction to be applied to the Sun's Declination.

Lat.	Hours from the Meridian.						Lat.	Hours from the Meridian.					
	1 H.	2 H.	3 H.	4 H.	5 H.	6 H.		1 H.	2 H.	3 H.	4 H.	5 H.	6 H.
° 10	56	33	24	20	18	17	° 36	94	82	71	64	60	59
12	63	39	28	24	22	21	38	95	85	74	67	63	62
14	69	45	33	27	25	24	40	95	87	77	70	65	64
16	74	50	38	31	29	28	42	96	88	79	72	68	67
18	78	55	42	35	32	31	44	96	89	81	74	71	69
20	81	60	46	39	35	34	46	97	90	83	77	74	72
22	84	64	50	42	38	37	48	98	91	85	79	76	74
24	87	68	54	46	42	41	50	98	92	86	81	78	76
26	89	70	57	49	45	44	52	98	93	88	83	81	79
28	90	72	60	51	48	47	54	99	94	90	85	83	81
30	91	74	63	54	51	50	56	99	95	91	87	85	83
32	92	77	66	57	54	53	58	99	96	92	88	86	85
34	93	80	69	61	57	56	60	99	97	93	90	88	87

For the construction of the above table, see **p.** 59.

Table III.

Inclination and Convergency of the Meridians.

Lat.	Inclination for one mile.	Inclination for six miles	Convergency for one township of 36 miles.	Lat.	Inclination for one mile.	Inclination for six miles	Convergency for one township of 36 miles.	Lat.	Inclination for one mile.	Inclination for six miles	Convergency for one township of 36 miles.
°	"	' "	LINKS.	°	"	' "	LINKS.	°	' "	' "	LINKS.
10	9.18	55	13.0	27	26.52	2 39	36.9	44	50.19	5 01	70.1
11	10.13	1 01	14.2	28	27.66	2 46	38.6	45	52.00	5 12	72.6
12	11.07	1 06	15.5	29	28.85	2 53	40.2	46	53.83	5 23	75.2
13	12.02	1 12	16.8	30	30.03	3 00	41.9	47	55.67	5 34	77.8
14	12.98	1 18	18.1	31	31.26	3 07	43.6	48	57.67	5 46	80.6
15	13.96	1 24	19.4	32	32.49	3 15	45.4	49	59.83	5 59	83.5
16	14.93	1 30	20.7	33	33.83	3 23	47.2	50	1 02.00	6 12	86.5
17	15.92	1 36	22.0	34	35.17	3 31	49.1	51	1 04.17	6 25	89.7
18	16.91	1 41	23.4	35	36.50	3 39	50.9	52	1 06.67	6 40	93.0
19	17.93	1 47	24.9	36	37.83	3 46	52.7	53	1 09.17	6 55	96.4
20	18.94	1 54	26.5	37	39.17	3 55	54.7	54	1 16.67	7 10	100.0
21	19.98	2 00	27.8	38	40.67	4 04	56.8	55	1 14.33	7 26	103.7
22	21.02	2 06	29.3	39	42.17	4 13	58.8	56	1 17.17	7 43	107.6
23	22.10	2 13	30.8	40	43.67	4 22	60.9	57	1 20.00	8 00	111.8
24	23.17	2 19	32.3	41	45.17	4 31	63.1	58	1 22.00	8 19	116.2
25	24.30	2 26	33.8	42	46.85	4 41	65.4	59	1 26.66	8 40	120.9
26	25.38	2 32	35.4	43	48.52	4 51	67.7	60	1 30.00	9 00	125.7

For the construction and use of the above table, see articles (13,) (14,) (15,) (17,) (18,) page 63.

For details of instruction in U. S. Government Surveying, see Hawes' System of "Rectangular Surveying," and Burt's "Key to Solar Compass."

To Find the Meridian from "Polaris."

The north star, Polaris, being out of the pole of the equator, is in the meridian but twice in a stellar day — once above and once below the pole — called the upper and lower transits, or culminations.

It is also at its extreme distance, east and west, twice in a stellar day, called greatest *elongations*, east or west.

At the time of a culmination, it would be only necessary to get the bearing of the star to have the place of the true meridian. But this would require an exact knowledge of the time, an element not usually possessed by surveyors. Moreover, the observation must be made with certainty, at the instant, which is not always practicable. On this account, this method is not in favor with surveyors.

At elongation, the apparent motion of the star is tangent to the vertical, and therefore, for a few minutes, with regard to azimuth, it appears to stand still, thereby affording ample time for deliberate observation.

The distance of this star from the pole—called its polar distance, was 1° 18′ 16″ on January 1, 1885, and is diminishing at the rate of about 19.06″ per year, whence its distance in following years may be known.*

The azimuth of the star, corresponding to any polar distance, is variable with the latitude. Thus, an observer at the equator would see this star — say at eastern elongation — in the horizon, and at the distance of 1° 18′ 16″ to the *right* of the pole, or true meridian.

If now the observer should go north, the azimuth of the star would increase with its altitude, till he should arrive at a latitude equal to the complement of the polar distance, when it would be N. 90° E. Between these limits, the bearing of the star, at elongation from the pole, would vary according to the following equation, in which $Z =$ the azimuth, or bearing:

$$\text{Sin } Z = \frac{\text{sin Polar Distance}}{\text{cosine Latitude}}$$

As the telescope of the surveyor's transit is not usually of sufficient power to show the star in the daytime, the observation must be made at night, in which case the cross-wires of the telescope must be illuminated by light reflected into the tube. A piece of stiff white paper, with an opening large enough to admit of seeing the star through it, and held obliquely in front of the telescope, will make a good reflector.

As generally but one of the elongations can be seen, on the same night, it is important to know, which one is observed. Also the latitude must be known, at least approximately.

The pole is nearly in line between *Polaris* and the star *Alioth*, which is the first star in the handle of the *Dipper*, reckoning from the bowl, so that when these two stars are in a line, nearly horizontal, and the Dipper is $\begin{Bmatrix} \text{east} \\ \text{west} \end{Bmatrix}$ of the pole, Polaris is at his greatest elongation $\begin{Bmatrix} \text{west} \\ \text{east.} \end{Bmatrix}$

In sighting to the star, the observer must be careful to keep his transit level transversely, for the star is so high that inattention to this might introduce a serious error into the resulting azimuth.

A satisfactory sight having been obtained, the telescope should be brought down to fix a mark on the ground, at a distance of 300 to 400 yards from the transit.

This mark should be something clear and definite, like a nail set in a hub, driven into the ground, which may be located by means of a plummet lamp, or by means of a common lamp in a box, having a vertical slit in one side of say ¼ or ⅛ an inch in thickness, with a plumb-line suspended from the slit, and manipulated by an assistant.

The direction of the star being satisfactorily marked, compute the azimuth from the above equation, and set the resulting angle off to the $\begin{Bmatrix} \text{right} \\ \text{left} \end{Bmatrix}$ of the mark for $\begin{Bmatrix} \text{western} \\ \text{eastern} \end{Bmatrix}$ elongation.

It may happen, that the resulting azimuth may have an odd number of seconds, or fraction of a minute, not convenient to be set off with a vernier graduated to

* Small corrections to the distances thus calculated are needed, but do not amount to more than 30″ in all; see a Nautical Almanac.

single minutes. In this case, find the distance carefully between the transit and the mark, and multiply this distance by the tangent of the azimuth. This result set off to the $\begin{smallmatrix} \text{right} \\ \text{left} \end{smallmatrix}$ $\Big\{$ for $\begin{smallmatrix} \text{western} \\ \text{eastern} \end{smallmatrix}$ $\Big\}$ elongation, will point out the place of the true meridian.

Meridian from Equal Altitudes of the Sun.

If the sun, like the stars, were stationary, with regard to declination, the meridian could be found by simply bi-secting the angle between the bearings of the sun, at two consecutive equal altitudes, at forenoon and afternoon. But the declination is constantly changing, so that a reduction is required, before meridian can be found from the daily motion of the sun.

The sun is going *north* in declination for six months in the year, or from the winter solstice, in *December*, to the summer solstice, in *June*, and during the next six months, from *June* to *December*, he is going *south*.

A table accompanying this paper gives the hourly motion of the sun in declination for every fifth day in the year, from which the change of declination, between two consecutive equal altitudes, may be readily computed.

To an observer, in north latitude, when the sun is going $\begin{smallmatrix} \text{north} \\ \text{south} \end{smallmatrix}$ $\Big\}$ in declination, the meridian, as deduced, from the bi-section of the angle between two consecutive equal altitudes, would be to the $\begin{smallmatrix} \text{right} \\ \text{left} \end{smallmatrix}$ $\Big\}$ of its true place.

The amount of error depends on the hourly rate of change of declination, the time elapsed between the epochs of equal altitude, and the latitude.

By multiplying the time elapsed, in hours, by the corresponding hourly rate of change, we have the change of declination during the time between the two equal altitudes of the sun. But declination is measured in a direction parallel with the earth's axis, which makes an angle with the horizon equal to the latitude. Whence, *the change of declination must be multiplied by the secant of the latitude* in order to find its value, as seen in azimuth in the horizon.

Again, the afternoon observation will not generally be made, when the sun is due west, but to the south of west, by an angle equal to half the supplement of the angle between the two points of equal altitude. Whence, the above corrected change of declination must be multiplied by the cosine of half the supplement of the angle above mentioned. The effect of this correction will be to reduce the corrected declination by from about 4 to 8 per cent., so that the formality of this calculation may be dispensed with, or may be omitted entirely, as only half of it affects the place of the meridian.

From the foregoing considerations, we have the following routine of work for finding meridian with a surveyor's transit:

(1.) Having set up the instrument and leveled it carefully, take a sight on some object, preferably to the left of the sun — the plate vernier reading zero — for an azimuth mark, and *clamp the plate firmly to the spindle.*

(2.) Having set the telescope to a convenient angle of altitude, *above* that of the sun, follow the sun in azimuth by the upper motion of the instrument, keeping the vertical wire of the telescope bi-secting the sun, by means of the plate tangent, till the sun's lower limb *touches the horizontal wire of the telescope.* Clamp, and *note the reading of the vernier,* keeping the telescope, also, carefully clamped in altitude, and *note the time of day.*

(3.) In the afternoon, turn the telescope, and bi-sect the sun's disc, as in the morning, with the vertical wire, and when the sun's lower limb comes down to the horizontal wire, *clamp the horizontal plate, note the reading of the vernier, and the time of the day.*

(4.) From the table of the hourly change of declination take out the rate for the day of the month, and multiply the same by the time in hours between the times of equal altitude, and by the secant of the latitude, or if there are no secants at hand, divide by the cosine of the latitude. This last result, diminished by about 5 per **cent.** of itself, as above explained, will be the *correction* to be applied to the last plate-reading, before finding the angle between the bearings of forenoon and afternoon observations.

(5.) If the sun is going $\left.{\text{north} \atop \text{south}}\right\}$ in declination, this correction must be $\substack{\text{subtracted} \\ \text{added}}$ $\left.{\text{from} \atop \text{to}}\right\}$ the plate-reading of the last observation.

Then, half the difference between this corrected plate-reading and the first reading, is the angle to be set off to the right from the bearing of the first altitude for the place of the meridian.

EXAMPLE.

Date of observation, April 10. Lat. 43° N.
Plate reading at first altitude, 13° 26′ (right of mark).
 "   "   " second   "   148° 13′.
Time between observed altitudes, 9½ hours.
Hourly rate of change of declination (see table), 56″.
Total change of declination north = 9½ × 56″ = 8′ 52″.
Multiplied by secant of latitude = 8′ 52″ × 1.367 = 12′ 12″.

This, 12′ 12″, is the amount by which the change of declination changes the sun's place in azimuth, and is *subtractive*, because the sun is going north, whence,

$$\frac{148° \; 13′ \mp 12′ \; 12″ - 13° \; 26″}{2} = 67° \; 17′ \; 24″ = \text{the angle between the bearing of the}$$

first altitude and the meridian. Or, from the assumed azimuth mark, the angle would be 64° 17′ 24″ + 13° 26′ = 80° 43′ 24″. It is best to use the double sign $\mp$ before the corrections of the last plate-reading, in order to avoid mistakes.

A "Davis' Screen," or a prism-attachment, to the telescope will add greatly to the facility of making observations.

With good appliances, this method of finding meridian will be found very efficient and satisfactory — indeed, one of the best in use. The computations are few and simple. Moreover, the resulting angle for the place of the meridian is affected by *only one-half* of any uncertainty in the correction for change of declination. The greatest objection to this method is the amount of time required.

Hourly Motion of the Sun in Declination.

Day of Month.	Jan.	Feb.	March.	April.	May.	June.	July.	Aug.	Sept.	Oct.	Nov.	Dec.
	″	″	″	″	″	″	″	″	″	″	″	″
1	+ 12	43	57	58	45	21	10	38	54	58	48	23
5	17	46	58	56	43	17	14	41	56	58	46	19
10	22	49	59	54	39	12	19	44	57	57	42	14
15	28	52	59	52	36	7	24	47	58	56	38	8
20	32	54	59	49	31	+ 2	28	49	58	54	34	− 2
25	37	56	59	47	27	− 4	32	52	59	52	30	+ 4
30	41	. .	58	46	23	9	36	54	59	49	25	10

Transit Solar Attachment.

For running Meridian or other lines by the Sun.

Written for this catalogue with special reference to the wants of Public Land Surveyors, for both common and mineral lands, by J. B. Davis, Assistant Professor of Civil Engineering, University of Michigan.

1. Remarks. The attachment herein referred to is the Davis and Berger solar screen, prism, and colored shade glass, used for direct solar observation. These inventions have been devised by the Mr. Berger, of the firm of C. L. Berger & Sons, and by the writer. They are simply for the purpose of enabling one to make an observation directly upon the sun's centre. This observation being secured by readings of the horizontal and vertical circles, is reduced so as to give the direction of the line of sight of the transit at the instant of the observation. Thus knowing the direction of the line of sight at a given instant it becomes simply necessary to turn off the angle which this line of sight makes with the meridian, to ascertain the position of the meridian. This angle is what is obtained by reducing the observation, as above mentioned. A brief reference to the history of these devices will best explain them. It occurred to the writer to see if an image of the sun could be formed behind the eye-piece of a telescope at the same time an image of the cross-wires was, and the latter image be made to quarter the former, by allowing the sun to shine into the object end of the telescope and thence directly through it. The experiment was made by holding a piece of white paper behind the eye-piece, and adjusting the focus of the eye-piece and object glass. The very first trial was readily successful. The next thing was to see if the position of the instrument could be located by this means as near as the circles would read. By the same simple means it was soon found that a motion given to the telescope by either tangent screw might be so slight that the eye could not detect it upon the circles, but evidence of it would be apparent in the position of the images with reference to each other. This fact at once settled the question of whether this would be a sufficiently delicate means of observation. It showed that the observations would be closer than the circles would read. After some trials and some months rest these facts were brought to the notice of others, and finally were submitted to Mr. Berger for his opinion. He made a screen which the writer exhibited at the first annual convention of the association of Michigan Engineers and Surveyors at Lansing. The matter was further studied by Mr. Berger. The screen was much improved, and the mechanical construction of it brought to the standard of the work done by this firm. Mr. Berger soon conceived the idea of making the screen of ground white glass in a brass frame, as shown in figs. **1** and **2,** so one might observe the position of the images directly upon it, and thus secure not only the comfort of an easy position in observing, but the consequent accompanying accuracy. The arm of attachment was perfected from time to time. The screen of ground glass is mounted upon an arm that admits of all adjustments of position, and is so attached to the side of the telescope tube that it can be turned up out of the way when not needed. The reflecting prism can be screwed on to the eye-piece cap for observing at high altitudes. This also is adjustable so as to look in any desired direction from the telescope tube. The diagonal eye-piece also has its movable colored shade glass as above stated. With these attachments observations on the sun at all altitudes may be made in two ways. By looking directly at it through the simple colored glass for low altitudes, or through the prism and its shade glass for high altitudes. The other way is to receive on the screen the images of the cross-wires and the sun and make the image of the cross-wires just quarter the image of the sun by means of the slow motion screws to the circles of the instrument. For this method the colored shade glasses are not to be used. With this complete outfit one may work whichever way seems best.

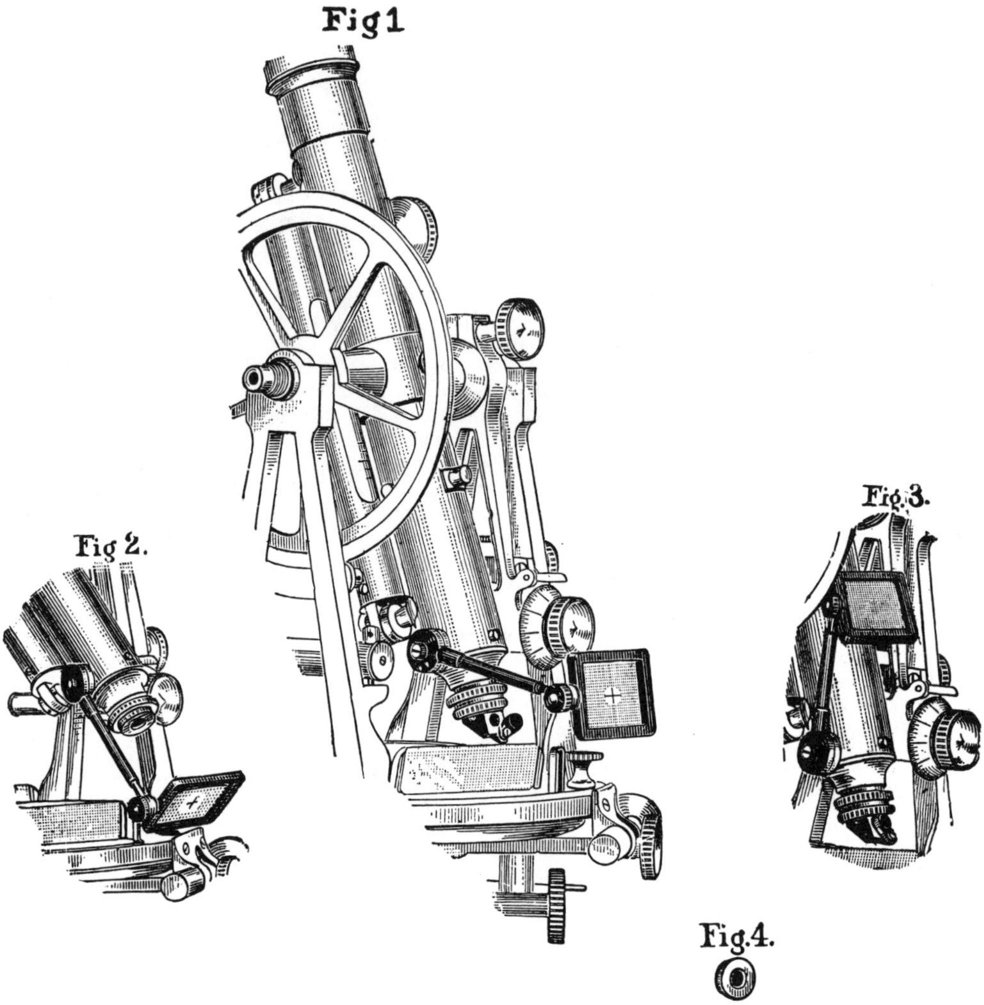

Fig 1

Fig 2.

Fig.3.

Fig.4.

Davis' Patent Solar Attachment.

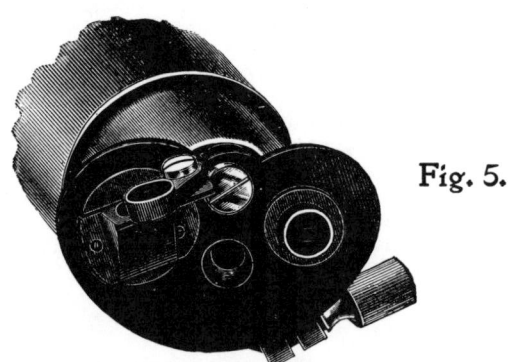

Fig. 5.

C. L. Berger & Sons' Improved Prism and Colored Glass Attachment for Solar Observations.

These devices are being more and more perfected, and will be protected by letters patent, and Messrs. C. L. Berger & Sons make and sell them exclusively.

2. Remarks. Certain precautions are necessary in the use of this method of finding the true direction of a line as well as in any other. It is not wise to observe the sun, read the circles, note down the readings and leave the instrument standing there while making the reductions. It will get out of place in some way, very likely. Therefore, as soon as the observation is completed and the readings of the circles noted, set the line of sight on some fixed point and read the plate again, noting this reading. Of course the two plate readings will give the horizontal angle from the sun to the line. This will enable the observer, after finding the direction of the line of sight when set on the sun, to readily ascertain its direction as set on the fixed point referred to, thus determining the direction of the line from the point over which the instrument is set to the fixed point. This line may be chosen before beginning the observation, and become the reference line for the work in hand.

3. Remarks. For the purposes of reduction the process by equations is used instead of one by rules. The introduction of symbols and signs is a much simpler matter than many suppose. It is nothing but this. We agree that a character of some sort or other shall represent a certain thing and nothing else. Whenever this character occurs, therefore, it simply means the thing we have set it for. That is all there is of symbolical representation. These very words here printed are all symbols. The method is universal. We here, as elsewhere in algebraic processes, make a special application of it. The rules for a case of this kind would be very cumbersome and give the user far more trouble than will be necessary for mastering the few equations given below. The record of the processes is hereby reduced to a few lines, and one has not to go searching through a page for a point here and there, but places his eye at once upon what he wants, where all will be found in a compact form. Of course one needs to read each word and each sign. Nothing must be slurred over or missed. The record as set forth below is exact, complete and reliable.

4. Remarks. All computations should be thoroughly checked, and check equations and devices are given. These should always be applied, without fail, as no one can implicitly trust a computation by a single process, unrepeated, even if simple. No one should who is a surveyor or engineer. Several checks are given. One *used* is sufficient, usually. If one distrusts the check because it shows the work to be wrong, it may be of some satisfaction to use another or more than one.

5. Remarks. The *directions* prepared below are intended for use, word by word, and step by step. It is hoped that they will prove in convenient form for use as a chart to direct the efforts of the observer in his first use of these attachments and this method. Therefore, it is thought that one may safely do as told, trusting the next step to the next statement. They have been prepared with this view.

6. Using the Screen.

a. **Directions.** Set the instrument so the sun can shine in at the object end of the telescope, and directly through it. Run out the eye-piece and adjust the screen behind it, by its sliding arm, so that a distinct image of the cross-wires can be seen on the screen within the lighted spot made by the shining sun, as shown in fig. **2.** Set the object glass so as to clearly define the image of the sun on the screen. Repeat these trials, and adjust the parts of the telescope and screen so that the clearest image of both the cross-wires and the sun will be obtained that the telescope will give. Mark the slide on the arm of the screen and the eye-piece, so they can be easily set thereafter for an observation.

b. **Remarks.** The eye-piece, when all is in exact position, will be found to be considerably farther out than for an ordinary sight. The marking of the sliding arm and eye-piece will save time in the future. These trials, when made with a new apparatus, should be conducted at leisure and with extra care, for the purpose of fitting the apparatus carefully to the telescope. A few trials may be needed at first in order to accustom the observer to recognize the best definition of the images.

This solar screen is especially adapted to the ordinary surveyors' and engineers' transit telescopes, with erecting eye-pieces. It is not adapted to be used with inverting or astronomical telescopes, unless during an observation the aperture of the objective is cut down to $\frac{1}{4}$ inch diameter, by means of a diaphragm placed in front of it, when the image can be seen as sharply defined as those of the erecting telescope; or the observations must be made with the shade glasses and reflecting prism alone.

7. Using the Colored Shade Glass.

a. **Directions.** Attach the colored glass shown in fig. **4,** to the eye-piece, to shield the eye from the sun and look directly at it, setting the cross-wires so as to quarter it.

b. **Remarks.** This will be found entirely satisfactory when the sun's altitude is so low as to enable the observer to bring his eye in apposition with the eye-piece of the telescope with ease.

8. Using the Diagonal Eye-piece.

a. **Directions.** Screw on the prism, as shown in fig. **3,** to the end of the common eye-piece. Look directly through the shade-glass, if observing in that way, turning the prism either way so as to make it convenient to look into it. If any trouble is experienced in finding the sun with it. let the sun first shine through the telescope, the colored shade-glass being turned aside, till the brilliant light perceived in the aperture of this eye-piece shows the telescope to be rightly directed. Cover the aperture with its shade-glass and proceed.

b. **Remarks.** By attaching the reflecting prism to the eye-piece of the telescope, the light is reflected at right angles to the the line of sight of the telescope, and it thus becomes what is termed a diagonal eye-piece.

This prism can be used for direct observation when the altitude of the sun is too great to allow the eye to be applied *directly* to the eye-piece of the telescope, and not so great as to bring the eye-piece too far over the plate, but through this range of altitudes the solar screen can be used without the prism, as shown in fig. **2,** and it will usually be found advantageous to do so.

Since the prism in effect withdraws the eye about half an inch further from the eye-piece of the telescope than its natural position, that being about the distance traversed by the light in passing through the prism, the high magnifying power used in C. L. Berger & Sons' transit telescopes makes the use of the reflecting prism for *direct* observation a little awkward, and it will usually be found more satisfactory when using the prism to use the solar screen with it.

9. Using the Reflecting Prism and Solar Screen combined.

a. **Directions.** Attach the prism, and direct the telescope as in **8.** Then, leaving the aperture of the prism uncovered, adjust the solar screen so as to receive the images of the sun and the cross-wires. as shown in fig. **1.**

b. **Remarks.** For observing the sun at high altitudes it will be found that in this, otherwise most difficult of all positions, the use of the solar screen combined with the prism will enable the engineer to make his observation with the greatest ease and precision.

10. Making the Observations.

a. **Directions.** Direct the telescope to the sun, and by means of the slow motion screws. cause the image of the cross-wires to exactly quarter the sun's image. Read both circles and record the readings. Refer the position of the instrument to some fixed line, and once, *after* the above work, by another plate reading. Also note and record the exact instant of time of the observation by the watch.

b. **Remarks.** This observation with the watch may be used as hereafter indicated to simplify and lessen the amount of work in making the reductions. A fair watch of ordinary accuracy is sufficient. The entire work can be carried on without a watch at all, but it takes some more figuring.

11. Use of the Nautical Almanac.

a. **Remarks.** In order to use the observations, made as above directed, it is necessary to find the sun's apparent declination for the time of observation. This is done as directed below.

b. **Conditions.** Let all the algebraic signs be carefully observed throughout the work. Use the watch time.

c. **Directions.** *For finding the Sun's apparent declination.* Look in the table of Washington Solar Ephemeris against the date of the observation, and take out the following quantities. First, the sun's apparent declination, with its sign, + when N., — when S., from its column. Second, the hourly change, with its sign, from its column. Find from a map or otherwise, the difference in longitude between the place of observation and Washington, as near as one-half hour, or seven and one-half degrees. This is + when W. and — when E. of Washington. Add to this difference of longitude the time of the observation from noon, this time being + when the sun is W. and — when E. of the meridian. Multiply the hourly change by this result, *in hours*, noting all the signs. Apply this product, regarding its sign, to the sun's apparent declination as taken, from the table, for the sun's apparent declination at the time of the observation.

d. **Example.** Date, 1881 — 6 — 14. Hour, 9h — 26m — 24s, A.M. Longitude about 40 minutes East of Washington, considered in time.

⊙'s apparent declination, 1881 — 6 — 14.
Washington mean noon, + 23° 18′ 15″
Hourly motion, + 7″

Time of observation from noon, —2 hours 30 minutes, about.
Longitude East of Washington, — 40 minutes.
Total time of correction, —3 hours 10 minutes, = 3⅙ hours.
Amount of correction = —3⅙ × 7″ = —22⅙″
⊙'s apparent declination from table, + 23° 18′ 15″
⊙'s apparent declination at time of observation, + 23° 17′ 53″ nearly.

12. Reducing Observations.

a. **Conditions.** Let h' = the sun's altitude, as observed.
Let ϕ = the latitude of the place of observation.
Let δ = the sun's apparent declination at the time of observation, found as above directed.
Let z' = the sun's observed zenith distance.
Let z = the sun's true zenith distance, always +.

Let k and k' be two auxiliary angles used in the reductions.
Let A = the azimuth of the line of sight of the instrument at the instant of the observation, reckoned from the N. point of the horizon, either E. or W. as the sun is E. or W. of the meridian.
Let t = the sun's apparent hour angle at the time of the observation, that is the local apparent time from apparent noon plus the change in the sun's right ascension between apparent noon and the time of the observation. This is + when W. and — when E. of the meridian, or + for P.M , and — for A.M. times. The mean or watch time is sufficient for use in **2.**
Let p = an auxiliary angle used in some of the reductions.
Let all signs be faithfully regarded. Let logarithms be used.

b. **Directions.** *For finding z from z'.* Use the following equations.

$$z' = 90° - h' \qquad (1)$$
$$z = z' + 55'' \tan z' \qquad (2)$$

c. **Directions.** *For finding A when ϕ, δ and z are given.*

Find $\tan \frac{1}{2}(k - k') = \cot \frac{1}{2}(\phi + \delta) \tan \frac{1}{2}(\phi - \delta) \cot \frac{1}{2} z$ (3)
When $\phi < \delta$ and of the same name find $k = \frac{1}{2} z + \frac{1}{2}(k - k')$ (4)
When $\phi > \delta$ and of the same name find $k' = \frac{1}{2} z - \frac{1}{2}(k - k')$ (5)
When ϕ and δ have different names find $k' = \frac{1}{2} z - \frac{1}{2}(k - k')$ (6)
Then find A from $\cos A = \tan k \tan \phi$ or $\tan k' \tan \phi$ (7)

Checks.

When (4) is used $\qquad \dfrac{\sin \phi}{\sin \delta} = \dfrac{\cos k}{\cos k'}$ (8)

or $\dfrac{\sin \phi}{\cos k} = \dfrac{\sin \delta}{\cos k'} = \cos p$ (9)

When (5) or (6) is used $\dfrac{\text{Sin }\phi}{\text{Sin }\delta}=\dfrac{\cos k'}{\cos k}$ . . . . (10)

$\qquad$ or $\dfrac{\text{Sin }\phi}{\cos k'}=\dfrac{\text{Sin }\delta}{\cos k}=\cos p$ . . . . . (11)

Find $\qquad\qquad$ Sin $p=\sin A\cos\phi$ . . . . . . (12)

Sin p and cos p are at the same place in the table.

d. Example. $\qquad \phi=42°\ 16'\ 30''$ N. $\qquad z=52°\ 43'\ 30''$
$\qquad\qquad\qquad\quad \delta=18°\ 13'\ 20''$ N. $\quad \frac{1}{2}\ z=26°\ 21'\ 45''$

$\qquad\quad \phi+\delta=60°\ 29'\ 50''$
$\qquad\quad \phi-\delta=24°\ \ 3'\ 10''$
$\qquad \frac{1}{2}\,(\phi+\delta)=30°\ 14'\ 55''$
$\qquad \frac{1}{2}\,(\phi-\delta)=12°\ \ 1'\ 35''$

$\qquad\qquad\qquad\qquad\qquad\qquad\qquad\qquad$ Checks.

Cot $\frac{1}{2}(\phi+\delta)=0.2342195$ $\quad$ Tan $\phi=9.9586273.$ $\quad$ Sin $\phi=9.8278148$ $\quad$ Cos $\phi=9.8691875$
Tan $\frac{1}{2}(\phi-\delta)=9.3284570-$ Tan $k'=9.2477939.$ $\quad$ Cos $k'=9.9933068$ $\quad$ Sin $A=9.9943079$
Cot $\frac{1}{2}\ z\ \ =0.3048785-$ Cos $A=9.2064212.$ $\quad$ Cos $p=9.8345080$ $\quad$ Sin $p=9.8634954$
Tan $\frac{1}{2}(k-k')=9.8675550$
$\qquad\qquad\qquad\quad 346 \qquad\qquad\qquad 3894.$
$\qquad\qquad\qquad\quad \overline{204} \qquad$ A $=99°\ 15'\ 22''.5$ $\qquad \overline{318.}$ $\qquad$ At same place in table.

$\frac{1}{2}\,(k-k')=36°\ 23'\ 45''$ $\qquad\qquad\qquad\qquad$ Sin $\phi=9.8278148$ $\qquad$ Cos $k'=9.9933068$
$\quad\ \ \frac{1}{2}\ z\ \ =26°\ 21'\ 45''$ $\qquad\qquad\qquad\qquad$ Sin $\delta=9.4951325$ $\qquad$ Cos $k\ =9.6606232$
$\qquad\quad k'=-10°\ \ 2'\ 00''$ $\qquad\qquad\qquad\qquad\qquad \overline{0.3326823}$ $\qquad\qquad \overline{0.3326836}$
$\qquad\quad k=\ \ 62°\ 45'\ 30''$

e. Remarks. Look out tan ϕ, cos ϕ, and sin ϕ, at one search. Use either check as may be preferred. This operation need not be performed oftener than the demands of the work require, the plate being used mean time.

13. Remarks.

The observations and reductions can be always made, according to the process given, without a watch, but the latitude of the place must be known. It must be carried on as the survey proceeds, by measurement, or an observation made to determine it with the instrument. If it becomes necessary to find the latitude it may be done as follows:

14. Finding the Latitude by the Sun.

a. Directions. *For Observations.* Near noon begin to observe the sun a little before it reaches its greatest altitude. By means of the slow-motion screws keep the sun's image exactly in place on the screen, or by direct sight keep the cross-wires exactly on the sun. As it moves upward just carefully follow it, recollecting that the object is to get its greatest altitude. Be careful to stop following it when it turns and begins to descend.

b. Directions. *For Reductions.* Find z, as in **12, b.** Find the sun's apparent declination, δ, as in **11, c.** Then

$\qquad\qquad\qquad z+\delta=\phi$, the required latitude. . . . . (13)

Be sure to observe the Algebraic signs, as δ may be $+$ or $-$.

c. Remarks. Having the latitude in this way, the observations and reductions may be conducted according to the processes above given. The latitude once carefully ascertained by this or some other method, may be preserved by the distance traversed *north* or *south* of the point of the last observation for latitude. It will at once appear that the measurement and observation may be made to check each other. The method of reducing the *change* in latitude by linear measurement may be as follows:

15. Finding the Latitude by Linear Measurement.

a. Conditions. The latitude of the point measured from, or reckoned from, must be known. The measurements must be reduced to the north and south direction from the reference point. Let reduced distances north be $+$, and those south be $-$. Let all signs be observed. Let the true bearings, or directions of all lines with the meridian of the reference point, be given. Let any number of courses be run in any direction.

b. **Directions.** *For reducing the north or south distances.* Multiply the length of each course by the cosine of its bearing, the results being given signs as above indicated, + for northerly courses, and — for southerly courses. Sum these results regarding the signs.

c. **Remarks.** This sum will be the distance north or south of the reference point.

d. **Directions.** *For reducing feet to minutes of Latitude.* Find the length of a minute of latitude for the place by this equation.

$$m = 6076.36 \left(1 + \frac{\text{Sin} \, 2(\phi - 45°)}{200} \right) \quad . \quad . \quad . \quad . \quad (14)$$

Then divide the traversed distance north or south of the reference point by the value of m found from this equation.

e. **Remarks.** The result will be the minutes and decimals of a minute of the new point from the reference point. This value of m will be in feet, hence the north or south distance must be in feet.

16. Remarks. The latitude may be dispensed with during a day's work after the first satisfactory observation. It may be for a longer period if the watch is to be depended upon. It will be well to find the latitude, and check the work occasionally, where the watch is used. In order to prepare the watch for this work, proceed as follows:

17. Correcting the Watch.

a. **Directions.** *For correcting the Watch by a Noon Observation.* Having ascertained the bearing of a line without the aid of the watch, as at first directed, near noon set the line of sight in a meridian. Set the telescope so the sun can be seen in it, or received on the screen as it passes the meridian. Note the time by the watch when the sun's west side comes in apparent contact with the vertical cross-wire. Note the watch time when the east side of the sun just touches the vertical wire. Find the time half way between these two noted times for the time of the meridian passage of the sun's center, or the time of apparant noon, by the watch.

b. **Remarks.** The time as above found should differ from exact noon by just the equation of time for that date and time as given in the Nautical Almanac. Observe the sign there attached to the equation of time. The watch may then be set to true time if not correct. That is, it may be set so that the time of the sun's meridian passage will be just the equation of time, with its sign, from exact noon.

c. **Remarks.** The watch may also be corrected directly from an observation, reduced as at first directed in **10** and **12.** Here it will be necessary to take the watch time of the observation, as directed in **10.** Having done so, and reduced the observation by **12,** proceed as follows:

d. **Directions.** *For correcting the Watch by an observation at any time.* Having found A and z, and knowing δ, find t by the following equation.

$$\text{Sin} \, t = \frac{\sin A \sin z}{\cos \delta} \quad . \quad . \quad . \quad . \quad . \quad (15)$$

This being in arc, reduce it to time at the rate of four minutes of time to one degree of arc.

e. **Remarks.** This result should differ from the watch time of the observation from mean noon, by just the equation of time, *with its sign.* If it does not, set the watch so it would have done so had the observation been made with the corrected watch.

18. Remarks. Having corrected the watch by the last method, the value of t in time may be found from the value of t at this observation by noting the time by the watch of another observation, and thence finding the elapsed time. This applied to the first value of t will give its value for the last observation. Thus the value of t may be carried forward as long as the watch runs true. Of course it will occur to many at once that the watch can just as well be used to measure the elapsed time without being corrected. This is too careless. The better way is to keep a careful oversight of the watch by correction. Thereby it may be known how much the watch is to be trusted. It is always best to establish a routine system in these matters, as soon as practicable, and adhere faithfully to it.

19. Remarks. When the watch is corrected by either method, it will give the value of t in time directly as follows: Note the time of an observation. Apply to this time the equation of time *with its sign*, as given in the Solar Ephemeris Table of the Nautical Almanac. The result will give the apparent time of the observation from apparent noon, $+$ when the sun is west of the meridian, and $-$ when it is east. This found is the required value of t.

20. Reducing Observations.

a. Conditions. Let the notation be as before.

Let $t =$ the sun's apparent hour angle at the time of the observation, that is the local apparent time from apparent noon. This is $+$ when W. and $-$ when E. of the meridian, or $+$ for P.M., and $-$ for A.M. times.

Let the value of t be found by **18** or **19,** and reduced to arc at the rate of one degree of arc to each four minutes of time, the work being carried out to seconds of arc.

b. Directions. *For finding* A *when* $\delta, t,$ *and* z *are given.* Find A from the following equations.

$$\text{Sin A} = \frac{\cos \delta \sin t}{\text{Sin } z}. \qquad . \quad . \quad . \quad . \quad . \quad . \quad (13)$$

Check.

$$\frac{\text{Sin } z}{\cos \delta} = \frac{\sin t}{\sin \text{A}} \qquad . \quad . \quad . \quad . \quad . \quad . \quad (14)$$

c. Example. $\delta = 18° \ 30' \ 20''$ N. $\quad z = 52° \ 43' \ 30'' \quad t = 55° \ 46' \ 32''.5$

$$
\left.
\begin{array}{l}
\text{Cos } \delta = 9.9776554 \\
\text{Sin } t = 9.9174225 \\
\hline
9.8950779 \\
\text{Sin } z = 9.9007700
\end{array}
\right\}
\quad
\left.
\begin{array}{l}
9.9231146 \\
9.9231146
\end{array}
\right\}
\text{Check.}
$$

$$
\begin{array}{l}
\text{Sin A} = \overline{9.9943079} \\
A = 99° \ 15'22''.5
\end{array}
$$

20. Remarks. The value of A as determined in these examples is greater than 90°, because the sun is south of the zenith. The value of t used in the second example was found from the first, hence the exact check. It may be noticed how much less figuring is required in the second example than in the first. It should be noted, however, that more than one check is figured out in the first example, and so more than the *necessary* figures shown. The value of A is carried out with exactness in order that the process may be fully illustrated.

21. Summary. Several courses are hereby opened to the surveyor. This is done that he may have the more checks at his command, and so make certain of his work. It may be well to indicate these courses in a catalogued form for easy reference. The courses are

The processes of **10, 12,** and **14** or **15.**

The processes of **10, 12,** and **14** or **15,** and thence
16, *a,* or **16,** *d,* and **18** or **19** and **20.**

22. Cautionary. Keep the levels and the vernier of the vertical circle in good adjustment. Also keep the adjustment of the axes of the instrument, the transit axis and the vertical axis, in good order.

23. General Remarks. It will be seen that in doing solar work with these attachments in the manner explained above, the observation of the sun depends on the ordinary line of sight of the telescope exactly as in all Geodesic work.

For this reason *no extra adjustments are required.* The accuracy of the observation in no way depends on these attachments, which are merely conveniences to enable one to make solar observations with the ease and precision of ordinary terrestrial work.

Other Solar Attachments are mechanical devices requiring special adjustments, and considerable care is necessary to keep these adjustments perfect, while they cause some degree of anxiety and doubt in the mind of the engineer as to whether they are quite perfect or not.

With this invention all these sources of anxiety are avoided, the solar observation being made with the telescope of the transit itself, while it has the advantage of being applicable to every surveyors' and engineers' transit, is so light as not to add appreciably to the weight of the instrument, so simple as to require no special provision for its care, and so cheap as to be within the reach of every surveyor.

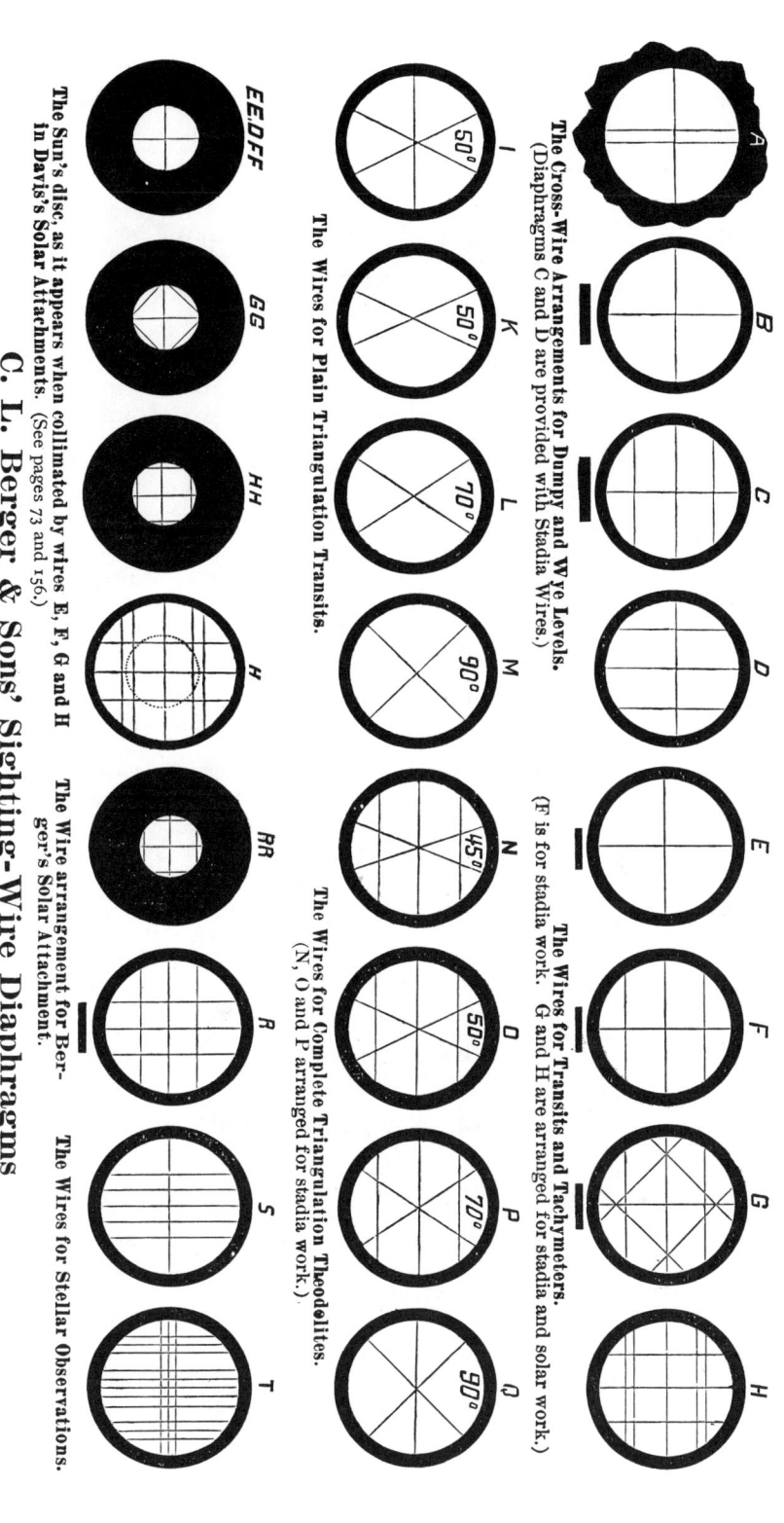

The Cross-Wire Arrangements for Dumpy and Wye Levels.
(Diaphragms C and D are provided with Stadia Wires.)

The Wires for Transits and Tachymeters.
(F is for stadia work. G and H are arranged for stadia and solar work.)

The Wires for Plain Triangulation Transits.

The Wires for Complete Triangulation Theodolites.
(N, O and P arranged for stadia work.)

The Sun's disc, as it appears when collimated by wires E, F, G and H in Davis's Solar Attachments. (See pages 73 and 156.)

The Wire arrangement for Berger's Solar Attachment.

The Wires for Stellar Observations.

C. L. Berger & Sons' Sighting-Wire Diaphragms

As they appear in the Telescopes of the various Instruments.

Patented.

NOTE.—Diaphragms marked thus ▬ are the only proper ones to use in the common Engineers' and Surveyors' practice. The other kinds are simply shown here to afford a full inspection of styles appropriate for special purposes and instruments, which latter are, however, made to order only. Spare diaphragms with arrangements of wires different from those supplied with an instrument can be furnished, if desired, at an expense of from $3.00 to $10.00, when ordered with the instrument, but it is not advisable to do so except in very rare cases. The difficulty to keep them intact outside of a telescope and to insert them without breaking is very great. As a rule, when for special work other styles of wires are desired, it is necessary to send the telescope, if not the whole instrument, to the maker, to enable him to place the wires in the relation fixed by the focal length of the object-glass, as when stadia and solar wires must be inserted.

In special cases a patented device of ours is to place two sets — for instance, the cross and the stadia wires — upon two different but parallel planes of the same diaphragm, where they remain in the same relation to each other, — which is not the case when they are mounted on two different diaphragms and controlled by two sets of adjusting screws. Then, by focussing the eye-piece for the cross wires or the stadia wires, they are seen one set at a time only. However, to arrive at correct results, the stadia wires should always be on the same plane with the cross wires to be seen simultaneously in the field of view, as without this the initial point in a telescopic measurement (the center wire and zero point of vertical circle) may not remain exactly the same during a measurement. — The above diaphragms are very strong and unyielding, and no amount of pressure by the capstan-headed screws can break them.

The Planimeter.

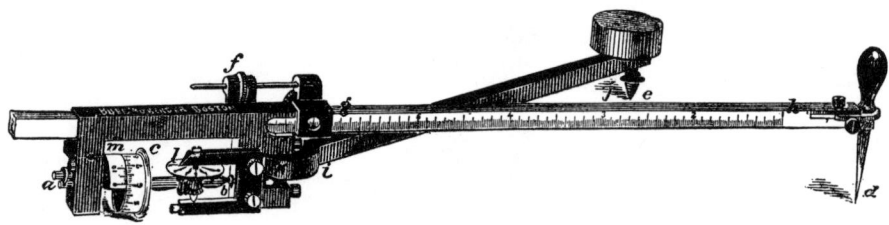

An instrument for measuring the areas of plane surfaces, by passing a pointer around their periphery. It is of great convenience to all classes of engineers, and practically applicable to a great variety of purposes. To measure the areas of figures that are bounded by irregular lines, such as :—drainage areas ; lots bounded by rivers or creeks ; contour lines of ponds, etc. ; to get the true average of observations taken at irregular intervals ; to measure indicator and other diagrams, and for many other portions of engineering work. As these instruments will not only give the area of any figure, but also any multiple of such area, and the sum of any number, or series of such multiples, at one operation, they may be used to very great advantage in the calculation of the cubical contents of solids ; as in the calculation of earth-work, etc. See on this point an article by Clemens Herschel, Esq., in the *Journal* of the Franklin Institute for April, 1874. The planimeters graduated by us are rated to read square inches of area, square centimeters of area, any multiple of these areas, and so as to give the cubic yards in any cut or fill, if used according to the directions that will accompany each instrument, Two consecutive measurements of the same area need never differ by more than 0.02 of a square inch ; and by repeating the measurement in the same manner that angles are repeated with a transit instrument, the error of observation may be reduced to but a small fraction of one hundredth of a square inch of area.

The above illustration represents the planimeter, as sold by us, ready for use. The total length of the instrument is about nine inches. The graduated bar $g\,h$ can be slid in and out in a socket formed at the top of the frame, the thumb-screw f being used for fine movements of this sort ; by this means, and by the sensible form of graduation adopted, the planimeter may be made to do the various operations spoken of above. Theory requires that the pointer d, which is moved around the periphery of the figure whose area is to be measured, the pivot k, at the junction of the two arms $g\,h$ and $i\,j$, and the main axis $a\,b$, upon which turns the measuring and counter wheel c should all be in one and the same straight line ; for this purpose, our instruments have both the pointer d and the rear part of the frame which carries the rear bearing of the axis $a\,b$, adjustable. Each reading of the instrument consists of a record of the number of revolutions of the counter-wheel c read to three places of decimals ; the whole revolutions are read on the wheel l, the tenths and hundredths on the wheel itself, and the thousandth on the vernier m. With such simplicity of construction and of operation, the accuracy of work done by this instrument is one of the most surprising things about it. The figures given above in relation to accuracy of work are, however, reliable ; being derived from the experience of several years in the use of the planimeter for many kinds of work.

A brass scale sent with our planimeters can be used to prove the correct working of the instrument. To use it drive a fine needle as an axis through one of the small holes of the scale into the paper, then put the tracer into one of the other holes and describe a circle. In this manner large and small circles of known areas can be circumscribed with perfect accuracy. This operation should be repeated in the opposite direction, and if the results agree the instrument is correct. However. if the results differ correction may be made by means of the two adjusting screws by which the tracer can be moved to right or left of the graduated bar, as the case may be.

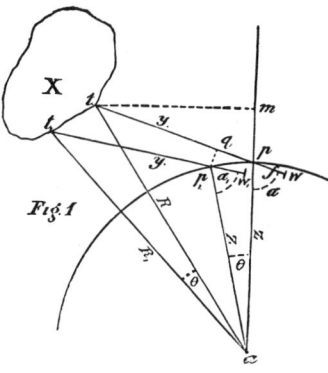

Fig 1

THE FOLLOWING DEMONSTRATION OF THE WORKING OF THE PLANIMETER IS FROM THE PEN OF WM. D. GELETTE, CIVIL ENGINEER, FORMERLY OF BOSTON.

Case I. When the anchor point is outside of the figure to be measured. Let X be the figure to be measured, and let a be the pole or origin, and R the radius of polar co-ordinates of the point t in Fig. 1. And let R_1 be the radius of a second point t_1 on the outline of the figure X, and let θ be the angle $t\,a\,t_1$—then the area of the triangle $t\,a\,t_1 = \dfrac{R\,R_1\sin\theta}{2}$. But if θ be taken so small, that for the small distance $t\,t_1$ the radius R may be taken constant, then area $t\,a\,t_1 = \dfrac{R^2\sin\theta}{2}$.

By polar co-ordinates the area $X = \Sigma\,\dfrac{R^2\sin\theta}{2}$ . . . . . . (1)

Let a be the anchor point, t the tracing point, and w the point of contact of the flange of the wheel of a polar planimeter, and call $t\,p = y$, $a\,p = z$, and $p\,w = f$, and the angle $a\,p\,w = \alpha$ then after the motion of the tracing point t to t_1 the point p comes to p_1, w to w_1 and the angle α changes to α_1. But when, as in this case, it is supposed that during the small motion $t\,t_1$ the radius R is constant, then for the same length of motion, α will be constant and $p\,a\,p_1$ will $= \theta$.

Expressing R in terms of y, z, and α, we have $R^2 = (t\,m)^2 + (a\,p + p\,m)^2 = (z + y\cos\alpha)^2 + (y\sin\alpha)^2 = z^2 + 2\,z\,y\cos\alpha + y^2\cos^2\alpha + y^2\sin^2\alpha$. But $\sin^2\alpha + \cos^2\alpha = 1$ and $R^2 = z^2 + 2\,z\,y\cos\alpha + y^2$. And the area $t\,a\,t_1 = \dfrac{R^2\sin\theta}{2} = \dfrac{z^2\sin\theta}{2} + \dfrac{2\,z\,y\cos\alpha\sin\theta}{2} + \dfrac{y^2\sin\theta}{2}$ and $X = \Sigma\,\dfrac{z^2\sin\theta}{2} + \dfrac{\Sigma\,2\,z\,y\cos\alpha\sin\theta}{2} + \dfrac{\Sigma\,y^2\sin\theta}{2}$. But in the summation, owing to the fact that the instrument returns to the same position from which it started, the $\Sigma\sin\theta$ must $= 0$ and wherever combined with constants only, in the above equation will reduce to 0, hence the first and last terms will disappear, but the middle term which contains the variable $\cos\alpha$ will remain, hence,

$X = \Sigma\,z\,y\cos\alpha\sin\theta$ . . . . (2)

It will be seen by reference to Fig. 1 that $z\sin\theta = p\,p_1$ when θ is very small, $\therefore$ $z\sin\theta\cos\alpha = p\,p_1\cos\alpha = p_1\,q$ which is the component of the motion of the wheel which is at right angles to its axis, and is therefore the part which represents the rotation of the wheel for a small motion $t\,t_1$ of the tracing point. And this component multiplied by the arm y gives $z\,y\sin\theta\cos\alpha$ which by equation (2) expresses the area of X after summation. But $\Sigma\,z\sin\theta\cos\alpha$ is the resultant rotation of the wheel after the tracing point has completed the circuit of X, hence the area $X =$ distance rolled by the wheel multiplied by the length of the arm y. Calling the circumference of the wheel c, and the number of resultant number of revolutions made during the measurement n, we have $X = y\,c\,n$. . . . . (3)

And if the instrument is graduated so as to record $y\,c\,n$, and we call the record of the instrument r, we shall have, $X = r$.

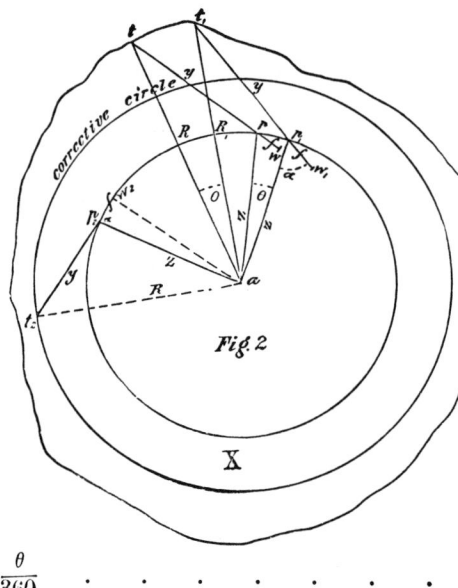

Fig. 2

X

Case II. When the anchor point is inside the figure to be measured,— and the rotation of the wheel is forward or plus with reference to the figures on its circumference,— it will be seen that there will be one position of the arms y and z, which, while the point t describes a circle around a as a center, (see fig. 2) will produce no rotation of the wheel. The condition under which this will occur will be fulfilled when the angle $t\,w\,a = 90°$, in which case $R^2 = z^2 - f^2 + (y+f)^2 = z^2 + 2\,y\,f + y^2$, and the area of the circle described under these conditions $= \pi\,R^2 = \pi\,z^2 + 2\,\pi\,y\,f + \pi\,y^2$ . . . . . (4)

Call this the corrective circle. In (fig. 2) the area of $t\,a\,t_1 = \dfrac{\pi\,R^2\,\theta}{360}$ the area of the whole figure $X = \Sigma\,\pi\,R^2\dfrac{\theta}{360}$; as before $R^2 = z^2 + 2\,z\,y\cos\alpha + y^2$ and hence $X = \Sigma\,\pi\,z^2\dfrac{\theta}{360} + \Sigma\,2\,\pi\,z\,y\cos\alpha\dfrac{\theta}{360} + \Sigma\,\pi\,y^2\dfrac{\theta}{360}$ . . . . (5)

But in the summation the instrument makes a complete revolution around a, the sum of $\Sigma\dfrac{\theta}{360} = 1$, and when combined with constants only, will not appear in the result as a factor. Hence we have

$$X = \pi\,z^2 + \Sigma\left(2\,\pi\,z\,y\cos\alpha\frac{\theta}{360}\right) + \pi\,y^2 \qquad . \quad . \quad . \quad . \quad (6)$$

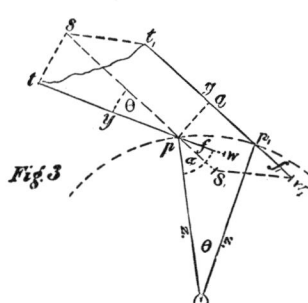

Fig. 3

Now follow out on the diagram in Fig. 3, the motion of the wheel, which corresponds to a motion of the tracing point from t to t_1 first dividing that motion into two parts $t\,s$ and $s\,t_1$ swinging the arm y around the point p until it becomes parallel to $t_1\,p_1$ while z remains fixed, produces the first motion and causes the wheel w to roll backward to s_1 and as the path of its motion is everywhere perpendicular to its axis, $s_1\,w$ will represent rotation or distance rolled during that motion, but as $s\,p\,t = \theta$ the distance $s_1\,w = 2\,\pi\,f\dfrac{\theta}{360}$ which is the backward or minus rotation of the wheel.

The second part of the motion is by moving the arm y from $s\,s_1$ to $t_1\,p_1$ during which the wheel moves from s_1 to w_1 and this motion is part sliding and part rolling, the rolling component is $p\,q$, and causes forward or plus rotation of the wheel, the value of $p\,q$ is $2\,\pi\,z\dfrac{\theta}{360}\cos\alpha$, and the resultant rotation of the wheel is on completing the circuit $= \Sigma\left(2\,\pi\,z\cos\alpha\dfrac{\theta}{360}\right) - \Sigma\,2\,\pi f$; and the area expressed by the wheel as by Case I is

$$\Sigma\left(2\,\pi\,z\,y\cos\alpha\frac{\theta}{360}\right) - \Sigma\,2\,\pi\,y\,f \qquad . \quad . \quad . \quad . \quad (7)$$

Comparing the area expressed by the wheel which we will call r with the true area of the figure as given by Eq. 6, we have

$$X = \pi\,z^2 + \Sigma\left(2\,\pi\,z\,y\cos\alpha\frac{\theta}{360}\right) + \pi\,y^2$$

$$r = \qquad \Sigma\left(2\,\pi\,z\,y\cos\alpha\frac{\theta}{360}\right) - 2\,\pi\,y\,f$$

$\mathbf{X} - r \qquad = \pi z^2 + 2\pi f y + \pi y^2$ which is seen to be identical with the expression in Eq. 4, for the area of the corrective circle. Hence, calling the area of the corrective circle C, we have for Case II

$$\mathbf{X} - r = \mathbf{C}; \text{ and } \mathbf{X} = \mathbf{C} + r \quad \cdot \quad \cdot \quad \cdot \quad \cdot \quad \cdot \quad (8)$$

It will be seen from the above equation that when $z \cos \alpha = f$, the forward rotation of the wheel just balances the backward rotation, and the result is 0 for a reading. This occurs when the tracing point moves on the arc of the corrective circle and the area passed around is $\mathbf{X} = \mathbf{C}$. When $z \cos \theta$ is less than f, then the backward rotation preponderates over the forward rotation, and the result is a minus reading; this is Case III.

Case III. When the anchor point is inside the figure to be measured, and the rotation of the wheel is backward with reference to the figures on its circumference. In this case r is negative, and instead of $\mathbf{X} = \mathbf{C} + r$, we have $\mathbf{X} = \mathbf{C} - r$. Hence, we have for Case I, $\mathbf{X} = r$, Case II, $\mathbf{X} = \mathbf{C} + r$, Case III, $\mathbf{X} = \mathbf{C} - r$.

Suppose it to be required to make the instrument record 100 for every square inch passed around by the tracing point. As the wheel is divided on its circumference into 100 divisions, and by the vernier can read tenths of these, then $\frac{1}{10}$ of a revolution will give a reading of 100, make $\mathbf{X} = 1$ sq. in., and $n = \frac{1}{10}$ then by Eq. 3, $y\frac{c}{10}$ $= 1$, and $y = \frac{10}{c}$: the length of the arm varies inversely as the reading, so for any other reading we may obtain the length of y by simple proportion. Suppose it is required to read v for every sq. in. passed around, we have

$$\frac{100}{v} = \frac{y}{\frac{10}{c}} \text{ from which } y = \frac{1000}{c\,v}.$$

The range of the arm renders it impossible to set it so that for one sq. in. of area it shall read more than 250, or less than 50, so the value of v must lie somewhere between these limits. Having determined the value of y for any particular scale, the value of C may be found by substituting in Eq. 4; the values of z and f being measured on the instrument.

A New Prismatic Stadia.

Devised by Prof. Robert H. Richards,
Massachusetts Institute of Technology.

In this prismatic stadia there is placed in front of the objective a prism or wedge of glass which half covers it.

If we hold up such a prism with a narrow angle, say 1° to 2°, and compare the transmitted image with the image seen above or below the prism, the former will be found to be thrown to one side by an amount varying with the angle of the prism. Speaking of the two rays as the direct ray and the bent ray, we may say that when the bisecting plane of the prism is at right angles to the line of sight, the angle between the direct ray and the bent ray will be constant for any given prism.

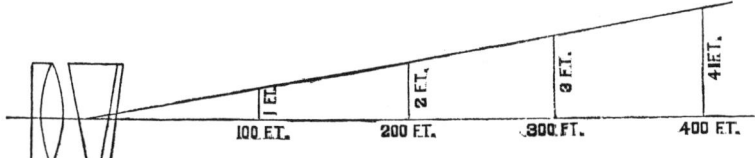

Fig. 1.

If now we place a prism in such a position that it half covers the objective of a telescope, as seen in Fig. 1, we shall obtain, upon looking through it, two images of every object seen — one image by the direct ray, which comes through the uncovered half of the objective, the other by the bent ray, which comes through the prism. The angle of divergence of these two rays will be constant and unalterable, whether the telescope is directed to a near or a distant object.

These two rays form the basis of this method of measuring, and the distance from the telescope to any given point will be proportional to the space between the lines at that point (see Fig 2).

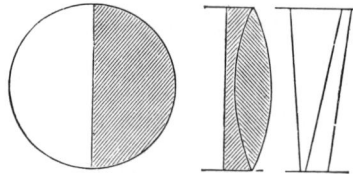

Fig. 2.

In the prismatic stadia the two spider-lines are dispensed with, and in their stead a prism is used in such a way that a double image of the target is formed. The observer, by means of a suitable adjustment, then brings the two images to coincide, whereupon the desired distance is read, either directly or indirectly, upon the rod.

Since the angle, Fig. 2, between the bent ray and the direct ray is constant, the employment of the prism obviates the variation to which the angle is subjected by the use of the spider lines.

The telescope for use with this stadia must be chosen to suit the purpose in view. For a hand telescope, 10 diameters magnifying power is quite powerful enough. On the other hand, if a substantial tripod be used, the higher the power, up to 60 diameters, the better. However, telescopes of 20, 30 or 40 diameters do excellent work. The telescope should give ample light, and upon a tripod an inverting telescope can be used quite as handily as a terrestrial one, provided an agreement is made with reference to the targets so as to avoid confusion.

The two images, produced by the direct and bent rays respectively, have a strong analogy to the two images in a sextant. The sea captain brings down the sun's image with one glass till he gets contact between the sun's disk and the horizon, and the angle is then read from the vernier of the instrument. With the prismatic stadia-telescope the lower image is brought into juxtaposition with the upper one, and we read the amount that the lower image has cut off upon the upper one. This reading gives us the displacement of the image, and from this we can compute the distance.

Prof. Richards found that a prism having an angle of 1 : 100 gives the best results for long or short distances, and that a telescope of about 30 diameters is the best power, although good results can be obtained with other powers.

The advantages claimed for this prismatic stadia-telescope as compared with the usual forms read by spider-lines in the telescope are :

1st. The starting-point for the measurement is within the prism, instead of being at an imaginary point some distance in front of the objective.

2d. Only one observation is required for measuring a distance, and there is therefore only one personal equation instead of two.

3d. The portion of the self-reading target which is to be read is the only part that is distinct. It is therefore unnecessary to hunt for the reading.

4th. The prismatic stadia-telescope can be used in the hand for short distances. This is impossible with the spider-line instruments.

5th. Long distances can be satisfactorily measured by the tape and sliding-targets, but not with the spider-line.

6th. The extreme variation of this instrument as found by Prof. Richards is from .2 per cent to .052 per cent. The extreme error will therefore be from ± .1 per cent down to ± .026 per cent, diminishing as the distance increases, while the error of the spider-line stadia may be as high as .4 per cent, and often is .2 per cent, and does not diminish with distance.

In making an observation we must first focus the two images, next rotate into juxtaposition, and finally read the distance.

The target is made in two forms, the self-reading rod and the tape-target.

Self-reading rods should be prepared for each prism according to its ratio or factor; but it appears probable that prisms can be made that will vary only between the ratios 1 : 100.2 and 1 : 99.8, and thus vary not more than 0.2 from the ratio of 1 : 100 Measures will, therefore, be taken for preparing rods with standard graduation of 1 : 100, which will be near enough for most purposes.

If it is desired to attach this prismatic arrangement to an instrument already made it will be necessary *to send us the telescope.*

For a pamphlet giving a detailed description and method of using apply to either Prof. Richards, Massachusetts Institute of Technology, Boston, Mass., or to us.

On Stadia Measurement.

Written especially for this Catalogue by GEO. J. SPECHT, C. E., San Francisco, Cal.

A transit or theodolite, which is provided with the so-called stadia wires and a vertical circle, furnishes the means to obtain simultaneously the distance and the height of a point sighted at without direct measurement, and with the only use of a self-reading rod, held at the point of which the horizontal and vertical position is to be determined in reference to the instrument-point.

Besides the ordinary horizontal and vertical cross hairs of the diaphragm of the telescope, two extra horizontal hairs are placed parallel with the center one, and equally distant on each side of it, which, if the telescope is sighted at a leveling rod, will inclose a part of this rod or stadia-rod, proportional to the distance from the instrument to the rod. By this arrangement we have obtained an angle of sight, which remains always constant.

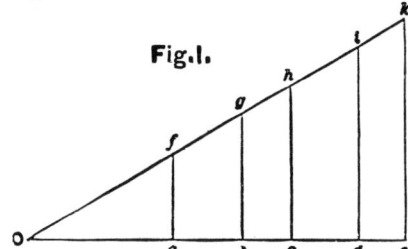

Fig. 1.

Supposing the eye to be in the point O (Fig. 1), the lines O e and O k represent the lines of sight from the eye through the stadia-wires to the rod, which stands consecutively at $k\,e$, $i\,d$, $h\,c$, $g\,b$ and $f\,a$. According to a simple geometrical theorem we have the following proportion:

$$O\,a : O\,b : O\,c : O\,d : O\,e = a\,f : b\,g : c\,h : d\,i : e\,k,$$

which means that the reading of the rod placed on the different points a, b, c, d and e is proportional to the distances O a, O b, O c, O d and O e.

The system of lenses which constitute the telescope do not allow the use of this proportion directly in stadia measurements, because distances must be counted from a point in front of the object glass at a distance equal to the focal length of that lens.

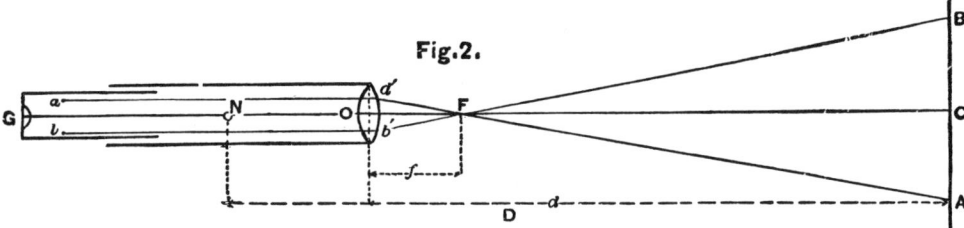

Fig. 2.

Fig. 2 represents the section of a common telescope with but two lenses, between which the diaphragm with the stadia-wires is placed.

We assume:

$f =$ the focal distance of the object glass.

$p =$ the distance of the stadia-wires a and b from each other.

$d =$ the horizontal distance of the object glass to the stadia.

$a =$ stadia reading (B A).

$D =$ horizontal distance from middle of instrument to stadia.

The telescope is leveled and sighted to a leveling or stadia rod, which is held vertically, hence at a right angle with the line of sight. According to a principle of optics, rays parallel to the axis of the lens, meet after being refracted in the focus of the lens. Suppose the two stadia wires are the sources of those rays, we have, from the similarity of the two triangles, a' b' F and F A B the proportion:

$$(d-f) : a = f : p.$$

The value of the quotient, $f : p$, is, or at least can be made, a constant one, which we will designate by the letter k; hence we have:

$$(d-f) = \mathrm{F\,C} = k\,a.$$

In order to get the distance from the center of the instrument N, we have to add to the above value of F C yet the value c.

$$c = O\,F + O\,N.$$

O N is mostly equal to half the focal length of the objectic, hence we have

$$c = f + \frac{f}{2} = 1.5\,f.$$

Therefore the formula for the distance of the stadia from the center of instrument, when that stadia is at right angles to the level line of sight, is:

(1) $$D = k.a + c.$$

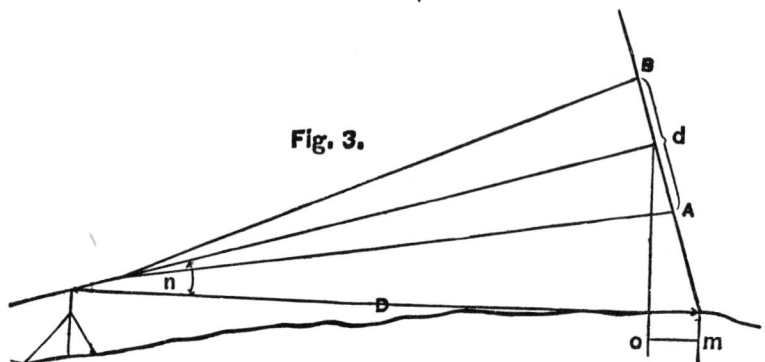

Fig. 3.

When the line of sight is not level, but the stadia held at right angle to it, the formula for the horizontal distance is:

(2) $$D = k.a.\cos n + c + om.$$

The member $\overline{om} = \frac{a}{2}\sin n$; for $a = 24'$, $n = 45°$ the value of $\overline{om}$ is but 8.4′, and for $a = 10'$, $n = 10°$ it is 0.86′; this shows that $\overline{om}$ in most cases may safely be omitted.

Some engineers let the rodman hold the staff perpendicularly to the line of sight; they accomplish this by different devices, as, a telescope or a pair of sights attached at right angle to the staff. This method is not practicable, as it is very difficult, especially in long distances, and with greater vertical angles for the rodman to see the exact position of the telescopes, and furthermore, in some instances it is entirely impossible, when, for instance, the point to be ascertained is on a place where only the staff can stand, but where there is no room for the man. The only correct way to hold the staff is vertically.

In this case we have the following: (Fig. 4)

$$MF = c + GF = c + k.C.D.$$
CD must be expressed by AB.
$$AB = a. \quad AGB = 2m.$$
$$CD = 2GF \tan.m.$$

And finally, after many transformations:

$$D = c.\cos n + a.k.\cos^2 n - a.k.\sin^2 n \tan^2 m.$$

The third member of this equation may safely be neglected, as it is very small even for long distances and large angles of elevation (for 1500′, $n = 45°$ and $k = 100$, it is but 0.07′). Therefore, the final formula for distances, with a stadia kept vertically, and with wires equi-distant from the center wire, is the following:

(3) $$D = c.\cos n + a.k.\cos^2 n.$$

The value of $c.\cos n$ is usually neglected, as it amounts to but 1 or 1.5 feet; it is exact enough to add always 1.25′ to the distance as derived from the formula

(3a) $$D = a.k.\cos^2 n$$

without considering the different values of the angle n.

In order to make the subtraction of the readings of the upper and lower wire quickly, place one of the latter on the division of a whole foot and count the parts included between this and the other wire; this multiply mentally by 100 (the constant k) which gives the direct distance D'.

In cases where it is not possible to read with both stadia wires, it is the custom to use but one of them in connection with the center wire, and then to double the reading thus obtained. With very large vertical angles, this custom is not advisable, as the error may amount to 0.50 %.

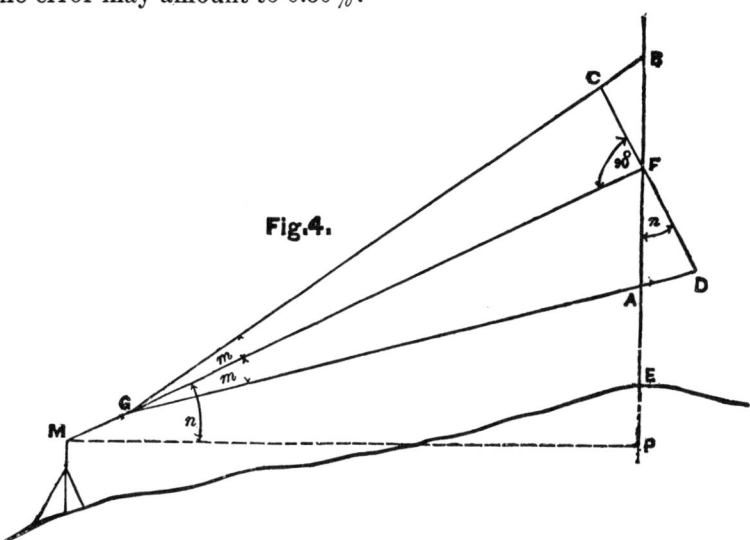

Fig. 4.

To find the height of the point where the stadia stands above that one of the instrument, simultaneously with the distance, we have the following:

We assume in reference to figure 4,

q = height of instrument point above datum.

$MP = D$ = horizontal distance as derived from formula (3).

n = vertical angle.

$h = FE$ = stadia reading of the center wire.

Q = height of stadia point above datum; it is
$$Q = q + D \tan n - h.$$

The substraction of h can be made directly by the instrument, by sighting with the center wire to that point of the rod, which is equal to the height of the telescope above the ground (which is in most cases $= 4.5'$); q will be constant for one and the same instrument point; then the formula:
$$Q = D \tan n;$$
this in connection with formula (3) gives
$$Q = c \sin n + a.k. \cos n. \sin n.$$

or
$$Q = c \sin n + a.k. \frac{\sin 2\,n}{2}$$

The first term of the equation can be neglected, when the vertical angle is not too large; hence the final formula for the height is

(5)
$$Q = \frac{a.k.\ \sin 2\,n}{2}$$

The position of the stadia must be strictly vertical.

The error increases with the height of m; (m = height of center wire on the rod). In shorter distances the result is seven-fold better when the center wire is placed as low as one foot than it is at 10'; in longer distances this advantage is only double.

It is always better to place the center wire as low as possible. If the stadia is provided with a good circular level, the rodman ought to be able to hold it vertically within 500″; that means, that the inclination of the stadia shall not be more than 0.023′ in a 10′ stadia, or 0.034′ in a stadia of 15′ length.

Determination of the two constant coefficients c and k. Although the stadia wires are usually arranged so that the reading of one foot signifies a distance of 100 feet, I will explain here, how to determine the value of it for any case. Suppose the engineer goes to work without knowing his constant, and not having adjustable stadia wires. The operation then is as follows:

Measure off on a level ground a straight line of about 1000′ length; mark every 100′, place the instrument above the starting point, and let the rodman place his rod on each of the points measured off; note the reading of all three wires separately, repeat this operation four times; the telescope must be as level as the ground allows; measure the exact height of the instrument, *i. e.*, the height of the telescope axis above the ground. Then find the difference between upper (o) and middle (m) wire; between middle (m) and lower (u) wire, and between upper (o) and lower (u) wire, from the four different values for each difference, determine the average value; then solve the equation for the horizontal distance (1) $D = k.a + c.$, with the different average values, and you find the value of k and c. In case the stadia wires should not be equi-distant from the center wire, there will be three different constants, one for the use of the upper and middle, one for the use of the middle and lower, and one for the upper and lower wire.

If the stadia wires are adjustable, the engineer has it in his power to adjust them so that the constant $k = 100$, or $k = 200$, which he accomplishes by actual trial along a carefully measured straight and level line.

The constant c, which is one and a half times the total length of the object-glass can be found closely enough for this purpose by focussing the telescope for a sight of average distance, and then measuring from the outside of the object-glass to the capstan-head-screws of the cross-hairs. This constant must be added to every stadia sight; it may be neglected for longer distances.

Stadia Measurements.

Written for this catalogue and manual by H. C. Pearsons, C. E., Ferrysburg, Mich.

In view of the great and growing interest in the subject of "*Stadia Measurements,*" the following solution of the problem is offered, as applied to inclined measurements.

This solution is made from a different geometrical consideration than that usually employed, and it effectually does away with the necessity for any subsequent corrections, as with most schemes in use for inclined distances.

In the following discussion, let

R = the reading of the stadia rod;
D = the horizontal distance from plumb line of transit to stadia rod, which must be vertical.
m = the angle of elevation or depression to the *smaller* reading of the stadia rod.
n = the same angle to the *larger* reading.

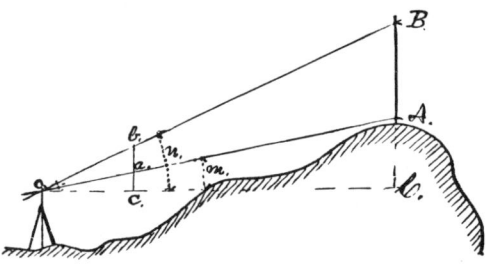

Through the point c, at the distance of unity from the centre of instrument, draw the vertical cb. Then the rod AB, being also vertical, the triangles aob and AoB are similar, as are also the triangles cob and CoB. But the reading, R, of the rod A B is the difference of the tangents of the angles of elevation, m and n. Also, the distance $a\,b$ is the difference of the tangents of these angles, m and n, to distance unity, as given in the trigonometrical tables.

Whence, to find the horizontal $o\,C = D$, we have simply to *divide the reading of the " Stadia Rod " by the difference of the tangents of the angles of elevation.* Or, by formula,—

$$D = \frac{R}{\text{Tan. } n - \text{Tan. } m}$$

If one of the angles should be a depression or —, we must then divide by the *sum* of the tangents, and the formula would be

$$D = \frac{R}{\text{Tan. } n + \text{Tan. } m}$$

Example.—If $n = 12° 16'$, nat. Tan $= .217426$
" $m = 10° 10'$, " " $= .179328$
The difference of the tangents $= .038098$

Then, if R $= 12.26$ feet,

$$oC = D = \frac{12.26}{.038098} = 322 \text{ feet.}$$

It may happen that our transit has no vertical circle, or that we have no trigonometrical tables at hand. In either case, introduce an auxiliary rod, $c\,b$ between the stadia rod and the plumb-time of transit, and at some known horizontal distance,— preferably 100 feet,— from the latter, and note the intercept $a\,b$.

This intercept is the analogue of the difference of tangents used in the former case, and must be used in the same manner, in dividing the reading of the stadia rod, when we shall have the distance, D, in terms of the distance of the auxiliary rod from the transit.

Example. — Suppose the intercept $a\,b$ on the auxiliary rod, at distance 100 feet, is .845 foot, and that the reading R, of the stadia rod is 12 feet, then

$$oC = D = (12 \div .845) \times 100 = 1420 \text{ feet.}$$

If the height, H, of the foot of the stadia rod, above or below the height of instrument, be wanted, it may be had from the following equation :

$$H = \pm\, D.\ \text{Tan. } m,$$

In which the $+$ sign must be used for angles of elevation, and the $-$ sign for those of depression.

Or if the auxiliary rod be used instead of the vertical arc, note the intercept $a\,b$ on this rod, between the level line $o\,C$ and the line of sight to the foot of the Stadia rod, and

Multiply this intercept by the ratio of D to o c.

Example — In the last case, if $o\,a = 1.06$ ft., $o\,c$ being 100 feet, and D $= 1420$ ft., then

$$CA = H = o\,a \cdot \frac{D}{o\,c} = 1.06\,\frac{1420}{100} = 15.05 \text{ feet.}$$

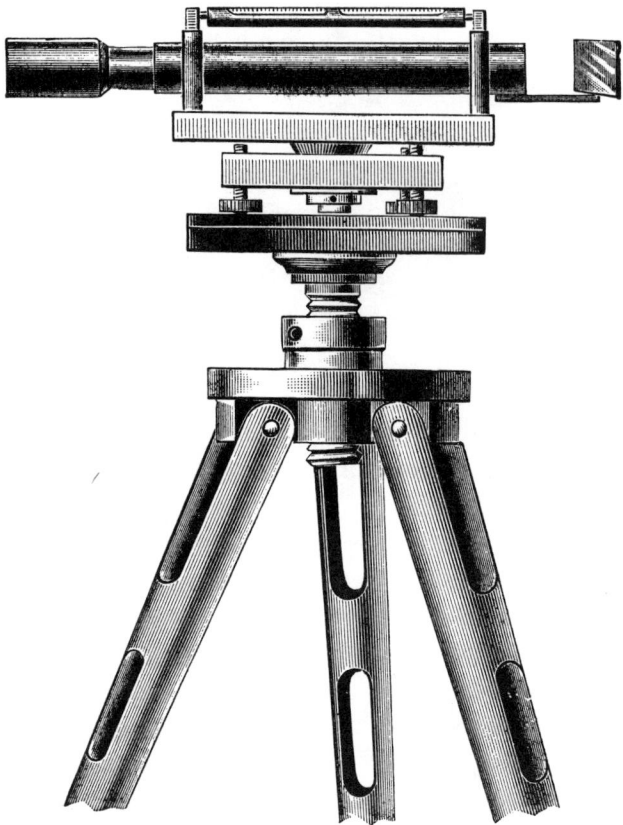

The Portable Collimator.

For use with Collimator A or B, for testing theodolites, by measuring and repeating angles in the manner these instruments are manipulated in the field. It can be set at any angle or height. It is a valuable aid in adjusting instruments with fixed angles, etc.; and also serves a variety of purposes where distant and fixed sights are a necessity in the shop.

Note : — The above procedure of measuring and repeating angles, by sighting at two collimator objects placed at any desired angle with the instrument, might be profitably adopted in the primary instruction of students at polytechnic schools before putting them on actual practice in the field. For this purpose alone the collimators may be of a most primitive construction. Any two cheap spy-glasses that can be bought in a store will answer, and the field of view can be simply scratched on the silvering of a mirror. The latter should be placed in the principal focus of the objective of the spy-glass. There is no need of a spirit-level. The whole apparatus can be improvised in any well lighted basement where it is not apt to become disturbed by sudden jars and vibration of the building. Where there is a well-adjusted wye level or transit at disposal for this purpose only one such collimator will be needed, and when they are placed opposite each other at a distance of about ten feet, so that an instrument can be readily placed and operated between them, they will then afford an excellent means of making the telescope adjustments by reversing the telescope of a transit upon a point, or successive points, in the opposite direction as explained on page 49.

In treating a similar subject (also on the same page) it is assumed that the geometrical and optical axes of the two telescopes are truly coincident with each other (a condition requiring some skill to make) in order to adjust the line of collimation in a transit telescope. By employing a cheap collimator of the above kind, so that the telescope can be reversed on the collimator object, this assumption becomes, however, invalidated, as the adjustments can be made precisely in the same manner as in the field. To test the adjustments for nearer distances it will then only be necessary to set the collimator object slide at places corresponding to these distances.

The use of such an apparatus will give the student clearer ideas of geodetic instruments and the optical principles upon which they are based, and every large engineering department ought to be in possession of an improvised apparatus of this kind. Much valuable time, which must otherwise be spent in the field, can then be saved.

C. L. Berger & Sons' Auxiliary Apparatus,
Used during the construction of their Instruments of Precision.

APPENDIX.

GIVING SOME METHODS, NOT ALREADY EXPLAINED, FOR ADJUSTING THE INSTRU-
MENTS ENUMERATED IN THIS CATALOGUE.

We feel that we owe an apology to our readers for the imperfect arrangement of our hand-book and illustrated catalogue. This has been caused by its gradual enlargement as our business has increased. It is our intention, should other pressing duties permit, to rearrange the descriptive part of the catalogue after the completion of several new types of instruments which are now contemplated, and to bring into their proper order the various topics treated in this book. — C. L. B.

Spirit-Levels on Metal Base. (See page 127.)

A level plane surface is required, upon which the adjustment of a level, mounted on a plane metal base, may be made. Such a level surface may be some portion of a field instrument, as the horizontal plate of a transit or the bar of a wye-level, which may be made horizontal by the leveling screws. If an instrument is not available, a plane surface upon any piece of machinery which may be suitably raised or lowered, may be used.

Place the level upon the plane surface with its edge coinciding with a straight line previously marked upon the surface, and bring the bubble to the center of the graduated lines or marks ruled on the glass tube, by raising or depressing the plane surface. Next reverse the level end for end, taking care to replace it precisely in the place that it previously occupied, and notice whether the bubble returns to the center of the gradua-tions ; if so, the level is in adjustment ; if not, correct one-half the error by means of the two opposing nuts, or by a capstan-headed screw, provided at one of its ends, and cor-rect the other half by raising or lowering the plane surface. Repeat these operations until the adjustment is perfect.

Locke's Hand-Level. (See page 127.)

There are several methods by which this instrument may be adjusted. (a) Select a sheet of water with an unobstructed view of not less than 200 feet, and sight through the level at an object whose height above the water has been made the same as the height of the eye. If the bubble is bisected at the same time that the object is bisected by the horizontal wire, the level is in adjustment ; if not, move the frame containing the wire, by means of the screw or screws placed at the ends of the mounting of the level, until this is the case. (b) If an adjusted wye-level is at hand, set it on its tripod at a height corresponding to the height of the eye above the ground, and direct it at some well-defined object of suitable size, as the target of a leveling rod, distant about 500 feet, and in the level plane given by the instrument. Now raise the hand-level to the eye and notice whether its wire bisects both the object and the bubble at the same time. If not, adjust the wire as explained above. (c) Select a tolerably level place where there are two trees, about 200 feet apart (two telegraph posts or two sticks fastened to a fence will answer the purpose). Find, by pacing or otherwise, the point midway between the two objects selected. Standing at this point, raise the hand-level to the eye, sight at each object in succession, and make distinct marks where the wire inter-sects them when the bubble is bisected. These two points will be approximately in a level line, although the instrument may be out of adjustment. Walking to the side of either of these objects, place the hand-level against the mark upon it and sight at the mark upon the other object. If the bubble is then bisected by the horizontal wire the level is in adjustment ; if not, adjust it as above.

Hand-Level and Clinometer. (See page 127.)

If the instrument is of the style shown in the cut, it may be adjusted thus. Clamp the index or vernier of the arc at 0° and test the adjustment by the methods described for Locke's hand-level. If the bubble is not bisected by the wire, the level must be ad-justed by raising or lowering the tube containing it by means of one of the adjusting screws at the end ; — or in some instruments by moving in or out a square tube to which the wire is fastened — until the adjustment is made. This latter tube should then be fastened by a screw or marked by a scratch, as the case may be.

If the style of the instrument permits of being placed with its base upon a plane sur-face, as is usual with an ordinary clinometer, the parallelism of the level with the base should be tested, when the index is at 0°, by the method explained above for adjusting the spirit-level with the metal base. If found to be out, the spirit-level should be ad-justed by one of the adjusting screws at the end of its tube. The line of sight must now be verified as explained for the hand-level. If found to be out, the adjustment must be made by moving in or out the square tube carrying the wire as described above.

The Adjustments of the Engineers' Transit for Leveling and for Measuring Horizontal and Vertical Angles.

These adjustments differ according as the telescope is provided with a fixed level, which is the more usual form in the engineers' and surveyors' transits, or is provided with a striding-level parallel to the telescope, as shown in the illustration of the transit-theodolite No. 11a.

1. A Fixed-Level Parallel to the Telescope.

This adjustment consists of two distinct operations. The first is to place the horizontal wire of the telescope in the combined geometrical and optical axes of the telescope, or the line of collimation (see "Some Facts Worth Knowing," Part I.); the second is to place the axis of the fixed-level parallel to the line of collimation.

The precision of the adjustments will depend in a measure upon the perfection with which the geometrical axis of the telescope has been made to coincide with the principal optical axis for all focussing positions of the object-slide. The more perfect the continual coincidence of these two axes, the more precise and accurate will be the measurements when the sights are of unequal lengths ; but in practical leveling, where sights of equal length are used, the effect of a non-coincidence of these two axes is mainly confined to "intermediate points," and is of minor consequence unless the non-coincidence is very marked.

The Adjustment of the Horizontal Wire. This may be done with sufficient accuracy in our transits having an erecting telescope, where the eye-piece is non-adjustable laterally, in other words, permanently centered with respect to the optical axis, by merely placing the horizontal wire in the center of the field of view of the eye-piece. The cross-wires of inverting telescopes, however, being stretched on a diaphragm which itself limits the field of view, they will appear in the center of the field whether the instrument is adjusted or not, and therefore another method of adjustment is needed. This can be done roughly in an emergency by merely placing the horizontal wire by the eye as nearly as possible in the center of the main tube of the telescope, the eye-piece having been first removed. Another method is that described by Gravatt, see pages 48 and 49. But the method which is the easiest in our shop practice, and sufficiently accurate for all practical purposes, is that followed by us, and as it might be followed with profit by the engineer, should circumstances favor, we give it here also. After the wires have been placed in a position perpendicular to the horizontal axis of revolution, the telescope is removed from the standards and placed in wyes where it can be rotated. Such wyes may be improvised by the engineer, by cutting the proper shapes out of thin wood and fastening a pair of them to a board in an upright position. The distance between them should be such, that the telescope may rest upon the outside of the mounting of the object-glass where the cap is placed, taking care to rest it against its shoulder on the inside of the wye, and upon the tube near the cross-wires, when practicable. The improvised wyes being placed on a firm support and fastened so that they will not move, the telescope may be revolved in them and the wires may be placed in the line of collimation as in a wye-level, using a distant point.

In some makers' transits the telescope is so arranged that it can be revolved in the center part of the transverse axis of revolution, after the level tube and the screw fastening it in the center of the hub which keeps the telescope and axis together have been removed. In this case the horizontal and vertical wires can be adjusted approximately for collimation by turning the telescope in its hub. This method, however, does not permit of close work, and besides being clumsy, generally leaves the telescope in a more insecure state of stability, than when the telescope and transverse axis are soldered together. On this account, whenever applicable, all our instruments are made in this latter manner.

There is another method which may be followed if the principal optical axis passes through the geometric horizontal axis of revolution of the telescope. Select a place for the transit alongside of a house or near a tree, and with an unobstructed view of at least 300 feet. Set the instrument and level up approximately. Clamp the telescope, and by means of its tangent-screw bisect a distant point with the horizontal wire. Then turn the instrument on its vertical center and find or mark a point, distant about 50 feet, on the house or tree, carefully focussing the telescope upon it. Now unclamp, reverse the telescope, clamp it again and bisect the nearer mark with the horizontal wire. Then turn the instrument again on its vertical axis and see whether the wire also bisects the distant point; the telescope having been carefully re-focussed. If the wire does not bisect it, correct the error by moving the wires, by means of the capstan-headed screws, in the direction explained in the foot-note on page 49.

The amount that the wire should be moved can be determined, by this method, only by successive approximations. If the horizontal wire is considerably out, move it an amount equal to two or three times the apparent error, and again test the adjustment. As the adjustment approaches completion, the wire should be moved less and less, until the correct position is attained.

As before stated, this method can only be relied on when the principal optical axis passes through the geometrical horizontal axis of revolution. When these two axes do not intersect, the amount of error is doubled and multiplied by reversing the telescope, and causes an eccentric position of the wire when the adjustment appears to be perfect. It is extremely rare that an instrument fulfills this condition exactly, the maker having to rely solely upon mechanical methods of construction; and it is not wise for the engineer to place much confidence in this method of adjusting the horizontal wire.

The adjustment of the vertical wire should now be made, as explained on page 49, by reversing the telescope through the standards, or over the bearings of the horizontal axis of revolution, as explained on page 100. The adjustment of the horizontal wire should then be examined again, and if found to be disturbed, the two wires should be adjusted again, in turn, until the position of each is found to be correct.

The Adjustment of the Fixed-Level of the Telescope. This is to be done by means of stakes set on opposite sides of the instrument, as described on pages 48, 50 and 54. By this method the axis of the spirit-level is placed parallel to the line of collimation directly, without depending on the accuracy of collars or other mechanical arrangements; thus ensuring great accuracy in the adjustment.

2. A Striding-Level Parallel to the Telescope.

The adjustment of a transit for leveling, when it has a striding-level parallel to the telescope, is similar in some respects to the telescope adjustment of the ordinary wye level. The telescope of the transit is provided with collars, or rings, upon which the striding-level rests. It may be placed upon them when the telescope is either in a direct or in a reversed position, and it can be reversed end for end upon them. In a theoretically perfect instrument it is assumed that the collars are cylinders of truly circular form and of exactly the same diameter; that they have a common axis which is parallel to or coincident with the geometrical and optical axis of the telescope, so that the line of collimation shall remain true for all focussing positions of the object-slide. It is also assumed that the combined geometrical and optical axis intersects the geometrical transverse axis of revolution of the telescope. The tests of these assumed conditions will be given below.

The following explanations will render the description of the adjustments more simple : —

The interior surface of the glass tube or vial is ground so that it has a slight curvature in the direction of its length, the radius of the curvature depending upon the degree of sensitiveness required for any particular instrument. This grinding gives a symmetrical barrel-shape to the interior surface. Consequently, if the tube be rotated slightly to one side, the bubble will still indicate a horizontal plane. The vertical plane which bisects the bubble longitudinally may be called the vertical axial plane of the bubble. The line tangent to the upper longitudinal element of the ground interior of the tube, at the central point of the graduations, is called the axis of the spirit-level. It is situated in the plane of curvature of the element, that is, in the vertical axial plane, and is horizontal when the bubble has settled normally in the center of its graduations.

The adjustment of such a transit for leveling consists in placing the axis of the striding-level parallel to the axis of the collars, and then, in placing the horizontal wire in the combined geometrical and optical axis of the telescope by means of the striding-level.

The adjustment of the striding-level consists of two separate operations, as follows : —

The lateral adjustment of the Striding-Level. This adjustment is the first to be made. Its object is to place the vertical axial plane of the spirit-level parallel to, or coincident with the axis of the collars, and is necessary to avoid errors in the position of the bubble, should the striding-level be slightly moved on its collars laterally during the manipulation of the instrument. The telescope should be clamped in a position nearly horizontal, and the striding-level should be placed with its wyes resting on the collars. The bubble must now be brought approximately to the center of its tube by the tangent-screw of the telescope. The striding-level should then be detached from its fastenings on the telescope, and, while resting on the collars, be inclined to one side of the telescope, say about 10° from its vertical position, and the reading of the bubble be noted. Next, it must be rocked on the collars an equal amount to the opposite side of the telescope, and the reading be noted. If both readings are the same, the adjustment is correct, but if the bubble moves towards one end of its tube when the striding-level is in one of these positions, and towards the other end of the tube when it is in the second position, it indicates that the vertical axial plane of the bubble is not parallel to the axis of the collars, and that the end towards which the bubble moves is elevated with respect to the axis of the collars, causing the movement of the bubble from the center of the tube. The adjustment to be made is to bring the bubble again to the center by means of the capstan-headed screws at the side of one leg of the striding-level. They are opposing butting screws, and the one towards which the tube must be moved to bring the bubble again to the center must first be slightly loosened. The opposite screw must then be tightened until it is perceived that a perfect metal contact has been established. This operation must be repeated until the reading of the bubble remains the same in both positions of the striding-level relative to its normal position. This will also indicate that the ground portion of the tube is truly barrel-shaped.

If the level be extremely sensitive, it is sometimes found that the bubble will run slightly towards the *same* end of the tube, as the lateral adjustment approaches completion, although the striding-level may have been inclined at the same angle when in each inclined position. This indicates that the ground portion of the tube is funnel-shaped, and then the adjustment is completed by making the bubble run to the same amount when the striding-level is rocked through equal angles laterally. The effect of this deviation of the glass vial from the barrel form is generally within the limits of error of observation with the telescope, but aside from this, it is imperceptible in most cases, as the striding-level can rock only one or two degrees laterally when it is fastened to the telescope. The lateral adjustment is necessary in order to avoid errors in leveling, should the striding-level be moved laterally on its collars during the manipulation of the instrument.

The horizontal adjustment of the Striding-Level. This, the principal adjustment, is to make the horizontal axis of the spirit-level parallel to the axis of the collars. To do this, bring the bubble to the center of its tube by means of the tangent screw of the telescope, then reverse the striding-level end for end, and note whether the bubble returns to the center. If it does, the adjustment is correct. If not, correct half the error by means of the vertical opposing capstan-headed screws at the other leg of the striding-level, moving them as has been explained for the lateral adjustment, and correct the other half by the tangent screw of the telescope. If, upon again reversing the striding-level end for end, the bubble remains in the center of its tube, all the adjustments of the striding-level are made. But it is well to verify the correctness of the lateral adjustment again, especially if the amount of correction required to adjust the horizontal axis of the level was large, and to repeat both adjustments in the order given above until assured of their correctness.

The adjustment of the horizontal cross-wire. After the striding-level has been fully adjusted, the adjustment of the horizontal wire of the telescope should next be made. This adjustment is to bring the horizontal wire into the combined geometrical and optical axis of the telescope, as follows. First make the vertical cross-wire perpendicular to the plane of the transverse axis by the process described on page 48. The striding-level is then placed upon the collars and is secured to the telescope by means of its fastenings. The instrument being leveled up, the telescope is clamped and the bubble of the striding-level is brought to the center of its graduation. Now turn the instrument on its vertical axis and find, through the telescope, a suitable and well defined object, distant about 400 feet, which the horizontal wire bisects when the bubble of the striding-level is in the center. Then remove the striding-level, and, if the telescope reverses through the standards, as is usual in these transits, unclamp the telescope and plates, reverse the telescope, turn the instrument on its vertical center and direct the telescope again towards the object previously bisected. Clamp the telescope, replace the striding-level on the collars and bring its bubble to the center. If the horizontal wire again bisects the same object, its adjustment is correct, but if not, move the reticule bearing the wires over half the error by means of the capstan-headed screws of the telescope, as explained in the footnote on page 49, and repeat the operation until the adjustment is correct. If the telescope is not reversible through the standards, as in Theodolite No. 14, (or as in most Plane-tables), the horizontal axis of revolution should be reversed end for end in its bearings, and the plates should not be unclamped. In other respects the process is the same as above described. The instrument is now in adjustment for leveling with the striding-level. To insure good leveling with the striding-level, it is necessary that its points of contact on the collars be free from dust, grit, and oxide.

The adjustment of the vertical wire should now be made as explained on page 49, by reversing through the standards, or if the instrument has low standards, over the bearings of the horizontal axis of revolution, as explained on p. 100, after which the adjustment of the horizontal wire should be examined again, and if found to be disturbed, the two wires should be adjusted in turn, successively, until the adjustment of each is found to be correct. See "Some Remarks Concerning Instrument Adjustments," Part I.

Tests of the assumed conditions.

The Collimation. It has been assumed in the above methods of adjustment that the line of collimation passes through the geometrical transverse axis of revolution of the telescope; otherwise, a change in its height would occur when the telescope is reversed, giving two *parallel* lines of sight, of which the distance apart is double the distance between the line of collimation and the geometrical transverse axis.

An error from this source in a well designed instrument of modern construction will be very small, and will be manifest only with short sights and in most cases will be within the limits of error of observation for long sights, since the space covered by the cross-wire is proportional to the distance to the object, and will soon amount to considerably more than this error. It would not, therefore, affect the adjustment of the instrument for long sights.

It is also assumed that the principal optical axis coincides with the geometrical axis of the telescope for all focussing positions of the object-slide; that is, whether the telescope be focussed on a distant or on a near object; otherwise the telescope would give incorrect readings for objects at different distances; more fully explained in Part I., "Some Facts Worth Knowing."

The existence of these two sources of error may be detected by repeating the test of the adjustment of the horizontal wire with the striding-level, using then a very short sight. If the horizontal wire does not bisect the near object when the telescope is both in a direct and in a reversed position, there is an error which affects readings on near points, caused by one or both of the above mentioned sources of error. This error may be eliminated with the striding-level, by the method used in precise geodetic leveling to remove errors of collimation, which is to take two readings of the rod at each point, one with the telescope direct, and the other with it reversed; the mean of the two being the true reading. This process may be used for distant points if it is suspected that the adjustment of the horizontal wire has been disturbed. This error cannot be eliminated with a fixed level; nor, as it might be supposed, with a reversion level, that is, a fixed level accurately ground and graduated upon the top and bottom (for an illustration see elsewhere) since both horizontal axes of such a level are adjusted to the line of sight for long distances, and because the level cannot be *reversed on the telescope tube.* The effect of an error of this kind in short distances can, however, be easily eliminated when backsights and foresights are equal in length.

The Equality of the Collars and Flexure of the Telescope. It has also been assumed that the collars are of the same diameter. This is of prime importance since inequality in diameter will cause large errors. If they are of unequal diameters the line of collimation will not define a horizontal plane when the bubble of the striding-level is in the center of its tube, although the axis of the collars may have been made parallel to the combined geometrical and optical axis of the telescope. The equality of the collars will always exist in our instruments when new, since the auxiliary and collimator apparatus used by our firm enables us to detect the least deviation of the collars from a circular form or from equality in diameter, and, consequently, the collars will be of unequal diameters only from the effects of wear and careless use.

A similar deviation from a horizontal line will exist, however, although the collars are of equal diameter, owing to the flexure of the telescope by its own weight, when in a horizontal position. This flexure of the telescope is greater in the telescopes of greater length. It is also aggravated when the telescope is not symmetrically mounted upon its transverse axis of revolution. This is the case to a certain extent in all geodetic instruments, owing to the changes in the position of the center of gravity of the telescope when focussing the object-slide for sights at different d s-tances, or when the sun-shade is removed, slight as the effect of the weight of the latter may be.

To test the combined effect of inequality of the collars, if any, and the flexure of the telescope, a method very similar to that used for adjusting the fixed telescope level, explained on pages 51 and 53, may be used.

After the striding-level and horizontal wire have been adjusted, as explained above, one may try the methods for adjusting the fixed-level, and see whether the line of collimation is in fact horizontal when the bubble of the striding-level is in the center of its tube. If the horizontal cross-wire bisects the target when the rod is held on the distant stake (see page 53), then the line of collimation is horizontal, but if there is any deviation, it may be owing to inequality of the collars or to a flexure of the telescope, or to both causes combined. To eliminate this error, the operator may move the telescope by the vertical tangent-screw until the horizontal wire bisects the target of the distant rod. The reading of the striding-level will then be noted by taking the mean of the readings of the two ends of the bubble, and the amount of displacement of the bubble from its central position will be the constant error of the instrument, which should be applied as a correction in fine leveling. If this error is caused solely by a flexure of the telescope, it may be corrected by actually making the collars unequal in diameter to an amount which will cause the line of collimation to be horizontal when the bubble is in the center of its tube.

It is thus seen that when the telescope is supported in the center only, a striding-level is no more accurate than a fixed level of equal sensitiveness, unless all of the above conditions are attained. Also, that the methods explained for adjusting the telescope with a fixed-level, being direct and avoiding the use of collars and eliminating the effect of flexure, etc., preclude the existence of these errors. The fixed-level has therefore its specific advantages and a legitimate place even in the best surveying instruments.

The Adjustments of the Telescope's Axis of Revolution of the Engineer's Transit, by means of the Transverse Striding-Level, so that the Line of Collimation shall describe a Vertical Plane.

These adjustments consist, firstly, in adjusting the striding-level; secondly, in making the transverse axis of the telescope horizontal, by placing it truly at right angles to the vertical axis of revolution of the instrument by means of the transverse striding-level; thirdly, in adjusting the vertical wire for collimation so that it shall describe a vertical plane.

The adjustment of the transverse striding-level. In our instruments for general work, where the telescope is reversible only through the standards, as is the case in the Tachymeters, the striding level will rest upon special collars of equal diameters on the transverse axis of revolution, as it may then be attached to the telescope and move with it; and besides being very compact and readily manipulated, it enables us to give the pivots of the horizontal axis a form by which greater lateral stiffness is secured for the standards, and without this the telescope of these instruments would not have the required stability.

The striding-level of the finest class of instruments, such as Transit-Theodolites No. 11b, 12, etc., used for triangulation wholly, and Mining Transit No. 8, used mainly for very steep sighting, will rest directly upon the cylindrical pivots of the transverse axis, at the circle of contact in the wyes. This latter method is preferred for the finest class of instruments, on account of greater simplicity and accuracy, although the striding-level is then more liable to injury, as it cannot be attached to the telescope and must be lifted off whenever the telescope is moved greatly in altitude or when it is to be reversed, and is liable to fall off if the instrument is taken up hastily.

The transverse striding-level is adjusted precisely as described, on page 96, for the adjustment of the longitudinal striding-level shown on page 180. To test the lateral adjustment, the striding-level may be rocked upon the transverse axis, if it rests upon its pivots; or, if mounted upon collars between the standards and attached to the telescope, the latter may be moved on its horizontal axis a few degrees up and down. The horizontal adjustment of the striding-level is tested by reversing it end for end.

The transverse striding-level is the most important level of the transit, when the latter is used for measuring horizontal angles and for ranging straight lines where objects differ greatly in height.

The adjustment of the transverse axis of revolution. The striding-level having been carefully adjusted, level up the instrument generally with the plate-levels, put the striding-level in position and bring its bubble to the center of its graduation by means of the leveling screws, then turn the instrument 180° on its vertical axis and note whether the bubble of the striding-level remains in the center of the graduation. If it does, the adjustment is correct. If it does not, correct one-half the error by means of the leveling screws, and the other half by means of the wye adjustment of the standard. Repeat the process until the adjustment is correct. Observe also, in adjusting the wye adjustment of the standards, that it will be best performed and more lasting when the last turns of the lower capstan-headed screw are always applied in an upward direction.

The transverse axis of the telescope is now adjusted for movement of the latter in a vertical plane. The striding-level, being very sensitive, is the best-known device for making the transverse axis of the telescope truly horizontal. The method assumes that the special collars, or the cylindrical pivots at the end of the axis, which rest directly on wye bearings in the standards, as the case may be, are of equal diameters, and if collars are used, that they are concentric with the pivots of the axis. These conditions are best fulfilled when the striding-level rests directly on the pivots of the horizontal axis, as in the finest triangulation transits.

Whenever exact vertical motion is desired, as when the objects sighted at differ materially in altitude, the striding-level in such instruments should be depended upon to the exclusion of the front plate-level. The latter is then entirely subordinate to the striding-level and should be depended upon merely for leveling the instrument approximately. The plate-levels are also useful in indicating quickly any large disturbances of position. When the objects sighted at do not differ much in altitude, the front plate-level is, in these instruments, sufficiently sensitive to give satisfactory results without using the striding-level. Where no striding-level is provided, the plate-levels are, of course, to be depended upon for all work, and as a rule, in our instruments the degree of sensitiveness of the front plate-level is such as to insure that the plane described by the telescope shall not differ from the true vertical plane by an amount larger than the smallest direct reading of the verniers of the horizontal circle.

The adjustment of the vertical wire for collimation. The object of this adjustment is to place the vertical wire so that the line of sight shall be at right angles to the horizontal axis of revolution. This operation presupposes that the optical axis is coincident with the geometrical axis of the focussing slide and also that this combined axis moves at right angles to the transverse axis, so that the line of collimation shall describe a vertical plane at all focussing positions of the object-slide when the transverse axis has been placed in a horizontal position. In which case, of course, the projection of the line of sight will not deviate from a straight line for intermediate points. See on this point "Some Facts Worth Knowing." It is also assumed that the point of suspension of the plumb bob is in the prolongation of the vertical axis of revolution of the instrument, and that the instrument is correctly set over the point chosen for its position.

The method of adjusting the vertical wire for collimation is sufficiently described on page 49. The result of this process is, however, affected by any variation of the transverse axis from a true right angle to the vertical axis of revolution, when the two objects sighted at are not precisely in the same horizontal plane, a condition somewhat inconvenient to meet at all times in the field. Therefore, in instruments not provided with a transverse striding-level, it will be best to select two points in opposite directions that are as near as possible in the same horizontal plane, then make the vertical adjustment as described on page 51, which, however, is a crude substitute for the more perfect methods of the striding-level, or two collimators arranged in a vertical plane. When this adjustment is completed it will be necessary in these instruments to again verify the adjustment of the vertical wire for collimation and also to repeat the adjustment of the vertical plane, as neither of them can be made at once correctly independently of the other.

The adjustment of the vertical wire of telescopes whose transverse axis is provided with cylindrical bearings resting in wyes, as in Transit-Theodolites Nos. 11 and 12, etc., may be made by the same process of reversing through the standards, or by the following process, which is the only one applicable to instruments having low standards, such as Theodolite No. 14.

The process is as follows. Level up the instrument approximately; clamp the plates; set the vertical wire in a vertical position by the method described on page 48. Then select a well-defined distant point and bisect it with the vertical wire. Next, lift the telescope out of its bearings, and reverse the transverse axis end for end, replacing the axis in the bearings with the telescope pointing in the same direction as before. If the distant point is again bisected by the vertical wire, its adjustment is correct. If not, move it in the direction stated in the footnote on page 49, by means of the capstan-headed screws, one-half of the distance between its present position and the point previously bisected, and repeat the process until the adjustment is correct.

Tests of the assumed conditions.

In the foregoing explanation of the adjustment of the transverse axis by means of the striding-level resting on special collars, it has been assumed that the special collars are of equal diameters and that they are concentric with the telescope's axis of revolution. Now to see whether the line of collimation moves in a truly vertical plane, after the horizontal axis of revolution has been adjusted to lie at right angles to the vertical axis of revolution of the Engineer's Transit, proceed thus:—

Having previously verified the adjustment of the striding-level and by it that of the wye adjustment of the standard as well as that of the vertical wire for collimation, set up the instrument as near to a church with a spire, or a building with a well defined object elevated sufficiently to require the telescope to move through an altitude of about 90° when it is elevated to the object and depressed to the base of the building. Now level up the instrument, bring the bubble of the striding-level to the center of its graduation, and with the plates clamped bisect the object at the top, then depress the telescope and find or make a mark at the base which is also bisected by the vertical wire. Now reverse the telescope, turn the instrument in azimuth and with plates clamped again bisect the elevated object and upon depressing the telescope see whether the mark at the base is also bisected. If so, the adjustment of the line of collimation for motion in the vertical plane by means of the striding-level is correct but, if not it will show that the collars are worn and that correction of the difference must be made by means of the wye adjustment of the standard, over a little less than one quarter of the error, taking care to make the final adjustment by an upward movement of the capstan-headed screws. Then repeat this adjustment, and find or make a new mark at the base upon each successive trial when the instrument is depressed, until correct.

Now note to what extent and towards which collar the bubble of the striding-level has moved and make a memorandum of it so that it may be applied as a correction in the most precise work. Such deviations, while never existing in our new instruments, are, as a rule, caused by unequal wear of the collars, or by an unequal distribution of weight of parts attached to the horizontal axis of revolution, or by an imperfect adjustment of the vertical wire for collimation — or by all three combined. Errors from this source can also be eliminated by the usual method of reversing the instrument and telescope and taking the mean.

The above test can also best be performed by the use of collimators arranged in the vertical plane, see "The Collimator Apparatus," when of course the *modus operandi* is the same as just described.

Instead of being obliged to find or make a mark at the base of a building in order to test the correctness of the vertical adjustment of the wyes a *true vertical plane* may be had by the use of an artificial horizon.

Artificial Horizon. This may be made of mercury placed in a shallow bowl of copper or iron (see page 194) or it may be made of molasses, or of oil mixed with finely powdered lamp black ; or it may be made of plane parallel glass suitably mounted on three leveling screws, so that it can be placed in a truly horizontal position by means of a sensitive spirit-level. In the absence of any of these a bucket full of water may serve the purpose, although the slightest air currents will cause a ripple in the water and make observations more difficult.

To test the vertical adjustment of the wyes with the aid of an artificial horizon proceed thus : —

Having previously verified the adjustments of the striding-level and instrument set it as near to an elevated object as can be conveniently observed, place the artificial horizon within 8 or 10 feet in front of the instrument ; level up carefully and bring the bubble of the striding-level to the center of its graduation. Now, if with plates clamped, the vertical wire bisects both the object and its reflection in the artificial horizon, it is proof that the line of collimation lies in a truly vertical plane ; but if not, remove half the difference by the wye adjustment and repeat until correct. Now make a memorandum of the deviation of the bubble of the striding-level as explained above in order to apply it as a correction in precise work.

While, with the use of an artificial horizon, it is not actually necessary, it will be well to repeat this test with the telescope in reversed position to see if the object and its reflected image remain bisected. If not, it will indicate that the instrument has not been properly adjusted before the test was made.

Test of the Equality of the Cylindrical Pivots of horizontal axis of revolution by means of the transverse striding-level resting at circle of contact in wyes. Carefully level up the instrument by the plate level. Next verify the adjustments of the transverse striding-level as explained on page 96 and then make the wye adjustment of the standards by reversing the instrument on its vertical center and, removing half the difference if any and repeating this operation until the bubble of the striding level remains in the center of its graduation. Now remove the striding-level, and upon reversing the telescope's axis of revolution end for end in the bearings and placing the striding-level again on the pivots, so that each end will be on the same side of the standard previously occupied before the telescope was reversed in the bearings, see if the bubble comes back to its center. If so, the pivots are of equal diameter, but if not, remove half the difference by the wye adjustment and note to what extent and towards which pivot the bubble of the striding-level moves and make a memorandum of it so that it may be applied as a correction in precise work.

The test of the equality of the pivots at the circle of contact in the wyes when the telescope is reversible over the bearings may also be made by any of the methods explained in the preceding paragraph for verifying the adjustment of the line of collimation for exact motion in the vertical plane. In this case, after the horizontal axis has been adjusted for the vertical plane with the telescope in the direct position, it is only necessary to reverse the telescope over its bearings, without unclamping the plates, and to repeat the test for motion in the vertical plane. If correct the pivots are of equal diameters ; but, if not, the inequality amounts to one-half of the deviation of the line of collimation from the vertical plane, and for best results in very steep sighting, it will be well to reverse the telescope for this purpose and to accept the mean of the two readings as the true result.

On the other hand, however, if the telescope is intended only for reversal through the standards and turning the instrument on its vertical center, as all our Transit-Theodolites do, then for good results after the cylindrical pivots are worn, the wye adjustment should be made only with the telescope in its direct and proper position in the instrument and by the striding-level alone, that is, if the latter rests directly on the pivots as shown in No. 11*b*. The telescope must then not be reversed over the bearings.

In Transit-Theodolites having no striding-level the wye adjustment of the standards, after the pivots are worn, must be made only with the telescope in direct position as explained above and as explained on page 51 for the Engineers' transit.

If the Transit-Theodolite has a striding-level resting on special collars (as most of these instruments are made by us for the sake of greater compactness and portability) and the pivots and collars have become unequal in diameter from wear and tear, then the instrument should also be treated as if it was non-reversible over the bearings and the wye adjustment must be made with the telescope in the direct position only by any of the methods given for the vertical adjustment of the wyes, and after completion of the adjustment this difference in the reading of the striding-level should be noted, etc., for correction as already explained.

Of course in making any of these tests for equality of collars, pivots and vertical adjustment of the wyes, it is of the utmost importance to have the instrument balanced as the maker intended it to be, as without this none of the above tests can be made. It will not do for instance, to detach the vertical circle after the wye adjustment has been made with it on the instrument, nor will it do, if a solar attachment or a side telescope, etc. has been attached to the telescope's axis of revolution, to suppose that the counterpoises made for each can be indiscriminately interchanged or left off at will without disturbing the equipoise of the instrument and its adjustments. All these parts bear a strict relation to each other and cannot therefore be changed at will without also deranging other existing conditions of the greatest importance for a successful working of the instrument.

Appendix to the Description of the Adjustments of the Transverse Striding-Level.

The above tests have been given in great detail to create a familiarity with this subject, on the part of the Engineer, that he may know the proper method to attain results consistent with his intentions. It is important to realize how much depends upon the degree of sensitiveness of the spirit-levels used, and upon the power of the telescope to reveal slight differences in its pointing. The utility of the striding-level in the finer instruments will be best understood when we consider that the method of reversion, so useful in detecting and eliminating errors of graduation and of collimation, is useless in detecting or eliminating deviations of the plane of motion of the line of collimation from the vertical plane passing through the center of the instrument. The adjustment of the vertical wire (described on page 49) causes the line of collimation to move in a plane perpendicular to the horizontal axis of revolution, and the vertical adjustment of the standards, described on page 51, renders it possible to place that plane in a vertical position as nearly as the degree of sensitiveness of the front plate-level or the striding-level will permit. The actual position of the plane of motion of the line of collimation being dependent upon the levels, a want of sufficient sensitiveness in them will cause the plane to deviate from the vertical plane passing through the center of the instrument. The angle between these planes measures the greatest angular amount of deviation that the line of collimation can make with the vertical plane. This occurs when the telescope is pointing vertically upwards or downwards. — The angular deviation of the line of collimation from the vertical plane is *nul* when the telecope is horizontal and it increases as the telescope is more and more inclined.

When a horizontal angle is measured between points so situated that the telescope does not need to be changed in altitude during the observations, each pointing deviates from the vertical plane to the same amount and in the same sense and the error is eliminated from the observation. This, however, is rarely the case, and when the telescope is changed in altitude during a set of observations, a deviation of the line of collimation from the vertical plane introduces an error which cannot be detected and which cannot be removed by repeating the angle, and which is greatest when one pointing is made with the telescope horizontal and the other with the telescope vertical, which can be done with some mining transits.

The above reasoning assumes that the deviation of the plane of motion from the vertical plane is constant or can be maintained constant during a set of observations. This, however, is far from being assured in ordinary instruments. The position of the plane of motion depends upon the accuracy of the adjustment and upon the degree of sensitiveness of the front plate-level, or of the striding-level. Their sensitiveness should therefore be such that the plane of motion shall not deviate from the vertical plane by an amount inconsistent with the degree of precision expected of the instrument. That is to say, levels whose sensitiveness is suitable for an instrument reading to minutes would be insufficient in sensitiveness for an instrument reading to ten seconds. We have seen cases where the more ordinary Engineers' Transits were provided with front plate-levels which would hardly indicate *five minutes of arc* in an instrument reading to *thirty seconds.* With such an instrument the plane of motion of the line of collimation cannot therefore be controlled within five minutes of arc, from a true vertical plane and all horizontal angles measured by it are influenced by the uncertainty arising from this defect.

The importance of knowing the degree of sensitiveness of the levels of the finer instruments is very great and on this account, it is usual for us to send with such instruments a statement giving the value of the divisions of the levels in seconds of arc. The sensitiveness of the transverse striding-level is made much greater than that of the plate-levels, and in such instruments the latter serve principally to level up approximately. Its greater sensitiveness is of especial importance in Mining transits, as steep sighting is the rule in mining surveying.

The Adjustments of the Arc, or of the Full Vertical Circle with Double Opposite Verniers.

In the Engineer's Transits provided with an arc or full vertical circle, the adjustment of its verniers to read zero when the telescope and plates are level should be made as described on page 50, in order to make the instruments serviceable for reading vertical angles.

When the vertical circle is, however, provided with double opposite verniers, as shown in No. 1c, Style 0, page 147, or as in Tachymeter No. 1g, the adjustment of the vernier zeros for a normal position should be made by the two opposing capstan-headed screws attached to the vernier frame; but if the telescope is reversible over the bearings also, requiring a separate tangent screw, as shown in our Universal Mining Transit No. 8, and in Transit Theodolites No. 11a and No. 11b, then the adjustment of the vernier zeros for position must be made by the vernier frame's tangent screw. See "Instructions for Using our Universal Mining Transit," page 177.

If the vertical circle has a separate level attached to the frame carrying the double opposite verniers, as shown in Mining Transit No. 8, and in Transit-Theodolites Nos. 11a and 11b, etc., then the adjustment of this level and verniers for position must be made as described on page 177, so of which we will repeat here much as properly belongs to this subject.

Place the telescope in the horizontal plane by means of its tangent screw, then move the vernier frames' tangent screw until the zero line of the double verniers, marked A, is in coincidence with the zero line of the vertical circle, and now raise or lower the adjusting screw of this level, as the case may be, until the bubble is in the center of its tube.

It is now supposed that the zero line of the double opposite verniers, marked B, is also in coincidence with that of the vertical circle. If not, the verniers marked B can be moved after releasing the capstan-headed screws, until both zero lines on that side of the vertical circle are also in coincidence. However, this is a very laborious proceeding for those uninitiated in this work, and as it cannot always be made quite exact, owing to the mode of mounting the vertical circle on the telescope's axis, it will be found easiest to eliminate errors of excentricity in the graduation of the vertical circle and verniers by reversing the telescope and taking the mean of the readings. The vertical circle is graduated from 0° to 90° and back, and the verniers are double, so that angles of elevation and depression can be read with ease and dispatch.

The Adjustments of the Auxiliary Telescopes of Mining Transits.

The Detachable Side-Telescope.

This telescope, illustrated on page 164, as ordinarily made, is attached to the transverse axis of the main telescope by means of a hub, which is screwed upon a prolongation of this axis beyond the standards. The hub contains an independent horizontal axis upon which the side-telescope may be revolved, and to which it may be clamped. The side-telescope is usually set parallel to the main telescope, and looking in the same direction, but it may be set so that it is inclined at a given vertical angle when the main telescope is horizontal. A counterpoise is attached to the other end of the transverse axis of the main telescope, so as to balance the weight of the side-telescope and retain that axis horizontal when the side-telescope is in use. The side-telescope is mainly intended as an auxiliary in measuring vertical angles, and it is on this account that the simple means of attaching it to the transit, here described, has been adopted by us as sufficient for the purpose, although it will be very difficult to place its line of collimation truly parallel to the main telescope for all focussing positions of the object-slide. When greater accuracy and greater permanency in the adjustments are desired, our Universal Mining Transit No. 8, page 172, should be chosen. This has duplex telescope bearings, one set of bearings being placed excentrically, so as to permit of vertical sighting up or down a shaft with the main telescope alone.

The adjustments of the detachable side-telescope are as follows : —

1. To place its vertical wire perpendicular to the transverse axis of the instrument. Attach the side-telescope and the counterpoise to the transverse axis. Clamp the side-telescope slightly to its hub, bisect a point by its vertical wire and move the main telescope on its horizontal axis of revolution. If the point remains bisected by the vertical wire of the side-telescope throughout its entire length this adjustment is correct. If not, loosen the capstan-headed screws and rotate the reticule bearing the wires, as explained on page 48, until the wire bisects the point throughout its entire length. Then slightly tighten the capstan-headed screws as explained in " Some Remarks Concerning Instrument Adjustments," page 24. Also see footnote on page 49.

2. To place the intersection of the cross-wires of the side-telescope in its line of collimation. This may be done in several ways.

(a) The side-telescope being detachable, it could be adjusted by rotating it in wyes, were any at hand. Such wyes, as we have shown before, may be improvised by cutting the proper shapes out of thin wood, and fastening a pair of them to a board in an upright position. The distance between them should be such that the telescope may rest upon the outside of the mounting of the object-glass and against its shoulder where the cap is placed, and upon the tube near the cross-wires when practicable. The improvised wyes being placed on a firm support and fastened so that they will not move, the side-telescope may be revolved in them, and the wires may be placed in the line of collimation as in a wye-level, using a distant point. The horizontal wire, being the more important one in the side-telescope, should be placed with some care.

(b) This adjustment for collimation may be made without removing the side-telescope, if for the adjustment of the horizontal wire, a small spirit-level* mounted on a metal base, similar to those described on page 127, is at hand. Then proceed thus :

Adjustment of the horizontal wire, First, level up the instrument by its plate levels. Then, placing the main telescope in a horizontal position by its level, find a well-defined object, such as the target of a leveling rod, distant about 300 feet. Now clamp the side-telescope when in a nearly horizontal position to its hub, and placing the auxiliary level, which has been previously adjusted, longitudinally on the side-telescope bring its bubble to the center of the tube by means of the tan-

---

* Such a spirit-level mounted in a cast-iron frame, and good enough for this purpose if carefully selected, can be bought in any of the better equipped hardware stores.

gent screw of the main telescope and now, by turning the instrument on its vertical center see if the horizontal wire of the side-telescope bisects the object or target also. If so, this adjustment is made, but if not, it must be completed by moving the vertical capstan-headed screws as explained in foot-note, page 49.

To verify this adjustment, the side-telescope may be reversed on its horizontal axis of revolution and clamped to its hub when nearly in the same level plane. Then turn the instrument a little more than 180° on its vertical center, place the auxiliary level on the side-telescope, same as before, and bring the bubble to the center of its tube by means of the vertical tangent screw. If now, when the side-telescope is in the reversed position the horizontal wire bisects the object also, this adjustment is completed, but if it does not then the horizontal wire must be moved again to a point half-way between the two readings.

This adjustment may also be made by the auxiliary level alone or by means of a striding-level without the aid of the main telescope. (See adjustment of the horizontal wire of the Engineer's transit telescope by means of the longitudinal striding-level, page 95.)

Adjustment of the vertical wire. Select a well defined object, as a church spire, distant 5 or 6 miles. Bisect it with the vertical wire of the main telescope, and without moving the instrument, look through the side-telescope and note whether the object is also bisected by its vertical wire. If not, make the adjustment by moving its vertical wire by the horizontal capstan-headed screws, until the object is bisected also. The distance between the two telescopes being only a few inches, the vertical wires will cover so great a width, if the object be sufficiently distant, that the effect of the excentricity of the side-telescope will be almost imperceptible and the same distant point may be used for each telescope.

(c) When a distant object is not available, measure with a pair of dividers the excentricity of the side-telescope, which is the distance between the centers of the two telescopes. Then transfer it to the face of a wall as far distant as practicable and make two marks whose horizontal distance apart is equal to this excentricity. Bisect one of these marks by the vertical wire of the main telescope and then look through the side-telescope and note whether the other mark is bisected by its vertical wire. If not, make it do so by moving the cross-wires of the side-telescope as described on page 49. The direction of the lines of sight should be at right angles to the surface upon which the two marks are made.

The position of the side-telescope with respect to the main telescope should be assured whenever the former is to be used. This may be done as follows: find a mark that is bisected by the horizontal wire of the main telescope. Then turn the instrument on its vertical axis and notice whether the horizontal wire of the side-telescope bisects the same mark. If so, firmly clamp the side-telescope to its hub. If not, gently tap one end of the side-telescope, which hitherto has only been loosely clamped, until its horizontal wire coincides with the mark and then clamp the side-telescope to its hub. The telescopes are now set to correspond with the zero of the vertical circle.

To place the telescopes at an angle with each other. Level up and fix a mark when the main telescope is level. Then raise or depress the main telescope the required angle and clamp the horizontal axis. Now move the side-telescope until its horizontal wire bisects the mark and clamp it firmly to its hub. During an extended operation with the side-telescope, the relative position of the two telescopes should be verified from time to time to detect any disturbance of the side-telescope.

Transits having the telescope mounted at the end of the horizontal axis of revolution are sometimes used in mines; or, as shown in the Alt.-Azimuths Nos. 15a and 15b, this construction is used in some instruments for geodetic and smaller astronomical work. The adjustment of such a telescope for collimation may therefore be explained in this connection. The following method is as simple as any: —

Select a well-defined object, as a church-spire, distant at least 5 or 6 miles. The instrument being leveled, bisect the object with the vertical wire and read the verniers of the horizontal limb. Then turn the vernier plate so as to read exactly 180° different from the previous reading, and revolve the telescope. If the vertical wire is adjusted for collimation it will again bisect the distant object, since the space covered by the cross-wires on an object at such a distance will be much greater than the change in the position of the telescope as caused by its excentricity from the center of the instrument. If it does not again bisect the object, correct one-half the error by means of the horizontal capstan-headed screws as explained in the footnote on page 49.

The adjustment of the horizontal wire for collimation may be made by selecting one of the methods best adapted for a particular design of telescope, as described on page 70 for a telescope provided with a fixed level. If the telescope has a longitudinal striding-level this adjustment should be made by the method explained on page 96.

These two adjustments should be repeated until both are correct.

To measure the excentricity of the telescope, set up the instrument as near to a wall or other vertical object as possible. Draw a horizontal line upon the wall at a convenient height. Point the telescope exactly at right angles to the wall, mark where the vertical wire intersects the line just drawn, and read the verniers of the horizontal limb. Turn the vernier plate exactly 180°, revolve the telescope and make a second mark where the vertical wire now intersects the line. The distance between these two marks will be *twice* the excentricity of the telescope.

When using an instrument of this description for short sights, it is very convenient to use sighting poles with excentric targets, or an offset at the foot of the pole corresponding to the excentricity of the telescope.

The Auxiliary Top-Telescope,
Now superseded by the interchangeable auxiliary telescope, see styles I and II.

This auxiliary, as previously made by us, was mounted in adjustable wyes on standards permanently fixed to the main telescope, so that both lines of sight could be made parallel. The weight of the top telescope was balanced by a counterpoise attached to a stem also permanently fixed to the cross-axis of the main telescope. When the top telescope was not in use it was kept in the instrument box, while the standards and counterpoise stem were permanently fixed to the main telescope so as to avoid frequent and tedious adjustments. This feature made the instrument troublesome and unwieldy for the more ordinary work in mines, and still less convenient for surface work.

This improvement when first introduced by us superseded all other forms of top telescopes as made by others whose main object seemed to be simply to straddle another telescope above the main one (a mere commercial article, not an instrument of precision) for the purpose of steep sighting. But since the line of sight of such a telescope can never be placed truly at right angles to the cross-axis, the line of collimation does not move in a truly vertical plane, therefore horizontal angles measured between points differing greatly in elevation or in distance are never correct.

It can also be readily seen that the telescope of a solar attachment as commonly made, having no means of lateral adjustment to the main telescope, is insufficient in this respect (even leaving aside its low power and aperture) and cannot meet the requirements properly. The adjustment by which the line of collimation of top telescope is placed in the same vertical plane as that of the main telescope is just as important as that of the main telescope itself.

A most convenient and practical device having all the advantages of that former style, i. e., means of adjusting the line of collimation parallel to that of the main telescope, so that *after having been removed it will retain its adjustments when again attached*, is our new mounting of the top telescope by means of threaded studs. This enables the engineer to read horizontal angles when the main telescope cannot be used, obviating the making of corrections for the eccentricity of the telescope.

Patent Adjustable Top Telescope.

This device consists of an adjustable trivet and an auxiliary telescope (see page 165) and an open central pillar, which latter screws to a threaded stud cast on or permanently secured to the cross-axis of the main telescope. When not needed, the auxiliary telescope and its counterpoise may be returned to the box and the instrument is free of incumbrances, save the stem for the counterpoise and the stud to which the central pillar carrying the auxiliary telescope is attached, and is ready for surface work. If desired, the top telescope may be entirely unscrewed from the central pillar, leaving the latter attached to the main telescope.

The Adjustment of the Auxiliary Telescope used as a Top Telescope: — It is assumed that all adjustments of the transit proper have been made, that is, that the plate and telescope levels, the line of collimation, the vertical plane, etc., have been verified and corrected, and that the verniers of the vertical circle read zero when plates are leveled up and that the bubble of the telescope level is in the center of its graduation.

The adjustment of Line of Collimation of Auxiliary Telescope : First examine the coincidence of the intersection of the cross wires with the optical axis. This may be done by rotating the telescope in improvised wyes of wood (see p. 104), or by rotating it in the socket of the pillar [as sometimes made by us] by unscrewing it about one turn, when the adjustment is made by moving the capstan headed screws as described in footnote, p. 49. The telescope must now be screwed to its bearing in such a manner that the cross-wires are parallel to those of the main telescope — to be verified as explained in "To make the vertical wire perpendicular to the plane of the horizontal axis," etc., p. 48.

To place the line of collimation of the auxiliary telescope in the same vertical plane with that of the main telescope. Bisect a distant object with the vertical wire of the main telescope; see if the vertical wire of the auxiliary telescope also bisects the same point. If not, move the auxiliary telescope by means of the pair of opposing milled-headed screws attached to its pillar nearer the eye-end until the distant object is bisected at the same time by both vertical wires. Now focus the main telescope on a near object and see if the vertical wire of the auxiliary telescope bisects the same point as the vertical wire of the main telescope. If not, make the adjustment by means of the pair of capstan-headed opposing screws on one side of the adjusting trivets of the pillar. Then re-examine both wires for coincidence with the distant object, using the milled-headed screws, and also repeat the adjustment for near object if necessary. The two lines of collimation are now in the same vertical plane.

To adjust the top telescope so that both horizontal wires bisect the same distant object. Bisect a distant object with the horizontal wire of the main telescope, and see whether the horizontal wire of the auxiliary telescope bisects the same point. If not, make the coincidence by means of the pair of opposing capstan screws in the trivets near the milled-headed screws. This being done, both these adjustments should be verified and repeated if necessary. These adjustments once carefully made assure the exact parallelism of both telescopes and will not require repetition except at long intervals, or after an injury.

The distance between the lines of sight of the two telescopes should be carefully measured by sighting at a vertical line on a wall — the telescopes being horizontal — when the distance between the intersections of the two horizontal wires on the line will be the eccentricity of the top telescope, for which every vertical angle measured with it should be corrected.

The adjustment of the extra level (if any is provided in place of a striding level, see footnote, page 158) near the eye-end of the telescope, as shown in the cut p. 165, is dependent on that of the telescope in the vertical plane. This latter must be verified, as explained on page 51, before this level can be adjusted by reversing, as in case of the plate levels.

Patent Adjustable and Interchangeable Auxiliary Telescope.

Style II with adjustable trivet, page 165.

In this device the auxiliary telescope is the same as described under "Patent Adjustable Top Telescope" (see cut, page 167), but it is so arranged that it can be attached (interchangeably) on top or at the side of the cross-axis of the transit and readily ranged into line with the main telescope. The excentricity will be the same in both cases. One counterpoise will be sufficient for both positions.

This improvement used as a side telescope cannot, however, be carried out with all instruments, since the cross-axis requires an extension ending in a threaded stud beyond the standards, and this to be enduring can be made only on the original instrument and cannot be attached to an old axis. From this it will be seen that this is only applicable to new instruments when so ordered.

The auxiliary telescope detaches from a stem permanently fixed to the cross-axis so that the excentricity is the same when it is placed on the side. It does not revolve on an independent hub, as before, with which style there was always danger of accidentally changing its position, but this device is simply ranged into line with the main telescope by use of the milled-head screws. However, to meet every want arising in a mine, it is provided with a clamp ring and capstan-headed screw, by which the auxiliary telescope may be changed on its hub and permanently secured in any position. When used as a side telescope the fine adjustment by the trivets is not so essential as is the case when used as a top telescope, where the accuracy cannot be too great and where the trivets are therefore essential, since the top telescope is then really the main telescope for measuring horizontal angles. To use it in measuring horizontal angles it is only necessary to bisect with the vertical wire of the main telescope as distant a point as can be found in a mine, then by means of the milled-headed opposing tangent screws, and by slightly revolving the transverse axis of the main telescope, the vertical wire of the auxiliary one must also be made to bisect the same point.

To use it as a side telescope for vertical angles the procedure is exactly similar to the above, and differs only in so far as the *now* horizontal wire of the auxiliary telescope must be made to bisect a point previously bisected by the horizontal wire of the main telescope. This is done by slightly turning the vernier plate on its vertical axis and by making use of the same milled-headed tangent screws.

Thus it will be seen that the ready interchangeability of the auxiliary telescope enables one to read horizontal as well as vertical angles when the main telescope cannot be used, obviating the making of corrections for the excentricity of the telescope in both positions.

Style I, with non-adjustable central post (see page 167).

The general design of this arrangement of attaching the interchangeable telescope and the method of manipulating it are quite similar to that described under style II. It differs from the latter only in so far as it is not provided with an adjustable trivet. In this device Style I, the auxiliary telescope screws direct upon an open central vertical post cast in one piece with the transverse axis to secure great rigidity, the degree of accuracy of the result depending in a large measure upon the degree of accuracy with which the center of the pillar, and the line of collimation of the principal (then vertical) wire of the auxiliary telescope are made to lie in the same vertical plane as the optical axis of the main telescope or parallel to it. With the care given to it and special machinery used for it, this condition, difficult as it is, is secured to an extent which leaves little to be desired for all practical purposes. As the auxiliary telescope is interchangeable from top to side there is really need of but one wire, which we will designate as the principal wire. This, when the auxiliary is mounted on top, is the vertical wire, and when on the side becomes the horizontal wire. Therefore it will be seen that when the auxiliary is mounted on top the line of collimation of its horizontal wire is immaterial, as no vertical angles will then be measured. When the latter are to be measured the engineer will then mount the auxiliary on the side, when in turn the vertical wire becomes immaterial. The auxiliary telescope is provided with two milled-headed opposing screws (same as in style II), for ranging in line with the main telescope.

Style I being more rigid, simpler and cheaper than style II, is now recommended. In neither of the two styles does the auxiliary telescope ordinarily revolve in a socket for the purpose of making the adjustment of collimation, and coincidence of the cross-wires and optical axis must be verified by the use of improvised wooden wyes (see above) should it become necessary.

The success which the interchangeable auxiliary telescope has achieved, both here and abroad, since first invented by this firm in 1895 is somewhat phenomenal. It shows that this combination is the most applicable one in solving the difficult problems arising in mine engineering. For this reason every preparation has been made to meet the demand and new improvements are added as experience may suggest. All our top telescopes are therefore now made interchangeable.

Credit is due, in working out the feasibility of using the top and side telescope interchangeably, to Mr. Dunbar D. Scott, mining engineer, for several valuable suggestions gathered from his experience and needs in mine work.

The Use of the Interchangeable Auxiliary Telescope for Astronomical Observations.

Besides its ordinary use for steep sighting in mines, the interchangeable auxiliary telescope, as described in the foregoing article, will at times be found very useful as an astronomical instrument. It is particularly advisable in making latitude observations by meridian altitude and in observing transits across the meridian for time. As a rule when the prism is attached to the eye-piece of the main telescope it is not possible with the engineer's mining transit to point the telescope at a greater angle of elevation than about 70°, consequently it would be impossible to make solar observations at a latitude lower than 40° when the sun is at its greatest declination or observation on stars near the zenith. However, by attaching the prism to the auxiliary telescope used as a top or side telescope, these observations may be made with ease and this difficulty overcome.

In making latitude observations the interchangeable auxiliary telescope should be attached at the side; and its horizontal wire is then, by means of the two opposing tangent screws, made to correspond to the line of collimation of that of the main telescope by bisecting with both telescopes some distant and well-defined object: then, if a meridian mark is used (which is not absolutely necessary), the transit should be set up in the meridian by the main telescope and the pointing on the sun or star may be made with the auxiliary telescope with or without the prism, as conditions may require.

In observing transits the auxiliary telescope should be mounted on top and ranged into line with the vertical wire of the main telescope by using the two opposing screws as explained.

In making solar and stellar observations with the main telescope and prism attachment, the telescope should always be reversed through the standards with the objective down instead of up.

C. L. Berger & Sons' Universal Level Trier.

A most valuable and indispensable Apparatus for testing the sensitiveness and regularity of curvature of Spirit Levels used in Engineers' Field Instruments, etc., etc.

Introduction. *The Spirit Level,* occupying, next to the graduations and telescopic measurements, so important a place on the Instruments of Precision used in Geodesy and for Scientific Research, it is absolutely necessary for those using such instruments to know the sensitiveness of a level as expressed in seconds of arc, and, whether it has been ground to a true curvature as indicated by a uniform run of the bubble, both of which are necessary in order to arrive at correct results in the use of instruments. To fully understand this we refer the reader to the various articles on this subject printed in our Handbook, notably those on pages 7, 38, 102 and 103. As the Engineer, Surveyor or Scientific Investigator has frequently no means of testing the character of these levels, we, as a rule, determine it, or its mean value, and send a statement, to this effect, with every important instrument issued by us. Indeed, so necessary is it to know the mean value of Spirit Levels, that many of the larger Schools of Civil Engineering, Physical Laboratories and Astronomical Observatories have been supplied by us with this apparatus for the acquirement of this knowledge so that the student might understand better the character of his instruments. However, as a more universal application of this apparatus is desirable at the centers of learning, and as there seems to be a growing demand for a simple, cheap and ready device, we have improved our ordinary apparatus and are now prepared to furnish the above-mentioned *Universal Level Trier.* By a more extended use of this apparatus at such schools we hope, in time, to abate the reprehensible practice of using spirit levels that are unfit for the character of an Instrument. We have seen cases where the more ordinary Engineers' Transits were provided with front plate-levels which would hardly indicate *five minutes of arc* in an instrument reading to *thirty seconds.* With such an instrument the plane of motion of the line of collimation cannot therefore be controlled within five minutes of arc, from a true vertical plane, and all horizontal angles measured by it are influenced by the uncertainty arising from this defect.

Description. *The Level Trier* as designed and made by us is exceedingly simple and of a more universal character than any we have seen before, and it is especially adapted to the wants of Engineering Schools and Laboratories. It consists of an iron base plate, upon which is mounted an iron bar having at one end two pivotal centers resting in receptacles provided for them in the base plate, and at the other end a micrometer screw carrying a disk graduated into 100 parts.

The bar is provided with fixed wyes in which levels to be tested may be placed, also suitable scales are attached; when the latter are not needed they may be turned back out of the way. At the pivotal end means are provided for supporting an entire instrument or the parts containing the levels to be examined. An adjustable fork is provided at the right end, which serves to steady the telescope when a wye level bubble is to be examined. If a plate, box, or telescope level of an instrument having three leveling screws is to be tested, the latter may be placed in the grooves provided for them, but if an instrument has four leveling screws a special plate can be attached so as to raise it above the wyes on the bar. The points of the pivots and micrometer screw are hardened, and the latter bears on the base plate on a hardened and polished surface. This bearing is often made movable excentrically with regard to the screw point, so that the point of bearing can be changed in case of wear so as to always present a smooth surface. The arm is about 18 inches long and the pitch of the micrometer screw one-sixtieth of an inch. The cut will give a clear idea as to the arrangements of the parts and the positions to be occupied by the specimens to be tested.

Mounting. The best place to mount the apparatus is upon a solid masonry shelf built into a wall at a convenient height (so that the operator can be in a sitting position), or upon a window sill where direct contact can be made with the stone. A well-lighted room free from sudden jars, direct sunshine or artificial light is desirable, and the apparatus should be protected by a suitable case of wood and glass from air currents and the heat of the body of the observer.

This case must be so arranged as to allow ready access to the micrometer screw and ready removal if the different instruments, telescopes, levels, etc., are to be put into position entire. Ordinarily, if the shelf is built into the wall, or if the apparatus is mounted on a "pier," the case may have three sides of wood and the top of glass and the left-hand end open for the introduction of the observer's hand when using the micrometer screw, or at a suitable height above the apparatus a plate of ordinary thick plate glass may be supported on brackets when in use, and when not in use it may be buttoned up against the wall. Another plate can be arranged to be raised or lowered to protect the apparatus from front and side currents whenever necessary. The screw end must be left open for the operator's hand. If the apparatus is mounted on a window sill the case had better be made wholly of wood, with open glass top.

Some such protection as is above suggested is indispensable when sensitive levels are to be tested. Of course, for the test of levels of low degrees of sensitiveness, such care may not, in general, be necessary.

Having secured a desirable place for the Level Trier, the next thing is to properly mount it. The heavy cast iron base plate is provided, on the bottom, with three projecting studs which should rest securely on the stone pier when the base plate occupies as near a level position as can be assured by the use of a common carpenter's level. Now carefully place the bar upon the base plate so that the pivots shall rest in their receptacles in the hubs on the right-hand end of the base plate, and the micrometer screw upon its proper bearing at the left-hand end, and the apparatus is ready for use.

The points of contact of the micrometer screw and the pivots should be kept free from dirt, grit or rust. The resting place for the micrometer screw will sometimes be made so that it can be raised or lowered by a screw arrangement so as to bring different parts into use and prevent local wearing. This resting place can also be moved excentrically and bring new points of bearing under the micrometer.

Mathematical Part. It is clear that the rotation of the micrometer screw raises or lowers the bar of the apparatus. Any point on this bar, rotating about the two pivots at the right-hand end, moves in the arc of circle whose radius is the distance from the pivots to the screw point. Knowing the pitch of the micrometer screw and its distance from the pivots, it is a simple matter to find how much arc is travelled by moving the divided disc through one or more divisions: in other words, to find the value of the angle of inclination subtended by this motion of the screw.

Example. Suppose the length of arm 17.9 in., the pitch of the screw one-sixtieth of an inch and the disc to be divided into 100 parts. Through how many seconds of arc will the bar move when the micrometer screw is changed 30 divisions?

There are 206,265 seconds in the arc whose length is equal the radius.

Let $x =$ the number of seconds required:

$$\frac{x}{206,265} = \frac{.005}{17.9}, \quad x = 57.6''$$

If we wish to know the radius of curvature of a level, it is necessary to measure the run of the bubble when the level has been changed a certain amount in arc, and then solve as follows:

Example. Observed run of bubble for 30 seconds change in arc is found to be one inch. Find radius of curvature of level.

Let $r =$ radius required:

$$\frac{r}{1 \text{ inch}} = \frac{206,265 \text{ seconds}}{30 \text{ seconds}} = 6875.5 \text{ inches,}$$

$$\text{or } 572.96 \text{ ft.}$$

Method of making Tests. The level to be tested is placed in the wyes on the bar, which is raised or lowered till the bubble is at one end of the scale, or, if the level has no scale, up to the point which will be the limit of the run of the bubble in practice. The micrometer disc is then turned over equal spaces and careful notes of the run of the bubble are then taken. Having moved the bubble over its course it should be moved in the opposite direction in the same manner and the whole operation be repeated several times, and, with very sensitive levels, at differing temperatures, to ensure accuracy in the results.

The mean value of all the observations may then be determined and the *value of one division of the level expressed in minutes or seconds of arc,* as the case may be. By this same process we may find, as already shown, the radius of curvature to which the interior surface of the glass tube has been ground.

For spirit levels having no scale graduated on the tube the german silver scale graduated to 20ths of inches and attached to the apparatus is to be used as shown in the cut. For testing the levels of an Engineer's Instrument as already indicated it is not necessary to remove them, but the instrument entire can be placed directly on the Trier.

The Instrument should then be leveled by its leveling screws, and the level to be examined should be brought parallel to the center line of the bar by the use of the tangent screws. (This saves much unnecessary wear on the fine micrometer screw.) This done, proceed in the same manner as explained for the unmounted spirit level. If we have the case of a wye level bubble that needs examination, the telescope and level tube may be removed from the wyes of the Instrument and placed on the wyes of the Trier, as shown in the cut. The forked arrangement at the pivotal end which can be clamped at any convenient place on the projecting rod, will tend to steady the telescope and prevent it from falling while the level is being tested.

This ability to readily and speedily examine the levels of an instrument without detaching them and thereby deranging instrumental adjustments will, we think, be quickly appreciated.

As already indicated above, a level should be tested forward and backward, and, in case of very fine spirit levels, such as are used for measuring angles, this should be repeated a number of times so as to get a considerable number of observations from which to deduce the mean value of one division in seconds of arc. In case of a level provided with an air chamber reading to single seconds of arc as a rule, it will be found necessary in making the test to keep the bubble of a constant length (a good length is two-fifths the length of the tube, excluding that of the air chamber.)

We here insert specimen sets of observations made upon levels taken from stock, and such as are used in our engineer's wye level.

Number of Trials.	Micrometer Readings.	Level Scale Readings.* A End.	B End.	Differences A End.	B End.	Length of Bubble.
1	7	9.8	51.8			61.6
2	17	14.2	47.2	4.4	4.6	61.4
3	27	18.5	42.7	4.3	4.5	61.2
4	37	23.0	38.5	4.5	4.2	61.5
5	47	27.0	34.4	4.0	4.1	61.4
6	57	31.5	30.0	4.5	4.4	61.5
7	67	35.8	25.7	4.3	4.3	61.5
8	77	40.2	21.0	4.4	4.7	61.2
8	77	40.0	21.2			61.2
7	67	35.5	25.5	4.5	4.3	61.0
6	57	31.1	30.2	4.4	4.7	61.3
5	47	26.7	34.6	4.4	4.4	61.3
4	37	22.3	38.8	4.4	4.2	61.1
3	27	18.2	42.8	4.1	4.0	61.0
2	17	14.0	47.1	4.2	4.3	61.1
1	7	9.5	51.5	4.5	4.4	61.0
				60.9	61.1	

Mean value of differences, 4.36 20ths of an inch.

*Level scale is graduated to 20th of inches, reading right and left from zero at the center.

Number of Trials.	Micrometer Readings.	Level Scale Readings. A End.	B End.	Differences A End.	B End.	Length of Bubble.
1	5	27.0	9.1			36.1
2	10	25.0	10.4	2.0	1.3	35.4
3	15	23.9	12.1	1.1	1.7	36.0
4	20	22.2	13.9	1.7	1.8	36.1
5	25	20.8	15.3	1.4	1.4	36.1
6	30	19.2	16.9	1.6	1.6	36.1
7	35	17.8	18.2	1.4	1.3	36.0
8	40	16.2	19.8	1.6	1.6	36.0
9	45	14.8	21.2	1.4	1.4	36.0
10	50	13.3	22.7	1.5	1.5	36.0
11	55	11.8	24.2	1.5	1.5	36.0
12	60	10.6	25.8	1.2	1.6	36.4
13	65	9.0	27.4	1.6	1.6	36.4
13	65	9.3	26.8			36.1
12	60	10.8	25.2	1.5	1.6	36.0
11	55	12.4	23.6	1.6	1.6	36.0
10	50	13.9	22.2	1.5	1.4	36.1
9	45	15.4	20.7	1.5	1.5	36.1
8	40	16.9	19.2	1.5	1.5	36.1
7	35	18.3	17.9	1.4	1.3	36.2
6	30	19.7	16.4	1.4	1.5	36.1
5	25	21.2	15.1	1.5	1.3	36.3
4	20	22.8	13.6	1.6	1.5	36.4
3	15	24.3	12.1	1.5	1.5	36.4
2	10	25.8	10.6	1.5	1.5	36.4
1	5	27.4	9.0	1.6	1.6	36.4
				36.1	36.1	

Mean difference, 1.5 20ths of an inch.

Ten divisions of micrometer screw correspond to 19.2 seconds of arc, therefore one division of scale $= \frac{19.2}{4.36} = 4.4$ seconds and one division on bubble tube (tenths of an inch) $= 8.8$ seconds.

Five divisions of the micrometer screw $= 9.6$ seconds of arc, therefore one division of level $= \frac{9.6}{1.5} = 6.4$ seconds of arc.

All levels will not, of course, give such results, owing to the difference of the products of different makers, or to change of curvature after grinding, or to mode of mounting, too tight packing, etc. A level which seems of good quality when being tested may be found irregular after it has been mounted in its brass tube, which shows that the level has been improperly mounted. We do not wish to be understood as saying that a level is necessarily a bad one if its irregularities are too small to be revealed by the telescope whose power is properly adapted to the character of the graduations and other fine features of the instrument.

If the Civil Engineer should wish to know the value of one division of the telescope level bubble, in minutes or seconds of arc, it may be readily found in the field as follows: Having set up and brought the level over a pair of leveling screws a reading is taken on a rod held perpendicular to the line of sight, and a distance of from one to two hundred feet and the position of the bubble noted. The inclination of the level to the horizon is now changed by moving the leveling screws and a new reading taken on the rod and the new postion of the bubble noted.

Knowing the distance of the leveling rod from the instrument, the difference of rod reading and the number of divisions and distance over which the bubble traveled, we may readily write the following: Let

$x =$ the change in inclination of the line of sight.
$r =$ the difference of rod readings.
$d =$ distance of rod from instrument.
$n =$ number of divisions passed over by bubble.
$a =$ distance passed over by bubble.
$R =$ radius of curvature of inner surface bubble tube.
$\phi =$ number of seconds corresponding to one division of bubble.

$$\tan. x = \frac{r}{d}$$

$$\phi = \frac{x}{n} \text{ in seconds.}$$

$$R = \frac{a\,d}{r}$$

Example. — Bubble divided to tenths of inches. Distance from bubble to rod, 200 feet.

Movement Bubble. Inches.	Rod Reading. Feet.	Difference in Rod Reading. Feet.
0.0	4.499	
0.3	4.519	.020
0.4	4.544	.025
0.5	4.572	.028
0.5	4.608	.036
0.5	4.580	.028
0.5	4.542	.038
0.4	4.515	.027
0.3	4.493	.022
Mean, .425		.028

For a movement of bubble over one division the difference of rod reading is .0066 ft.

$$\frac{.0066}{200 \times .000005} = 6.6 \text{ seconds.}$$

(.000005 is the natural tangent for one second.

Note — For testing levels of the highest grade as required for astronomical purposes only, Prof. C. A. Young suggests the following modification: To make the above described Level Trier a repeating instrument. The base plate of our Level Trier will be lengthened and, at a small additional expense, an auxiliary screw of reasonable fine pitch and having a large but coarsely graduated head for convenient settings, will pass through it and rest against the masonry direct. This will make the Trier a repetition instrument; after running the bubble from one end of the level to the other by the micrometer screw, the bubble can be sent back by the auxiliary screw, and then the micrometer screw may be used again, getting a new reading on a different part of the screw. If this is repeated until the reading of the graduated disc of the micrometer screw is the same as that *started with*, it will eliminate an error of excentricity, if any exist, of the graduated disc, but what is more important, that of the *screw point* also, which in a measure always may exist after careless handling of the instrument, and then it may describe a little circle on the resting surface. If this surface is perfectly plane and at right angles to the axis of the measuring screw, this little excentricity of the screw point is harmless; but since it seldom happens that the bearing surface is truly at right angles to the screw point, a serious periodic error results. To make this arrangement complete, a small level will also be attached to the base plate to indicate when the latter occupies its normal position.

Why do different Magnetic Needles not always point in the same direction, though observed at the same place and time?

Paper by C. Louis Berger, read at Annual Meeting of Michigan Engineering Society, at Lansing, Mich., Jan. 15-17, 1895.

A close observer of the compasses used in surveying instruments must have noticed that the exterior shape of the magnetic needles, forming part thereof, frequently differs as much as do the styles of the instruments themselves. In some cases the needle consists of a bar of thin steel, oblong in shape, which rests flatwise on its supporting pin, so that its greatest superficial area lies in the horizontal plane as shown in Fig. 1. In others, as represented in Figs. 2, 3 and 4, the shapes are modifications of the above in a greater or less degree, inasmuch as their longer transverse dimensions — whether at the ends or center — also lie in the horizontal direction. Some magnetic needles carry a graduated circle or verniers, as the case may be, as seen in Figs. 5 and 6, whose zero points are supposed to lie in the geometrical axis of the needle; some are placed edgewise, so that the greatest superficial area lies in the vertical plane. There are also dipping needles to measure the vertical intensity of the earth's magnetism; short and stubby needles with aluminum extensions for galvanometers; complex needles, made in sections, used in marine compasses, and cylindrical or tubular forms used in scientific research. (Needles of the last mentioned type are not read by observing the ends, but by means of mirrors attached at the middle similar to those on a magnetometer.) Figs. 9, 10 and 11 show the principal types of the last mentioned needles.*

Since these latter forms are never met with in surveying instruments, we shall not consider them here, except to show the different styles of needles in vogue — suffice it to say that some of the shapes are simply selected by their propounders for the larger superficial area which they have, compared with others, according to the purposes for which they are intended; for it is well known that the larger the superficial area, combined with a minimum weight, the more delicate will the magnetic needle be; thus, a needle made of a very thin steel tube will be capable of receiving and retaining a greater charge of magnetic force, and, also being very light — preventing wear of the cap and pin — it is easily influenced to assume the direction of the magnetic meridian in azimuth.

However, we wish to remark here that it is not so important that the needle of a surveying instrument should have a great magnetic intensity up or nearly to saturation — which it may receive according to its superficial area and degree of hardness — as it is that it should have as constant an amount as possible, be it great or small. Of course such a needle should not have a surcharge, which it is apt to receive if improperly hardened, because its intensity is liable to be diminished and the needle thrown out of balance thereby, from time to time requiring a readjustment of its counterpoise for the same latitude in which it is used.

It is all-essential that a well constituted needle should have a proper symmetrical form as regards the longitudinal axis, and that it should be supported on a pivot as free from friction as it is possible for human ingenuity and skill to contrive. Besides the necessity of proper form and suspension, the quality and degree of hardness of the steel, length of the needle, and lastly the strength of its magnetic force, whether imparted by an auxiliary magnet or the more powerful electric battery and coil, have a most important influence on the behavior of a needle.

Returning to Fig. 1 we see exemplified the flat bar, and in Fig. 7 the edge bar needle as used in surveyors' compasses, but as all the intermediate styles of needles are simply modifications of the flat oblong form in order to be light in weight, partaking of the same principle that is involved, we shall consider these two forms almost exclusively. It remains therefore, for us to show in how far the exterior shape of a needle — whether of a faulty design or imperfect construction — may affect its reading, inasmuch as its geometrical axis, that is, the line passing through its ends and the center

---

* The largest and most delicate magnetic needle with which the writer is acquainted is that of the large magnetometer in the University of Marburg, made of a solid bar of steel two feet long and weighing about twenty-five pounds. It was suspended from the ceiling by a strand of silk fibers, and was provided with mirrors so that any oscillation could easily be read by the use of a scale and theodolite. This instrument was made after the style of Gauss' instrument, and with it many observations were made to determine the diurnal and annual changes in the magnetic meridian at that place during the younger days of the writer.

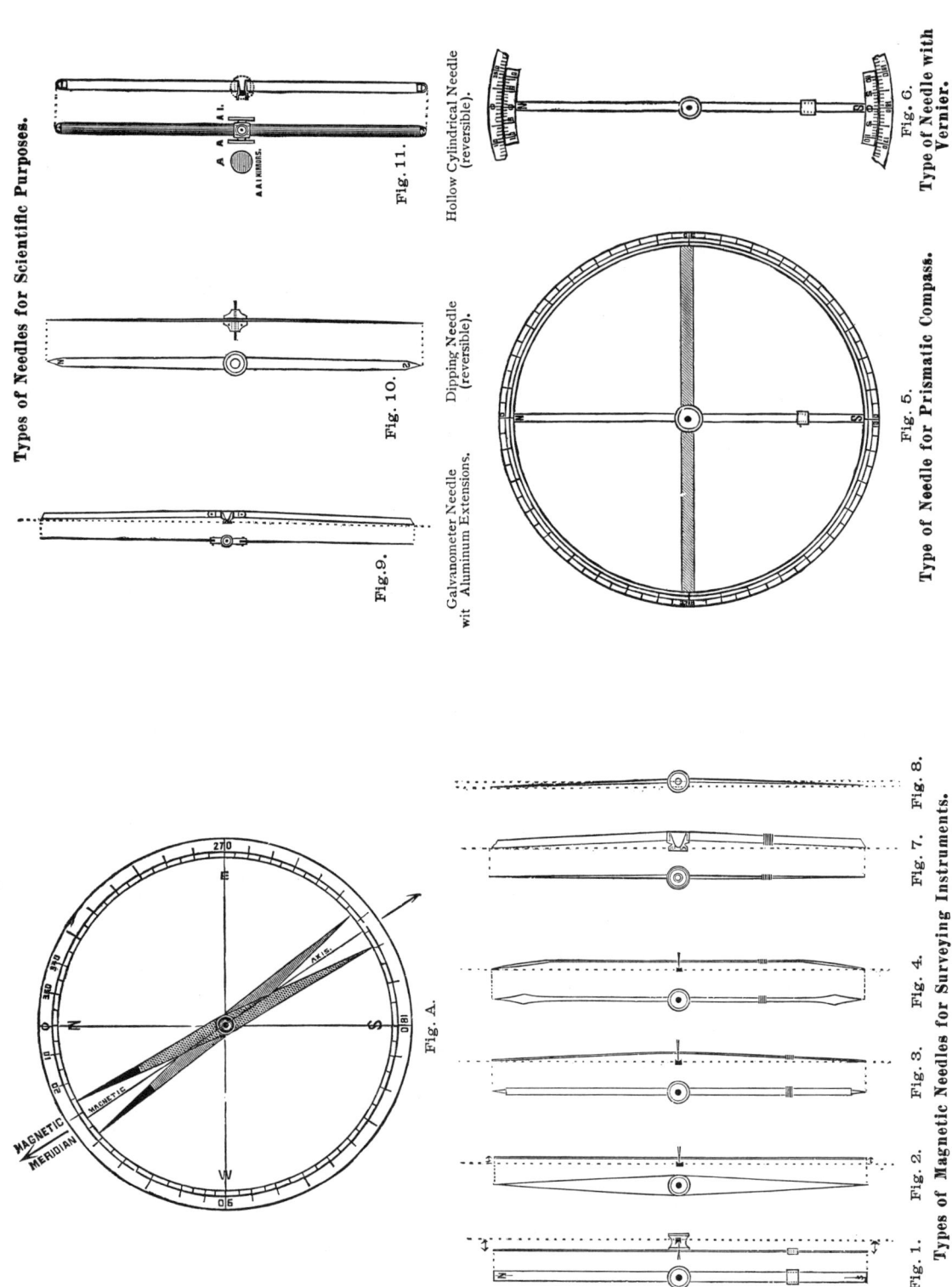

Types of Needles for Scientific Purposes.

Fig. 11.

Hollow Cylindrical Needle (reversible).

Fig. 10.

Dipping Needle (reversible).

Fig. 9.

Galvanometer Needle wit Aluminum Extensions.

Fig. 6.

Type of Needle with Vernier.

Fig. 5.

Type of Needle for Prismatic Compass.

Fig. A.

Fig. 1.   Fig. 2.   Fig. 3.   Fig. 4.   Fig. 7.   Fig. 8.

Types of Magnetic Needles for Surveying Instruments.

point in the needle cap, may not coincide with its magnetic axis, which lies in the plane passing vertically through its poles and in which the magnetic meridian is contained.

However, to get a complete understanding of the matter, we must go back to the ingot from which the steel of the needle was produced. If we remove a portion of the outer crust of a steel ingot or other casting and carefully examine the surface laid bare with a magnifying glass, we shall find that what seemed to the naked eye as a solid mass is now a spongy one, with numberless small and large cavities or blow holes. It is only by forging and rolling into bars and sheets, at the mill, that the metal becomes at all homogeneous. Some of the cavities not being welded, are elongated in the direction of the rolling, and form veins, flaws or blisters according to size, thus imparting to the bar or sheet a structural grain or fiber, which, while always traceable in poor steel, can only be detected in degree when the ingot has been converted into fine steel. In the absence of information to the contrary it is therefore but natural for the writer to believe that when a needle is being magnetized its resulting magnetic axis will in all likelihood tend to run in the direction of the general trend of this grain without strict regard to the geometrical axis, with which it may then be at an angle. The above divergence of these two axes may, therefore, be considered as one cause of the observed fact — *that two needles, though of the same size and shape, other things being equal, may not read exactly the same when observed at the same place and time.*

If, on the other hand, these two axes (magnetic and geometrical) do coincide, there will be no deviation, and two or any number of needles will read alike under the conditions above mentioned, if tested in the same compass.

Notwithstanding the fact that it is, in practice, very difficult to attain the desired result on account of the difficulty of tracing the grain, the writer would and does insist that the steel used in the construction of needles for surveying instruments shall not be forged, but shall be cut from fine shear steel, in the direction in which it was rolled, and be treated in such a manner as to ensure as near a parallelism of the structural grain with the geometrical axis as possible. Yet in spite of the utmost care and skill consequent minor magnetic poles are probably present, and there is almost certain to be a deviation of these two axes, and this deviation will vary in magnitude with the quality of the material and grain, the design, width and length of the needle, and the mechanical skill with which its outlines in azimuth have been made, as also with the degree of accuracy with which the point of suspension has been located in the geometrical axis.

While it may seem that just here the instrument maker's real work should begin in the testing and adjusting of these axes by grinding off a little from one side or the other, as the case may be, to insure a coincidence of the two axes, he, as a rule, will stop here either from lack of understanding or from want of necessary apparatus and a secluded space free from iron and provided with well defined meridian marks established by means of collimators or natural objects.

His chief reason for not doing this, however, may be said to be the fact that neither he nor the surveyor wish to bear the added expense.

For this and other reasons the scientist, not wishing to depend upon mechanical skill for good results, adopts the method of reversion, by means of which errors from this source can be entirely eliminated.

Therefore, the needle used in his investigations (See Fig. A) can be reversed by simply turning the needle upside down (the cap having been changed from the top to the bottom side of the needle) on the center bearing point, so that he can use the mean of the readings of each end in both the direct and reverse position as the result sought.

If the surveyor is desirous of making some such test he can do so readily, not by changing the center cap as above noted, for he cannot do so with American instruments, as the center cap is firmly fixed to the needle, but, by first taking a careful reading of both ends of the needle, he can, by means of a strong magnet, reverse the poles of the needle,* and, after rebalancing, read again the two ends, thus obtaining a reading correct in a measure and approximately finding the constant of his needle, which he can use when needed for important work.

The constant of the needle is therefore *the angle that the magnetic axis makes with the geometrical axis.*

If a needle has aluminum extension arms, as in galvanometers, Fig. 9, or a circle attached as in the case of prismatic compasses, Fig. 5, or if it is provided with verniers as in some compasses, Fig. 6, then the constant of the needle, or the index error is the

---

*Accidental changing of the poles by carrying the instrument on an electric car near the motor has been noticed by the writer, and has explained what, at first, seemed to be very queer behavior on the part of the needle.

angle that the zeros of the graduations or extensions (supposed to be in coincidence with the geometric axis) makes with the magnetic axis. As a rule, needles of the latter type always do have such an index error in addition to those arising from eccentricity of graduation and the difficulty encountered by the maker to determine the magnetic axis.

While the methods of reversion for eliminating errors of eccentricity and non-coincidence of the principal axes just described are good for scientific research, they are ill adapted for the work of the surveyor.

A surveyor's compass is ordinarily graduated to single degrees, and a fine instrument is seldom divided to read to less than one-half degree directly or from six to ten minutes by estimation; therefore it is useless to try to arrive at results closer than these. A needle will serve its purpose well if the two principal axes coincide within the degree of accuracy obtainable with its length and with the compass as a whole.

The following question may now properly present itself: What is the best form for a magnetic needle for the surveyor in order that it shall be so nearly free from errors due to the above causes as not to require reversion and yet to come within the degree of accuracy obtainable in practice? If now we have recourse to Fig. 1 and such modifications of the shape there shown as is delineated in Figs. 2, 3 and 4, we shall see that since the greatest superficial area lies in the azimuthal plane, deviations of the two axes cannot be infrequent, and, therefore, these shapes are not well adapted for surveying instruments. Fig. 4, in particular, presents the curious combination of a large surface at the ends and a small one at the hub. This increased spread of surface by the arrow-shaped ends renders the needle especially liable to this defect without so much as compensating for it in greater magnetic intensity as was probably the supposition, since the arms connecting it with the central hub are quite narrow and limit the magnetic intensity. This faulty desig.. is probably based on the supposition that since in a well hardened and well constructed needle the magnetic intensity is greatest at or near the extreme ends, gradually weakening towards the center where it becomes quite indifferent, that if the ends are made big the needle will the more readily assume the magnetic direction.

On the other hand, the edge bar needle as made by C. L. Berger & Sons and shown in Fig. 7, with its greatest superficial area in the vertical plane, reduces the chances of a non-coincidence of the two principal axes in the azimuthal plane to a minimum. In order to add strength and stability to this needle it is made thicker at the middle than at the ends, which are quite thin.

If in the manufacture of a needle of this form the arms are bent so as to be symmetrical to the axis passing through the center of the needle cap and at right angles to the line connecting the ends of the needle, as shown in the exaggerated diagram, Fig. 8, the magnetic axis will be contained in a plane parallel to the vertical plane passing through the ends of the needle and a little removed from it.

The error in the reading introduced by this small distance will be very small compared with the error due to the eccentricity caused by the above mentioned bending. But, supposing the case that in a badly constructed needle of this class one arm is bent more irregularly than the other, so as to unsymmetrically distribute the mass of the metal of the needle, then the resultant polarity consequent upon the irregular distributoin of masses of the metal may be slightly at an angle to the line passing through the ends of the needle, causing an error due to the divergence still very small as compared with the error due to the eccentricity such a needle would have.

So far in the foregoing explanation we have dealt with the magnetic needle alone. It will now be necessary to treat this task in relation to other functional parts that go to make up a surveying instrument, since there are other causes, which, singly or combined, may conduce to produce the differences mentioned in the reading of different needles.

However, before dealing with this subject in its further complexity in relation to the compass and instrument, we first of all take it for granted that there be no local attraction in the instrument itself or by any iron concealed on the person of the observer.*

---

*To determine whether an instrument itself has any iron in it to disturb the needle, it is a good plan, after setting the transit so that both compass needle and the vernier read zero, to go around the circle, setting the vernier ahead ten degrees each time, and noting whether the compass needle also describes an arc of precisely ten degrees. If it does not, there is some local attraction. Before making this test it will be well to first test the needle as to sharpness of pivot — mentioned later on — and to breathe on the glass cover of the compass and on the rubber frame of the reading glass so as to remove any electricity which may be present. Both of these articles being insulators are very easily electrified by the process of cleaning in a dry atmosphere, thereby affecting temporarily the reading of the needle.

Next we assume that the graduation of the compass is a good one and that the pivot on which the needle rests is in the center of the graduation, and that both ends of the needle read precisely 180° apart.

Next, that the needle will be so sensitive that when deflected from its pointing by the outside attraction of a piece of iron held about a foot or so away from it, it will settle to its original position several times in succession. This sensitiveness depends on the form and sharpness of the pivot, strength of its magnetism and its bearing upon the finely polished jewel or steel cap. (*It is generally owing to the dulling of the point and the scratching of the cap that a needle becomes sluggish and refuses to return to the same point.*)

Then the extreme ends of the needle should come close to the graduation and, together with the point of suspension, lie in the same plane with it, so as avoid parallax in reading. It is also important that the center of gravity of the needle be as far below this plane as possible in order that the quivering of the needle, so necessary to insure the proper settling of the needle on the pivot, shall not be annoying. If the extreme ends of the needle and the point of suspension are in the plane of the graduation, the quivering motion will not be annoying, since the extreme ends lie in the axis of quivering and consequently are stationary, as shown in Figs. 3, 4, 7 and 9. In Figs. 1 and 2 the plane in which the needle and its ends are contained is much below the axis of quivering which lies in the point of suspension and in consequence such needles can only be read when they have ceased quivering.

With a compass constructed as above we see no reason why its needle, or any number of needles of the same shape as shown in Fig. 7, should not give the same reading as long as the compass and its immediate surroundings remain undisturbed.

Our task of showing that in a theoretically perfect compass the needle will always assume magnetic North and South, — debarring erratic oscillations due to earth currents — would, therefore seem to be ended were it not for the fact that in a surveying instrument a sighting arrangement, telescopic or otherwise, must be provided, by means of which natural objects can be viewed and their relative positions in azimuth determined with regard to the stationary pointing of the needle in the magnetic meridian, affording a means for measuring angles and tracing lines. Seen from this standpoint, it is therefore all-important that the plane passing through the slits of a surveyor's compass, in which the line of sight is contained, be truly in line with the zero points of the graduated ring and at right angles to the plane of the same. With the transit, the line of sight, as defined by the optical axis of the object-glass and the cross wires of the telescope, must not only revolve in a vertical plane in which the line of collimation is contained, but this vertical plane must also either cut the zeros of the graduations or at least be parallel to the plane passing through them. This condition is, however, so difficult of attainment in instruments fitted with a telescope, and in cases where it may exist it is so very apt to become deranged, that the writer is sure that right here the principal cause can be found for most of the differences observed in the reading of the needles of different instruments, though observed at the same time and place. To illustrate this: suppose we bisect a distant natural object with the cross wires of different instruments whose lines of collimation are in perfect adjustment, then the readings of the needles — assumed to have both of their principal axes in coincidence — would be the same if the zero points of the compasses are contained in the vertical plane of the line of sight. But, if on the other hand the initial points of the compass are not contained in the vertical plane of the line of sight the reading must differ by an amount equal to the angle of this divergence in any one of the different instruments. This index error, while often of no account in ordinary instruments, will exist nevertheless, and must always be looked for even in the best class of instruments, since in the field use of an instrument and by rough handling it is very liable to change.

This is particularly the case where a telescope is involved, since in its construction we have to deal with the optical axis, which, of itself, is of a very complex nature and subject to changes affecting its line of collimation in relation to the compass, and because of the changes occurring by rough handling in the position of the standards carrying the wye bearings of the telescope's axis of revolution in the customary instruments. Therefore, in order to eliminate all the errors due to the change of the line of sight with regard to the zero points of the compass, the instrument should be tested before any important work is undertaken with the needle.

Before proceeding to determine the index error or constant of the compass of a transit, it will be well to verify the adjustment of the line of collimation as well as its motion in the vertical plane. Then, having previously established a meridian line by three points in line, point the telescope to the mark at the North, read the needle at both ends, and, in order to eliminate all errors of eccentricity of the graduation or needle and thereby arrive at correct results, the instrument should be reversed on its vertical axis and the telescope again pointed to the mark at the north end as before and another reading taken from both ends of the needle.

The average of the four results will be the combination of both the index and magnetic errors and that of the declination of the needle — the latter being the angle the magnetic meridian makes with the astronomical meridian at the place of observation. Naturally, then, in a theoretically perfect instrument the reading of a needle at a place where East and West declinations join, would be zero when the cross wires bisect the meridian mark. However, as in the field use no instrument will retain its fine qualities and adjustments, it will be best not to depend on them, but to determine the index error and constant of the needle combined simultaneously, from time to time, by an observation on a meridian as above described, in order to apply it as a correction in precise work. In instruments fitted with a variation plate permitting the declination for any particular locality to be set off, the combined errors of the instrument and needle can at once be added to or subtracted from the declination, as the case may be. But, as these constants differ in every instrument it furnishes the explanation to the question propounded in the title of this paper: Why do different magnetic needles not always point in the same direction, though observed at the same time and place? *Unless these constants have been previously determined and applied in every case.*

There are other reasons, such as the daily variations of a needle, which, according to temperature, latitude and season of the year, may amount to from six to sixteen minutes alone, showing why reliance on the reading of a magnetic needle can be placed only within certain limits and that it should be depended on only for general direction as required in filling in details. For exact work meridian lines run by solar or stellar observations should be depended on exclusively, to which end the modern solar attachments give the fullest satisfaction.

In conclusion, we would say that such tests on a meridian mark, as recommended above, should be repeated often, and that in order to facilitate the work such meridian marks should be located at some convenient place so as to be within easy reach of the surveyor at all times.

In the selection of such a place care should be taken that there is no iron in the vicinity and that there are no electric wires, particularly those used for arc lighting or the conveyance of power within a radius of three to six hundred yards, or else a needle is apt to be affected by these currents as would be the needle of a galvanometer. Such a place, if properly selected and provided with permanent meridian marks, could at the same time be made available for verifying the telescope and level adjustments of the transit so necessary to its proper use. For the benefit of surveyors, the City of Boston, many years ago, provided such meridian marks on stones deeply set into the ground on Boston Common. Unfortunately, electric wires now pass all around these grounds, and even traverse them, so that reliance can no longer be placed on observations there made.

The Berger Short Focus Lens Attachment.

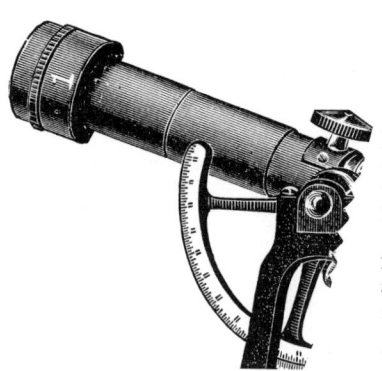

A very valuable addition to the engineer's outfit is found in the short focus lens attachment which has been brought out. The contrivance is simple, but, like many simple devices, is very effective in overcoming a practical difficulty. Probably every engineer has been annoyed by being obliged to sight a point a little too near for the telescope to focus. Most transit telescopes will not focus on a point much nearer than 5 or 6 feet (levels not nearer than 7 or 8 feet) away from the instrument, while it is frequently necessary to sight a point on the ground nearly under the transit, at a distance which is usually less than that.

In mine surveying as well as inside of factory buildings, one frequently needs to sight a point overhead or on the walls and very near the transit. Ordinarily the only way out of the difficulty is to focus as nearly as possible and do the rest by a guess. As a further instance, one often finds in leveling, that it will be necessary to take a reading on a point very near the instrument, and has to resort to various means (all of them inaccurate) of getting around the difficulty. The attachment mentioned consists of a small aluminum tube containing a simple lens, which is attached in front of the objective. The lens is so placed in the tube that it can be accurately centered by means of 4 adjusting screws. The effect of this lens is of course to bring rays to a focus nearer to the objective, and thus enable the observer to focus a nearer object than would otherwise be possible. When the telescope will focus no nearer than 6 feet, the attached lens, marked 1, is ground so that it will focus objects 6 feet away *when the objective tube is drawn away in.* This allows the entire motion of the focusing slide for distances between 6 and 4 feet. For distances nearer than 4 feet a second lens may take the place of the first and will focus up to about 2½ feet. If the two are used at once the distance is reduced to about two feet.

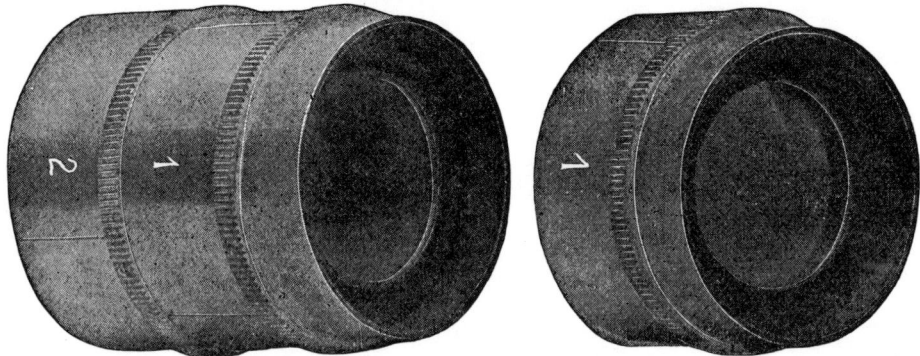

With this pair of lenses there is no distance between two feet and infinity at which objects cannot be focused. The accuracy of work done with this attachment is in no way affected by the centering of the attached lens itself, as this is capable of perfect adjustment. The only way in which error can occur is through the imperfection of the objective tube. If the cylindrical surface of the object-head of the telescope on which the attachment is placed is not concentric with the optical axis of the telescope this error will enter into the adjustment of the attached short focus lens. This error, however, is never large on an instrument sent out by our firm. But even admitting that there may be some error here, it must be rememembered that this lens is never used for objects more than about 6 feet away ; consequently the resulting error on the point is entirely negligible, and the convenience of the attachment in many cases is so great that it entirely outweighs any such consideration, since the work done at this distance will be entirely consistent with the work done with the instrument on the longer distances. The attachment fills a want that has long been felt by engineers and is certainly a step in advance in the perfection of instruments of precision.

To attach this device to their old instruments it will be necessary to send the instrument to them, as every lens attachment must be specially fitted and centered. However, it can be supplied with any of their new instruments, either Transits or Levels, made since 1899.

When attached to transits, No. 1 permits focusing objects to about 3½ feet, No. 2 permits focusing objects to about 2½ feet : both permit focusing objects to about 2 feet from center of instrument.

This is so important a feature that one trial will convince one that it is indispensable to the outfit of an engineer The device is patented. The Messrs. Berger are also prepared to attach it to their Wye and Dumpy level, for focusing nearly as close as stated above for transits. For prices see catalogue, page 169.

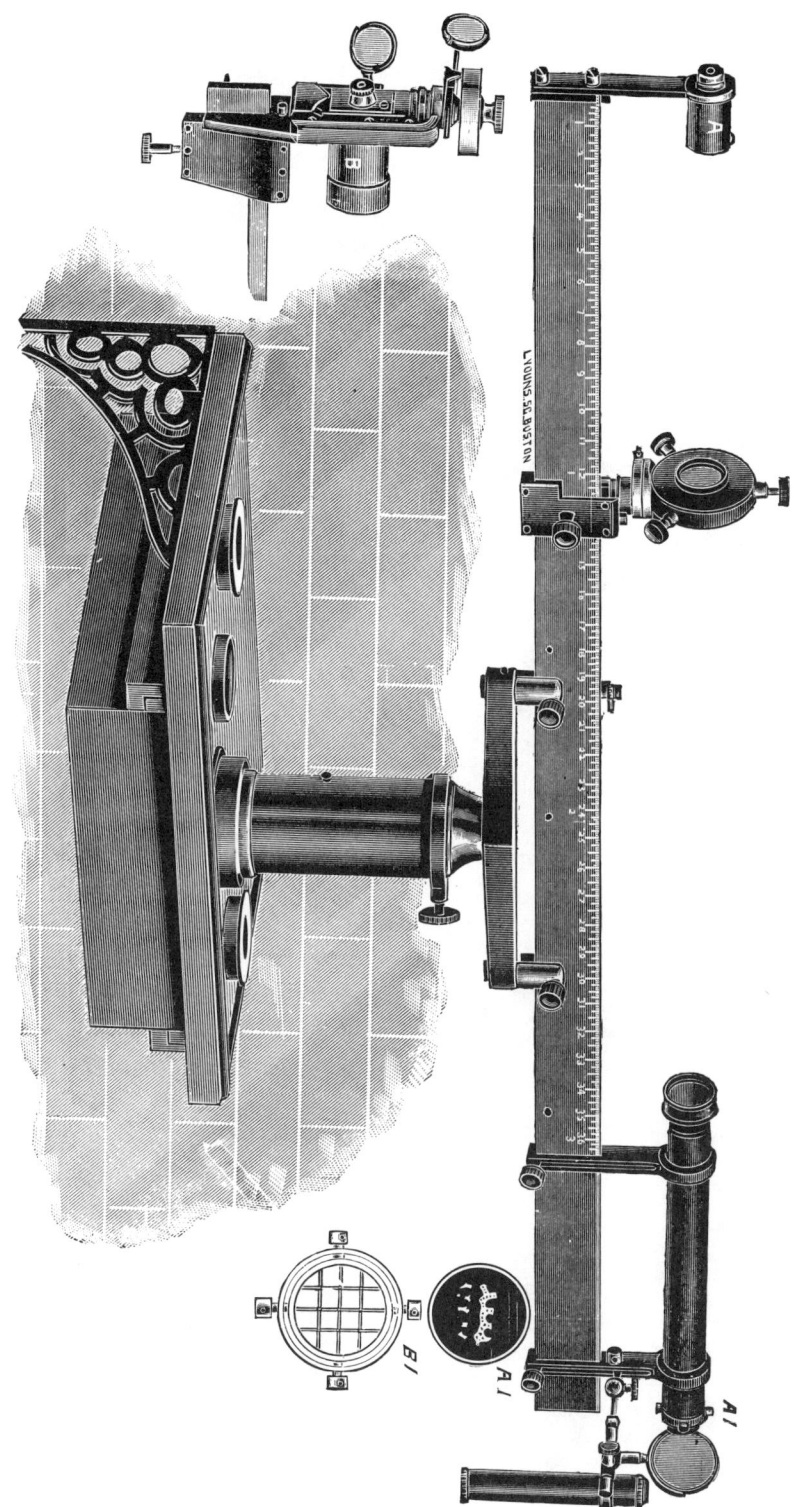

C. L. Berger & Sons' Focalimeter.

As made for the U. S. Naval Observatory. (In two styles.)

Style A, shown in principal cut, measures the focal length of an object-glass from the optical center on the scale.

Style B determines the focal length from the *second principal point* by using in place of the scale the micrometer eye-piece *B*, and measuring the interval 1 : 100 of the spider-line micrometer *B* 1.

Prices given on application.

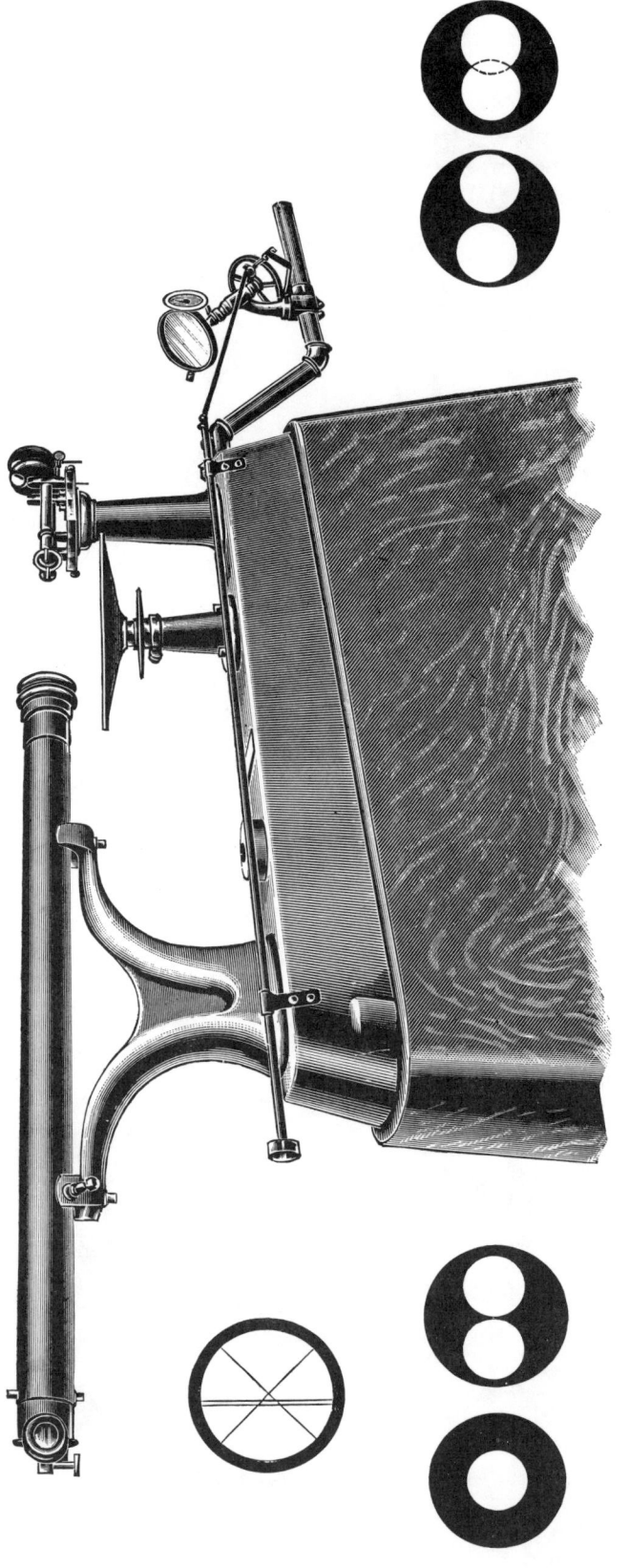

C. L. Berger & Sons' Sextant-Testing Apparatus.

As made for the U. S. Naval Observatory.

For testing the quality of mirrors and colored glasses used with Sextants.

Prices on application.

PART II.

Catalogue and Price List

OF IMPROVED

Engineers' and Surveyors' Instruments,

WARRANTED FIRST-CLASS,

MANUFACTURED BY

C. L. BERGER & SONS, No. 9 Province Court, Boston, Mass.

NOTICE.

In selecting instruments from catalogues, engineers should not be led so much by a simple comparison of prices, as by the advantage offered in *superior merits, working capacity, and preservation of fine qualities in case of severe treatment.* We can cite instances, where transits and levels of our manufacture had severe falls, resulting without injury to any part of instrument — not even disturbing the adjustments.

A greater outlay of $10 or $20 in the purchase of a superior article is a greater saving in time and expense in the end.

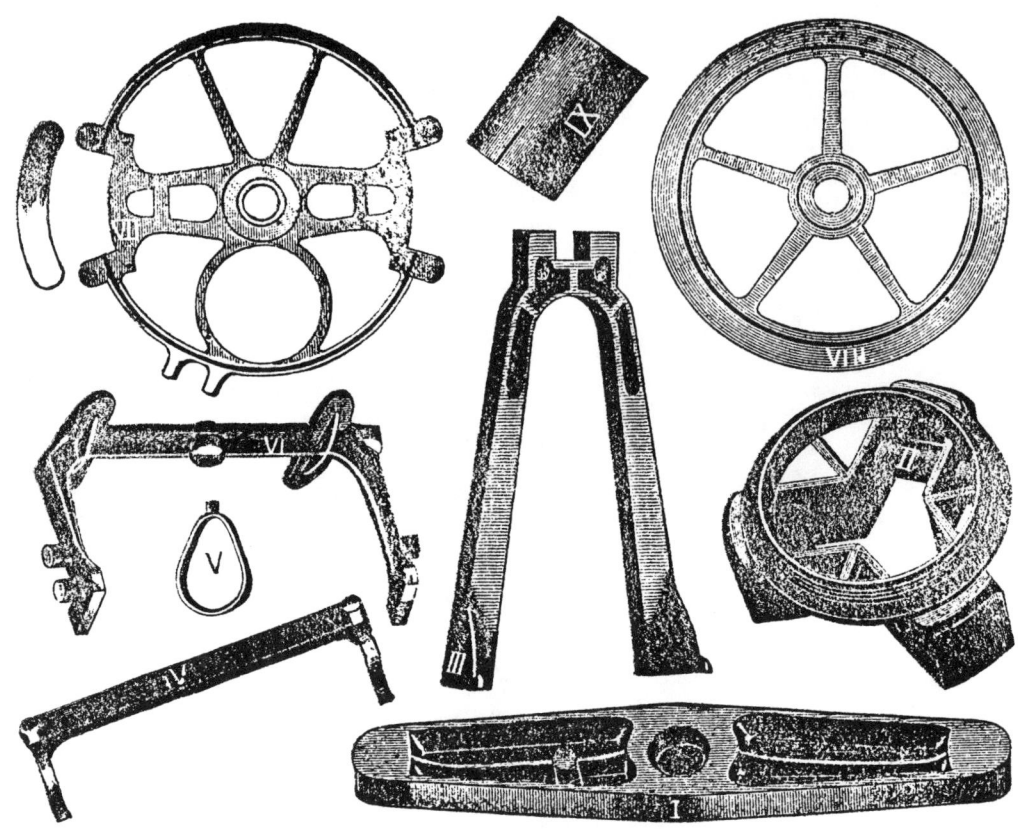

Parts of Instruments,

As made by C. L. Berger & Sons.

Showing the extent to which Aluminum bronze (containing 90% copper), or Aluminum alloyed with only a small percentage of copper or silver, is being utilized in the manufacture of our Instruments.

(Before ordering any of the above parts please read "Aluminum for Instruments of Precision," page 27. also notes following). Parts of Aluminum which are not of general adoption are indicated by the extra charge made below.

DESCRIPTION OF PART.	METAL USED.	ADVANTAGES.	DISADVANTAGES.
1. Crossbar* for 18-inch Wye Level	Aluminum Bronze	Greater tensile strength	None.
2. Tripod Head for Level or Transit	" "	" " "	"
3. Telescope Standards† for Transit	" "	" " "	"
4. Striding Level Frame for Plane Table	Aluminum	Saves a few oz. in weight	Softness.—Level is apt to lose [its adjustments.
5. Handle for Striding Level	"	None	None.
6. Compass Stand for Mining Transit	"	Saves one half pound in weight	"
7. Verniers and Vernier Frame for Vertical Circle	"	" one quarter " " "	"
8. Vertical Circle for Mining Transit	"	" " " " " "	Softness.
9. Sunshade	"	Saves fraction of ounce in wt	None.
10 Tripod Head for Plane Table	"	" four pounds in weight	Bolts and leveling screws may wear loose. $8.00 extra.
11. Alidade for Plane Table	"	" one and one half pounds in weight	Softness. — Edge may wear concave. The paper becomes discolored; when nickel-plated to prevent wear and discoloration, nickel is apt to peel off in time. Price extra, $3.00.
12. Standard Frame for Duplex Mining Transit	"	" two and one half pounds in weight	None.
13. ‡¶ Standard Frame for Transit No. 11	"	" one pound in weight	None. — Will be cloth-finished. — If ordered, can be finished bright or black. Price, $5.00 extra.
14. Tubes for various purposes.	"	" less weight	
* 1. ‡ Cross bar for 18 inch Wye Level	"	" 1½ lbs. in weight.	Softness. — $5 00 extra.
† 3. ‡ Telescope Standards for Transit	"	" 1 " " " "	Softness. The screws by which they are connected to the vernier plate may wear loose, endangering the stability and reliability of the whole instrument. Cloth Finish. Price extra, $5.00.

‡ Will be made to order only.
¶ Will be cloth finished. If ordered highly finished, $5.00 extra.

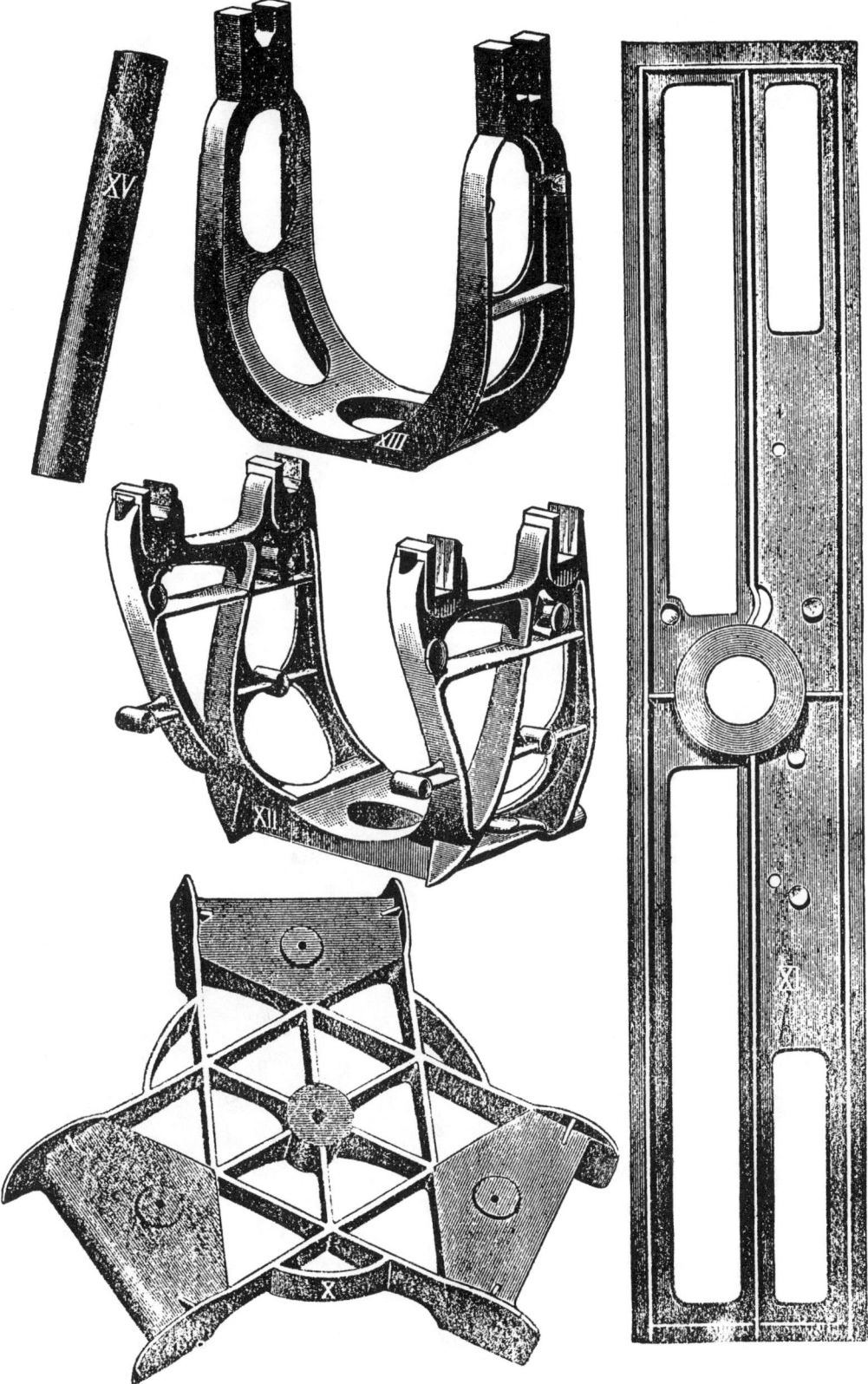

Parts of Instruments which can be made of Aluminum.

As made by C. L. Berger & Sons.

See Notes on preceding page.

Divers Parts of Instruments,

As made by C. L. Berger & Sons.

Showing method of ribbing, etc., in order to make Instruments light and stiff without resort-
ing to the use of the objectionable aluminum. (See "General Construction," page 9.)

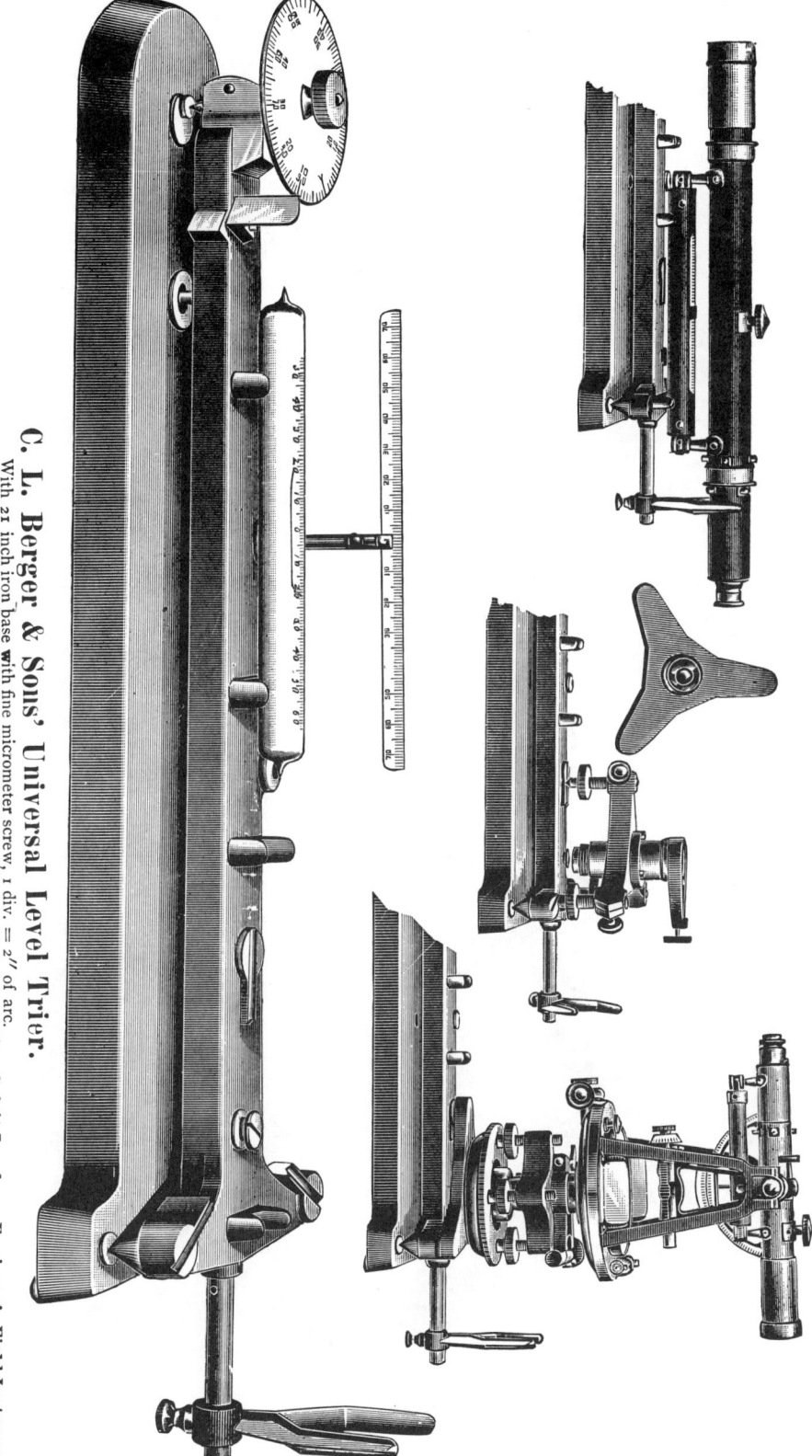

C. L. Berger & Sons' Universal Level Trier.

With 21 inch iron base with fine micrometer screw, 1 div. = 2″ of arc. Specially adapted for testing Spirit Levels on Engineer's Field Instruments without taking them apart.

For use in Engineering Schools, Physical Laboratories and Astronomical Observatories.

Price of this Instrument, complete, as above - - **$38.00.**

For a full description of this instrument, method of using, etc., see p 109 *et seq.*

Price of this Instrument with auxiliary screw and level to iron base as described in footnote, p.112, **Extra, $10.00.**

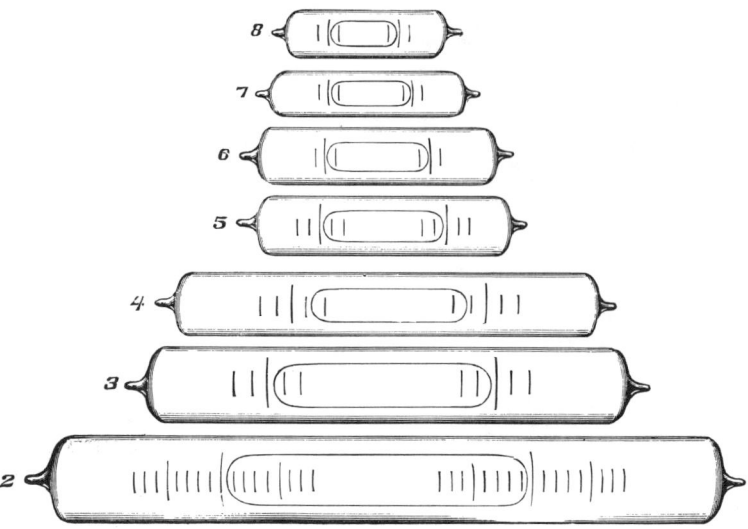

Spirit Levels.

For the benefit of our patrons we enumerate below the principal Spirit Levels we are prepared to supply at short notice. They are made by us, and are of the same superior quality as those furnished with our instruments. In the list below we give length, diameter, and degree of sensitiveness. They are graduated, as a rule, as shown above. — Levels different in size from this list can be made to order only, and will be furnished only when order is accompanied with *the tube or mounting for which one is intended, and also stating the kind of instrument it is for, and the degree of sensitiveness desired.* We will positively not make any levels upon written dimensions only, but require the tube to be sent in all cases, as otherwise we will not be responsible for any failure in that respect. Please read pages 7 and 18.

No.	Length from tip to tip in Inches.	Diameter in Inches.	Sensitiveness.			Price each, sent to us mounted in Tube.
2.	6.50 to 6.60	0.75 to 0.80	One div. (0.10) = 10" to 20" of arc			$4.50 to 5.00
3.	4.75	0.65 to 0.68	" (0.15) = 15" to 20"	"		3.50 to 4.00
4.	4.10	0.58 to 0.60	" (0.15) = 15" to 20"	"		3.80
5.	2.50	0.51 to 0.53	" (0.10) = 20" to 25"	"		3.20
6.	2.40	0.51 to 0.53	" (0.10) = 60"	"		1.80
7.	2.00 to 2.25	0.41 to 0.43	" (0.10) = 70"	"		1.80
8.	1.68	0.41 to 0.43	" (0.10) = 75"	"		1.80

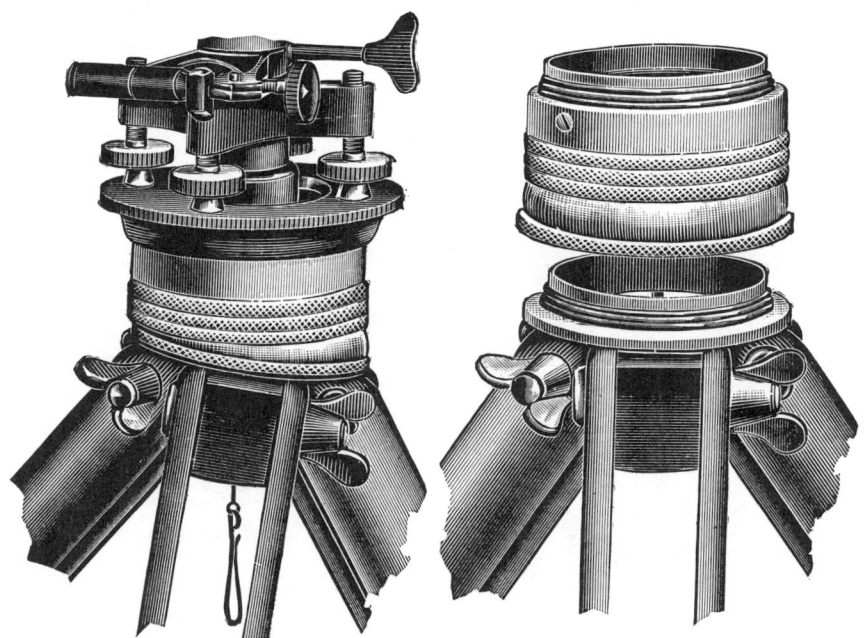

C. L. Berger & Sons' Quick Leveling Attachment.

Shown as applied to Levels and Transits.

(See page 39 of Manual.)

Price, $8.00

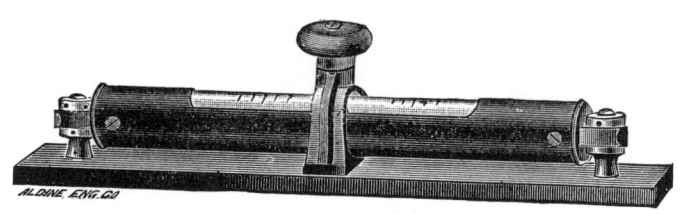

Spirit-Levels on Metal Base.

NOTE. — For use in setting weirs, fine machinery etc., and for leveling up apparatus in observatories, and physical and chemical laboratories, etc.

Ground Spirit-Level, one division of level about 1 min. of arc; mounted on 3½ inch metal base, provided with a handle. Level adjustable. In case.

Price $7.00

Ground Spirit-Level, one division of level about 30 sec. of arc; mounted on 8-inch metal base, provided with a handle. Level adjustable. In case.

Price $13.00

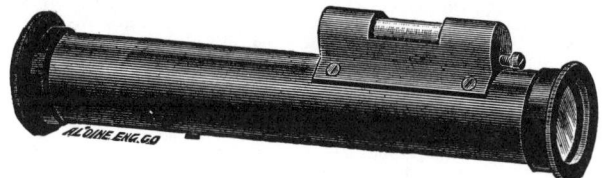 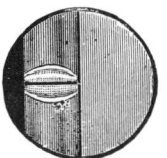

Locke's Hand-Level.

Brass or nickel-plated. In case . . . . **Price $9.00** *#8.*

NOTE.—This consists of a brass tube 6 inches long, with a small level mounted on its top to the left of its center near the object end. Underneath the level is a horizontal wire stretched upon a frame. This frame is made adjustable by a screw and a spring working against each other, or by two opposing screws placed at the ends of the level mounting. In the tube directly below the level is placed a totally reflecting prism, acting as a mirror set at an angel of 45° to line of sight. The images of the bubble and wire are thus reflected to the eye. The prism divides the aperture in two halves, in one of which is seen the bubble and wire focussed sharply by a convex lens placed in the draw tube, while the other permits of an open view. Putting the instrument to the eye and raising and lowering the object-end until the bubble is bisected, natural objects can be seen through the open half at the same time, and approximate levels can then be taken. To prevent dust and dampness from entering the main tube, both the object and the eye ends are closed up with plain glasses. In preliminary work this is a very useful instrument.

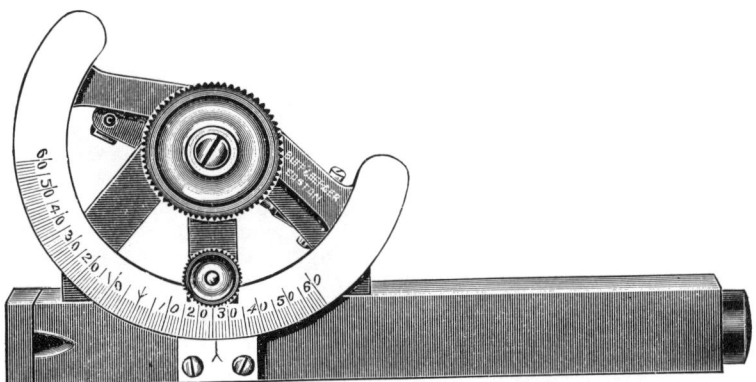

Hand-Level and Clinometer.

Abney Level and Clinometer.   **Price $14.00.**

NOTE. — This instrument is similar to the Locke's hand-level, but the small spirit level mounted on top can be moved in the vertical plane and clamped to a dial graduated in single degrees, thus the angles of slopes, etc., can be measured also.

Pocket Leveling Instrument with Divided Circle.*

In the ordinary practice of the Builder, Agriculturalist, Gardener, etc., etc.; it is often necessary to obtain relative positions and heights, within practical limits. For such needs we offer this instrument at once well made, reliable and low in price.

This instrument, shown on the opposite page, has the unusual advantage that the level bubble, cross wires and images can all be seen at a glance, since the bubble of the level can be seen from within the telescope and not from without. The bubble is somewhat magnified by the power of the eye-piece, and on this account the efficiency of the instrument is considerably raised and very much surpasses the old-fashioned water level. At the same time the handling of it is so simple that those who do not know how to work with the ordinary level and telescope may readily become adept in its use. In the sectional view opposite, the inner arrangement can clearly be seen. The telescope has a magnifying power of 5 diameters, the objective is achromatic, but the eye-piece has only a single lens.

The instrument can be mounted on any convenient support, while for approximate results, with a little practice, it may even be held in the hand. In order to obtain a stable mounting, it can be placed on a tripod or jacob-staff, or can be screwed to a bench, window-sill, or other convenient support. By means of the ball and socket joint above the fastening screw, the instrument can be approximately placed in a horizontal plane.

In order to see the image and cross-wires of the telescope distinctly, the eye-piece slide must first be moved out or in till the wires are very sharply defined, then by turning the focussing screw move the main tube out or in till the image of the object sighted at also becomes very distinct. For distances of 60 feet or so the main tube should be moved out till the mark on the tube just shows. For shorter distances the tube must be moved out farther, and for longer distances moved in *slightly*, till in each case the image is distinct. This last process can best be done by taking the instrument from the center and holding it in the hand. As shown by the cut opposite, there is a mirror with a hole in the center, placed at an angle of 45° inside the main slide tube. Below this mirror is the level, which has two adjusting screws. Through the opening in the bottom of the level tube light passes, illuminating the mirror, and when the instrument is horizontal, by looking through the eye-piece, the bubble can be seen in a vertical position and at the same time the image and cross-wires can be seen.

In order to obtain a horizontal position of the instrument it is necessary to operate the milled-head nut under the object-glass until the image of the bubble shall appear to the eye as much above the cross wire opening as it does below the same, as illustrated by the small cut opposite. Of course for every change in position in the horizontal plane, the bi-section of the bubble must be verified. The eye should be placed very close to the eye-piece so as to be able to observe the whole mirror surface. Near sighted persons, who, on account of their glasses, cannot bring the eye near to the eye-piece, should take the cross-wire opening as the mark and watch for the bubble above and below the same. In case the instrument is in a building where the light is poor, the bubble may be sufficiently illuminated by a piece of white paper or the bare hand held beneath the level tube.

Price, as above, with terrestrial telescope, horizontal circle reading to minutes, in Morocco Pocket Case, and light tripod, ... **\$32.00**

Pocket Leveling Instrument with Reversion Level.*

This instrument, fitted with a reversion level, has as compared with other designs provided with a striding level, the important advantage that it dispenses with the inconvenience of adjusting the level, also with the reversing of the telescope in its wyes.

The telescope magnifies 12 times; its definition is such that less than one-half an inch can be read off at a distance of 300 feet. The degree of accuracy is $\frac{1}{20000}$ of the distance.

The reversion level is encased and protected from external influences; the bubble is directly visible in the field of the eye-piece, the whole is mounted on a light tripod with a ball and socket joint.

Note. — This instrument, which is intended for accurate measurements, can be used without preliminary adjustment. The manipulation is very simple, and is done thus: set up the instrument with the level on the left and take a reading; unscrew the milled-headed screw by which it is fastened to the horizontal bar carrying the micrometer screw; turn the instrument upside down with the level on right and fasten in place; take another reading and the arithmetical mean is the correct value.

Price, including Pocket Case, \$42.00

The Road Builder's Dumpy Level

has been designed by us to meet a growing want for the road builder as well as for the drainage engineer. The degree of accuracy obtained with it is commensurate with the work required. It will not be as accurate as our 15 or 18 inch Dumpy Level, but we believe it will be quite satisfactory in the above cases, where a lighter and cheaper instrument is desirable.

The telescope is 12 inches long, is erecting, and has an aperture of 1¼ inches and a power of 24 diameters. The eye-piece is provided with an improved *screw arrangement* for the accurate focussing of cross-wires; field of view large and flat; objects erect; *telescope balanced each way from the center when focussed to a mean distance with sun-shade attached to it;* the center is very stout, long and of the hardest bell-metal; 5½ inch very sensitive spirit level; instrument does not detach from tripod above leveling screws; it packs whole and stands in the case erect. Mahogany case, provided with straps and hooks, contains sun-shade, wrench, screwdriver, and adjusting pin.

Note. — The above instrument, being of the Dumpy level type, has to be adjusted by the two peg method as described in our manual (p. 54) for the Dumpy Level.

Weight of instrument 7 lbs., weight of tripod from 6½ to 7 lbs.

Gross weight of instrument, packed securely for shipment in two boxes, about 40 lbs.

Price, *including protection to the object-slide,* **\$75.00**

Extras: Stadia wires, ... 3.00

Gossamer bag, ... 1.00

Bottle of fine oil, ...25

* Not of our manufacture.

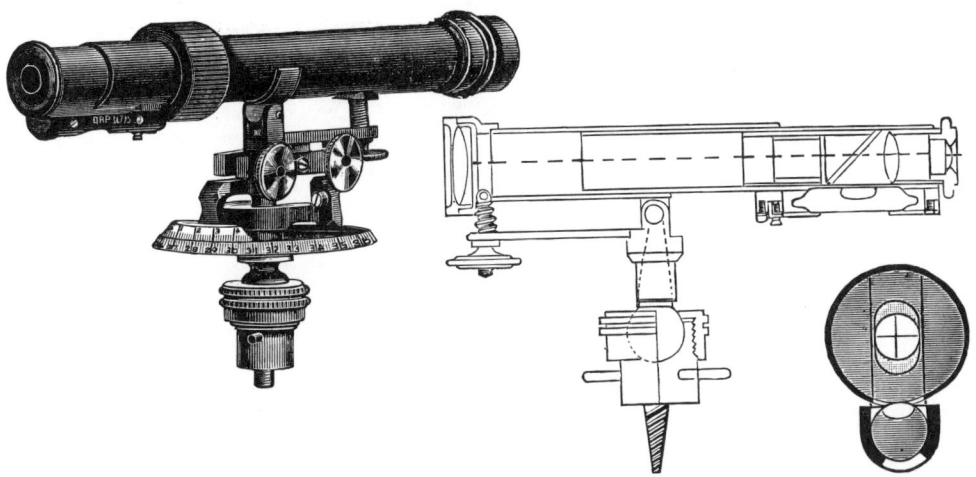

Pocket Leveling Instrument with Divided Circle
for laying off Horizontal Angles.
Suitable only for approximate work, such as reconnoissance, use in drainage work, for farmers, etc.

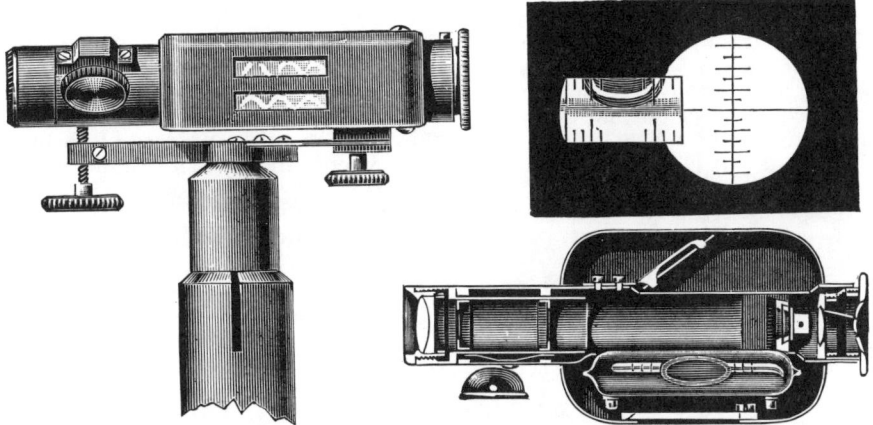

Pocket Leveling Instrument with Reversion Level.
For the use of engineers in reconnoissance work and filling in topographical details; capable of close results.

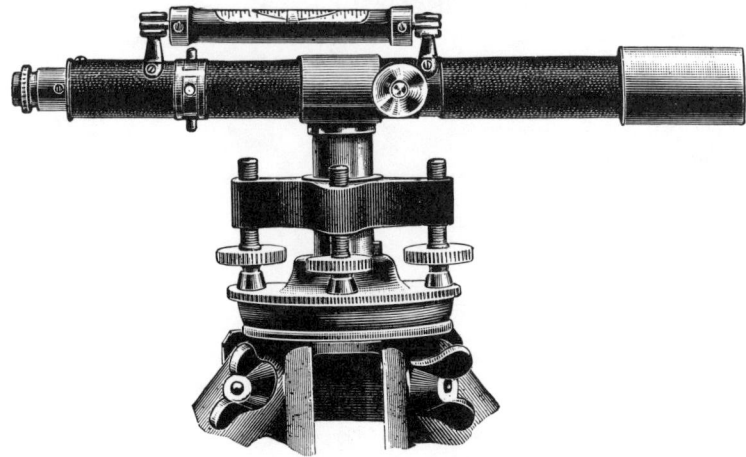

The Road Builder's Dumpy Level.
Suitable for ordinary road building, drainage work, and for filling in details between benches.

ENGINEERS' DUMPY LEVEL.

On account of the greater compactness, our dumpy level is best adapted for railroads, water works and reconnoissance, etc., permitting of high accuracy by greater simplicity, and is less liable to derangement of all parts. As regards size, it has all the advantages of the larger engineers' levels, but as it consists of a lesser number of pieces and screws, is superior to these in point of durability and permanency of adjustments. With a properly adjusted dumpy level of our make, (*see e adjustment of dumpy level,*) an engineer can perform as high a class of work as he is generally enabled to do with a good wye level, depending, as he does, not so much on *mechanical* perfections, as on his own superior skill and sense of accuracy in making adjustments. The upper part of this instrument is entirely cloth-finished.

The instrument is packed in a mahogany box, containing a sun-shade, a wrench, a screw driver and adjusting pin.

Weight of instrument 10 lbs., weight of tripod from 7 to 7½ lbs.

Gross weight of instrument, packed securely for shipment, in 2 boxes, about 45 lbs.

Price $100.00.

Extras to Engineers' Dumpy Level.

Center of instrument made of steel, and hardened, ... $10.00
Stadia Wires, fixed, ... 3.00
Gossamer water-proof bag, to protect the instrument in case of rain or dust, 1.00
Bottle of Vaseline, to lubricate the level center, ... 0.25

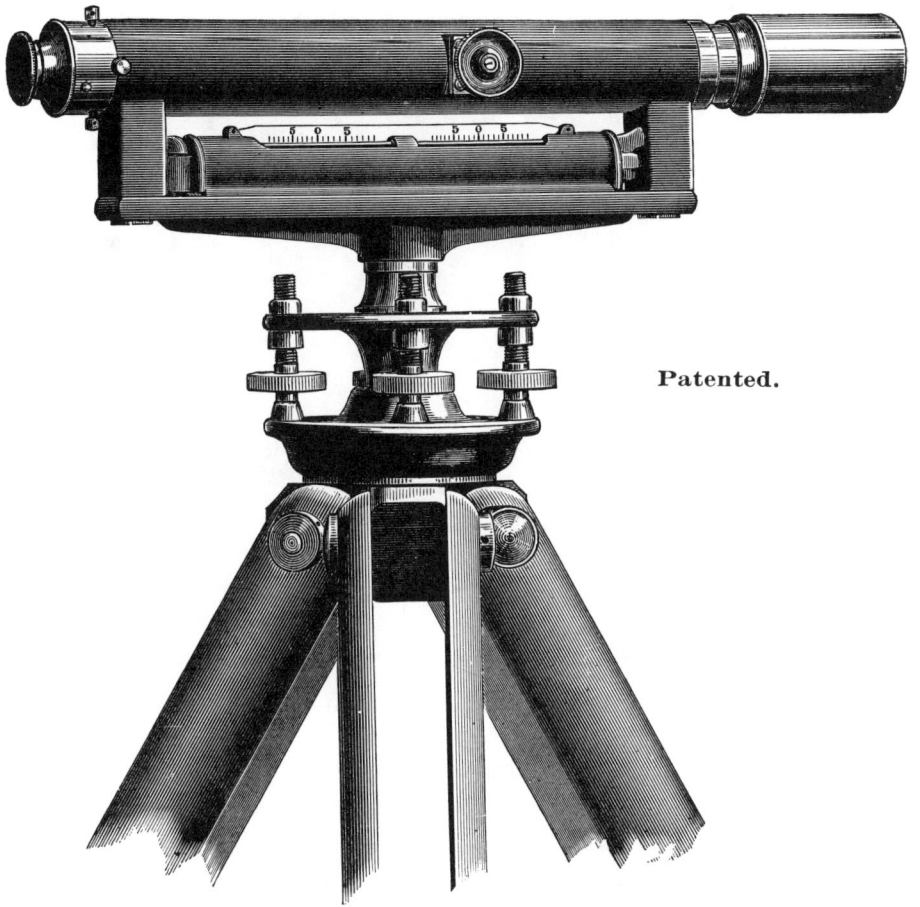

15-inch Dumpy Level.

As made by C. L. Berger & Sons.

NOTE.—The cut represents this instrument with a telescope, showing objects inverted. If ordered with an erecting telescope its length will be 17½ inches. Unlike the round nuts, shown in the cut, the tripod is provided with thumb-nuts with which to fasten the legs against the tripod head, and the instrument has a leveling head similar to those represented in the cuts of our Engineers' Transits and Levels.

ENGINEERS' 18 inch WYE LEVEL.

Leveling Instrument of Precision.

Eighteen or seventeen-inch powerful telescope; aperture of object-glass 1⅜ inches in diameter; eye-piece provided with an improved *screw arrangement* for the accurate focussing of cross-wires; field of view large and flat; telescope provided with an *adjustable stop* to readily set cross-wires horizontal and perpendicular; *line of collimation true on all distances;* objects erect; *telescope balanced each way from the center when focussed to a mean distance with sun-shade attached to it* to secure the highest accuracy attainable; telescope rings and the center are very stout, long and of the hardest bell-metal; cross-bar is cast hollow and provided with ribs; 8-inch very sensitive spirit level; instrument does not detach from tripod above leveling screws; it packs whole and stands in the case erect. Mahogany case, provided with straps and hooks, contains sun-shade, wrench, screw driver, and two adjusting pins.

Weight of instrument 11 lbs., weight of tripod from 7 to 7½ lbs.

Gross weight of instrument, packed securely for shipment in two boxes, about 48 lbs.

Price, *including a protection to the object-slide,* . . . . **$140.00.**

Telescope and level tube will be cloth-finished, unless ordered to the contrary when an extra charge of $5.00 will be made. However, we strongly advise the cloth-finish. (*See cloth-finish.*)

Extras to Engineers' Wye Level.

Center of instrument made of steel, and hardened, and running in a socket
of cast iron, improved style ‡ (See cut on opposite page) . . . $15.00
Stadia wires, fixed 3.00
Sort Focus Lens (see pages 118, 169) one pair 16.00
Metal-mirror with universal joint. (This is readily attachable to the instrument and facilitates the reading of the bubble on soft ground without stepping aside.) 10.00
Extra Sun-shade with smaller aperture, for use with the telescope when the sun's rays are too bright for accurate work, ... 1.50
Instrument provided with three leveling screws, as shown in cut of Hydrographer's Wye Level,* 18.00
Gossamer water-proof bag, to protect the instrument in case of rain or dust, 1.00
Bottle of fine watch oil, to lubricate the level center, ... 0.25

* Four leveling screws commend themselves in the more ordinary class of instruments for the greater rapidity with which an instrument can be leveled up approximately and that (no matter how much the levelling screws may be worn) when brought to a true bearing on the lower leveling plate, all such looseness is taken up.

NO. 2. ENGINEERS' 14 inch WYE LEVEL.

No. 2.—The essential features of this instrument are like those *enumerated above*, and shown in cut of eighteen-inch Engineers' Wye Level, with the exception of size and weight. It is designed to be used in cases *where a lighter instrument is desirable.* It is provided with a fourteen-inch telescope which has an aperture of 1¼ inches in diameter and a power of 27 diameter; six-inch sensitive spirit-level; steel center; four leveling screws.

Weight of instrument, 9 lbs.; weight of tripod, from 6½ to 7½ lbs.

Gross weight of instrument, packed securely for shipment in two boxes, about 40 lbs.

Price, *including a protection to the object-slide,* **$136.00**

‡ **Steel Centers.**—Although the centers of our Wye Levels, which are long and unyielding, to afford steadiness, are giving high satisfaction, inasmuch as they are made of very hard bell-metal (nearly equal to soft steel), in order to revolve with a minimum of friction, we also make them, when so ordered, of *steel* to run in a socket of *hard cast-iron.* In using these two latter metals whose co-efficient of expansion is nearly alike, the same condition as to a free motion in all temperatures obtains, whether below zero or 100 degrees above as with our hard bell-metal center running in a socket of brass composition, with the difference, however, that the steel ones will retain their precise fitting qualities so well that the bubble of the telescope's spirit level will hardly show any displacement upon revolving the instrument when leveled up, even after years of constant service. In order to make our superior steel center and cast iron socket construction a standing feature in our Wye Levels, we placed the cost at a nominally higher price only, although the cost of production to us is more than double that of the customary style of steel center. We strongly advise to order it where first greater outlay is not considered as important as greater wearing qualities.

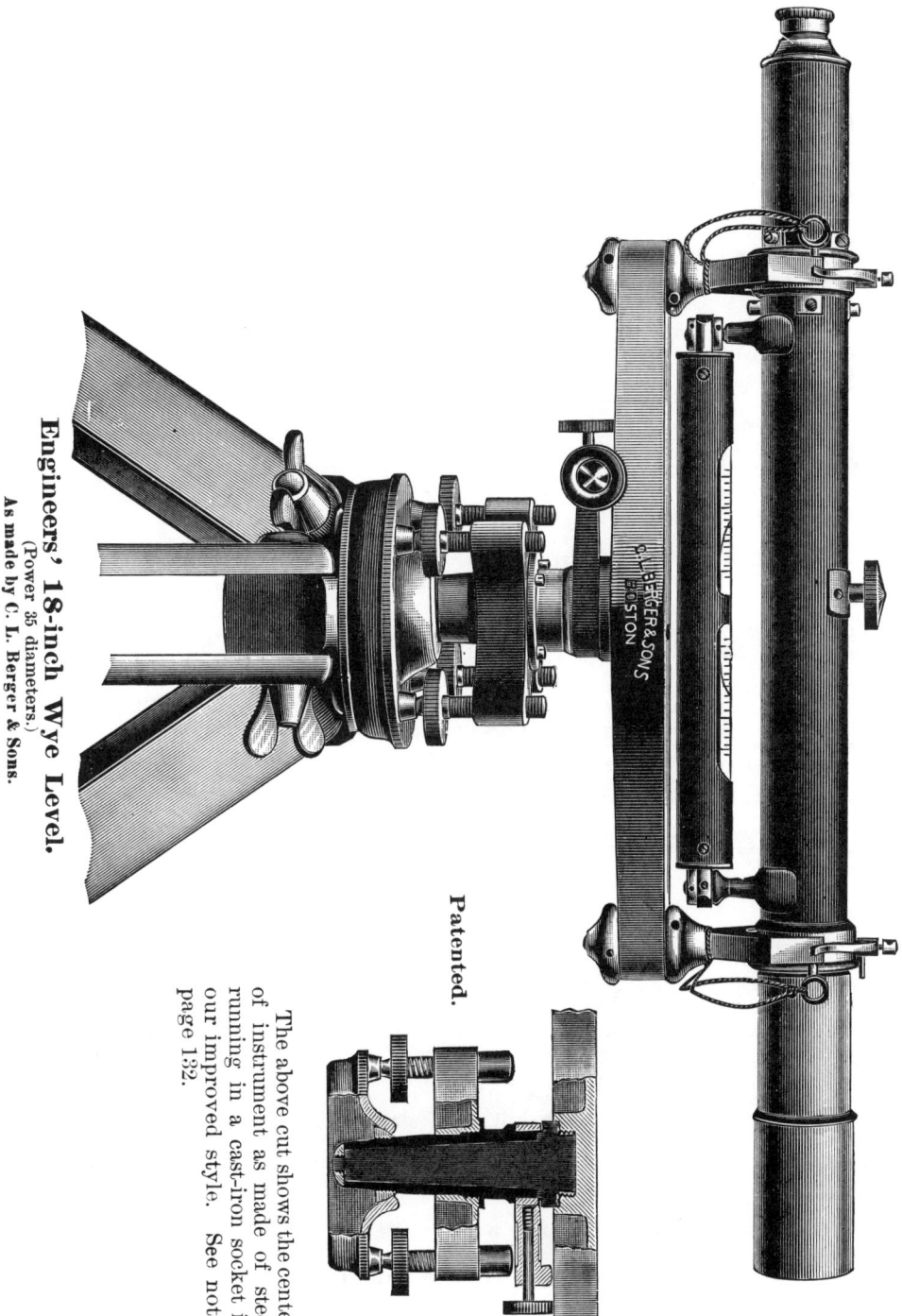

Engineers' 18-inch Wye Level.
(Power 35 diameters.)
As made by C. L. Berger & Sons.

Patented.

The above cut shows the center of instrument as made of steel running in a cast-iron socket in our improved style. See note, page 132.

Hydrographer's Wye Level.

With three Leveling Screws.

Please read: Three leveling screws versus four, p. 38. As regards mode of fastening this instrument to the tripod by means of the center piece or fastener, see cut and description, p. 46, in article "Shifting Center for," &c., also note below.

This instrument, as shown on opposite page, is exactly similar to our Engineer's 18-inch Wye Level (see p. 132), except the telescope, which in this Instrument is of the *inverting kind.* Unless otherwise specified, the sensitiveness of the spirit level will be such that one division ($\frac{1}{10}$ inch) will correspond to 8 to 10 seconds of arc.

The weight of this instrument is about the same as that of our 18-inch Wye Level. The box is about 1 inch wider and on this account is about one pound heavier. The tripod legs are spread as far apart at the tripod-head as the leveling screws, in order to ensure the proper degree of stiffness to such an instrument, and on this account the weight of the tripod is increased about 2 lbs.

This instrument will be made to Order only.

Price, as above, with cloth-finished telescope and level-tube, . . **$158.00**

Extras to Hydrographer's Wye Level.

Center of instrument made of steel, and hardened	$10.00
Stadia wires, fixed	3.00
Metal mirror with universal joint. (This is readily attachable to the instrument and facilitates the reading of the bubble on soft ground without stepping aside	10.00
Extra sunshade with smaller aperture, for use with the telescope when the sun's rays are too bright for accurate work	1.50
Gossamer water-proof bag, to protect the instrument in case of rain or dust	1.00
Bottle of fine oil to lubricate the level center	0.25

NOTE.—The advantage derived from the use of three leveling screws * in the Engineer's Wye Level, when mounted on a base or circle of larger diameter, consists in the greater ease and precision with which the bubble of a most sensitive spirit-level, and thereby the line of sight, can be controlled, in bench leveling and in work of a very close character.

This will be more readily understood when we mention that these levels are frequently made to read to single seconds of arc for every one-hundredth part of an inch on the bubble scale, as shown above.

After an approximate leveling of the instrument, to prevent a change of height of instrument, it is advisable to clamp one of the leveling screws by its clamp screw at the side, and to level up by the other two screws alone. This should be done in like manner also, to correct for slight changes in the level caused by the settling of the tripod-legs.

* Four leveling screws commend themselves in the more ordinary class of instruments for the greater rapidity with which an instrument can be leveled up approximately, and that (no matter how much the leveling screws may be worn) when brought to a true bearing on the lower leveling plate, all such looseness is taken up.

———————•◦•———————

Intermediate base-plate to make the wye level with three leveling screws interchangeable on tripods made for any of our four-screw transits or levels.

As a connecting link between the instruments with 4 and 3 leveling screws we formerly made an arrangement as shown in side cut on opposite page and described below, and are still prepared to supply it if desired. It, however, seems hardly desirable to be used, as the greater steadiness of tripod and simplicity and compactness as exemplified by the Hydrographer's Wye Level with 3 leveling screws (shown on page opposite) are not attained. While we do not recommend the adoption of this device we retain it as a feature in our manufacture simply to meet the desires that some of our customers may yet have. As can also be seen by referring to the side cut, our Quick Leveling Attachment (see p. 39) may in this case be applied, although it will detract still more from the steadiness, and in a measure the advantages of a finer spirit level become lessened.

With the intermediate base plate arrangement the three leveling screws rest in grooves provided for them in the base, or lower leveling plate. The instrument is attached to the tripod by means of a screw in the same manner as the ordinary wye level, inasmuch as the base plate is a part of the instrument proper and is not detachable. The upper part of this instrument is held firmly in the grooves of the lower base-plate by means of a strong spiral spring in the central socket, which permits the use of the leveling screws for their entire length. The leveling screws of this instrument should be kept, as far as possible, in the middle of their run to secure the best action of the spiral spring.

Price of instrument with this device is the same as above for the regular Hydrographer's Wye Level. If provided with a *Quick Leveling Attachment,*

Price, extra, $8.00.

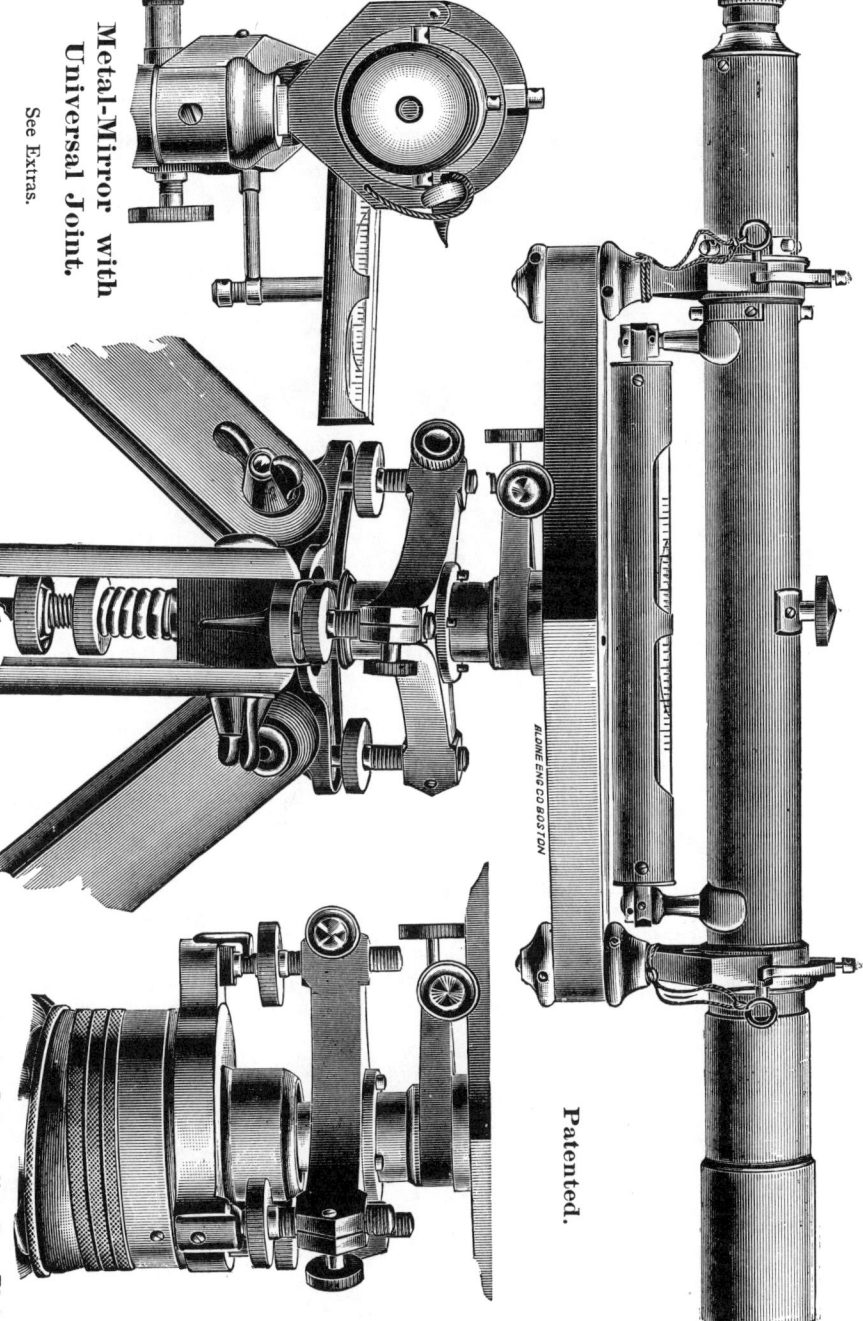

Metal-Mirror with Universal Joint.

See Extras.

Hydrographer's 18-inch Wye Level,

With three Leveling Screws. As made by C. L. Berger & Sons.

(For size, power, and other particulars of this instrument, see page 132).

Device showing Intermediate Base-Plate for attaching an Instrument with Three Leveling Screws to an ordinary Tripod supplied with our Instruments having Four Leveling Screws. (See page opposite). Also showing our Quick Leveling Attachment, which may also be supplied. (For particulars see page 39).

ALOINE ENG CO BOSTON

Patented.

Reversion Level.

Applicable to any of our Engineers' Wye Levels.

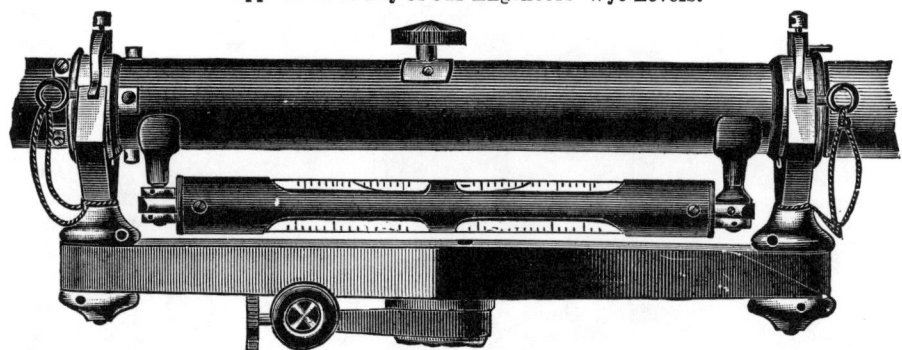

The spirit level used in this feature differs from the ordinary one in that it is ground to the true shape of a barrel so that the tangents to the level bubble curves, at the **zero** points of the scales, are parallel and diametrically opposite. By the use of this reversion level, attachable to any of our Engineers' Wye Levels, in place of the single reading level, it is possible to do good leveling, though the adjustments of the spirit level and cross wires are entirely deranged and the collars worn (see remark below), by first making the level bubble central and taking a reading, then by revolving the telescope 180° in its wyes, which point is indicated by an adjustable stop,* making the bubble again central and taking another reading. The arithmetical mean is the correct result.

This device will, in an emergency, be appreciated when it is known that by the use of the method above the work will average as good as that done with an ordinary good wye level, in adjustment. The adjustment of an instrument provided with a reversion level is made in precisely the same manner as if the spirit level was of the single reading kind, since the adjustment of the level when it is reversed will take care of itself.

The reversion level is guarded by a revolvable outer tube (*Patented, not shown in cut*) leaving a space of air, as a non-conductor of heat between it and the ordinary level mounting tube. This exterior tube serves both as a protection against breakage and sudden changes of temperature, and, as its inner surface is painted white, it also acts as a reflector which facilitates the reading of the bubble.

Remark : The inequality of worn collars cannot be eliminated in a strict sense by using the reversion level, yet for ordinary good work it may be said to be. Nor can the test for the equality of the collars be directly tested in this way but should be done as in the case of the ordinary wye level ; viz., by the two-peg method described under the adjustment of the Dumpy Level (pages 54 and 55 of our handbook). The following modification is to be noted: After the line of collimation has been adjusted for distant objects by rotating the telescope in its wyes and the spirit level has been adjusted by reversing end for end and adjusted laterally, (the telescope having the sunshade attached, as it serves to balance the telescope when the object slide is drawn in), the instrument is set up close to the near target, and a reading is taken with the level tube in the direct position. In order to eliminate the error of collimation for nearer objects, should any exist, another reading is taken with the telescope rotated 180° in the wyes, and the mean taken as the true reading. If, now, the horizontal wire also bisects the distant target and the bubble remains central in each position of the telescope, the collars are of equal diameter. Should the latter not be the case, the error may be corrected thus : Bisect the distant target with the telescope in its direct position, and adjust the level till the bubble is central. Rotate the telescope 180° in its wyes, indicated by the stop, and note the number of divisions through which the bubble moves in order that the distant target remains bisected, so that a correction can be made when most precise work is required.

It is assumed that in making this test the temperature of the two collars has been alike and that the telescope has been in proper balance by being focussed for a distance of about 300 feet with sunshade attached. A scratch on the telescope or object slide indicates the focus which the maker used in equalizing the collars. An apparent error found as above may be due to a change in the shape of the level tube which may occur in time (for which the maker, of course, cannot be held responsible), as well as to a worn condition of collars, or these causes combined.

In order to trace the error to its source the only sure test is made with a striding level. (See Engineer's Precise Level.)

Price, as above, *if ordered* with our Wye Level in place of the single reading kind . . . . **$20.00.**

* It is an extremely difficult matter to grind a level of this kind so that the bubble will remain central at all positions during this rotation through 180°. The stop just mentioned is so adjusted, however, by the maker. that when the level has been turned exactly 180° it gives a correct reading.

C. L. Berger & Sons' Engineers' Precise Level.

Patented. (*For cut see page 139.*)

With micrometer screw for close setting the spirit level.

For use in cities in establishing benches, etc., also for all work requiring speed and the highest degree of accuracy in spirit leveling.

It is a well-known fact that, satisfactory as it may be on account of its great simplicity and compactness, the ordinary wye level (pp. 132, 133) will fail in degree of accuracy or in rapidity of manipulation when the closest results are required. It often happens when precise work is required, the time spent in leveling up and keeping the level bubble of an ordinary good wye level in the center of its graduation by means of the four leveling screws is often very considerable and, when the course is over swampy or frozen ground, the vexation attending the work is apt to be great, and the results vitiated by the numerous readjustments required to keep the bubble in its place. This manipulating of the leveling screws is very apt to lead to a change in the height of the telescope, varying in magnitude according to the style of the instrument. (It is here to be noted that this change in the height of the telescope is less in our levels, or transits with leveling attachments, than is the case with the instruments of other makes).

To aid the Engineer in the prosecution of exact work, avoiding the errors caused by the readjustments above referred to, we have designed and are prepared to furnish the instrument shown on page 139.

By referring to the cuts it will be seen that this instrument is mounted on three leveling screws, and that the center about which the instrument revolves is unusually long and unyielding. Two small spirit levels attached to arms extending from what we may call the cross-bar (since the center of the instrument is permanently secured to it as in the ordinary style of levels) serve to put the center in a vertical position, thus securing at once a nearly horizontal position to the cross-bar. These small levels are adjusted the same as the ordinary plate levels of a transit.

At the eye end this cross-bar carries a micrometer screw by which the telescope and its level can be raised or lowered at will independently of the leveling screws. A strong spiral spring on the same side holds the wye-bar down upon the micrometer screw. This arrangement provides a most delicate motion up and down, and enables one to set the bubble accurately at every sight and in a very much better manner than can be done by the leveling screws alone. The head of the micrometer screws is divided into one hundred parts, and as a rule its pitch will be such that 250 to 252 parts of revolution of the screw will make a change of one foot in the reading of the rod held at a point 100 feet away from the center of the instrument. It may be seen that the instrument can be very advantageously used for making grade measurements. The graduated disc, when reading zero on the index-bar, brings the instrument at once within one or two divisions of its normal position. The disc can also be readily turned on its hub by taking hold of the milled head (the disc is held on its arbor simply by friction), so that, for convenience, a reading may always start from zero, though the cross-bar be not leveled up. This instrument, as above stated, is provided with three leveling screws, which give a firm support on the tripod, and allow a closer setting of the bubble when the instrument is run as an ordinary wye level, without making use of the micrometer. (*See p. 38.*)

The Chief Feature of the Instrument, however, consists in the fact that the pivots * on which the wye bar can be raised or lowered, are in the middle of the instrument and within a fraction of an inch of the plane of the line of collimation, thus securing to the telescope a motion in altitude free from any change in height of the line of collimation, though the telescope were to move throughout the entire range of the micrometer screw during an extended leveling operation. As a rule, the working range of the micrometer will be limited to a few revolutions each way from its normal position in order to keep the instrument as compact as possible. The instrument is also arranged so that, whenever desirable, it may be used as an ordinary wye level. For this purpose, it is provided, at the object end of the cross-bar, opposite the micrometer screw, with a milled-head screw and check nut, by means of which, and by the micrometer screw, when set at zero (see cut), the wye-bar may be set exactly at right angles to the vertical center. However, for the fine settings of the bubble in bench leveling or pointing of the telescope, etc., the micrometer screw should be used exclusively.

A clamp and tangent screw motion is also provided and so arranged, that it can be readily reached from the eye end of the telescope. The cross and wye-bars are cast hollow and the former fits inside the latter.

*Note.—It will be noticed that in instruments of a similar character, having pivot screws acting in and below the wye opposite the micrometer screw, as for instance, in the U. S. Coast Survey geodesic levels, designed after Stampfer (see Report 1879), any motion of the telescope in altitude will also change its height. By an injudicious use of the micrometer screw our own hydrographic wye level (see page 104a, catalogues 1888–1891), partook of this same error, and this together with the marked wear on the collars due to this same motion led us to the abandonment of it. We note, however, that other firms who are in the habit of copying our styles and patterns have since brought it out as a detail of a precise level.

The Telescope will be invariably *inverting* in order to admit of as large an aperture and as high a power as is possible. Thus : its aperture will be $1\frac{1}{4}$ inches, the total length is about 17 inches, and it will have a magnifying power of 40 diameters. It will be provided with fixed stadia wires, in the proportion of 1 to 100, the distance to be measured from a point in front of the objective equal to its focal length.

The Spirit Level is of the single reading kind, and is generally made so that one division (of $\frac{1}{10}$ of an inch) equals from 8 to 10 seconds of arc. The sensitiveness of the level will, however, *be adapted to the particular requirements*. It is not necessary, however, to have it any more sensitive than is required for a fine field instrument, as an over-sensitive level is apt to give more trouble than benefit in its use.

A Reversion Level of same sensitiveness can be applied instead of the single reading level, if desired, *as a convenience* (see Reversion Level, p. 134), when the highest precision is not needed. Of course in fine work the reversion level must be used in the direct position as with a single reading level.

The Auxiliary Striding Level, if one is ordered, is generally made so sensitive that one division (of $\frac{1}{10}$ of an inch) equals from 8 to 10 seconds of arc; but this sensitiveness† may be made a little greater if desired for exceptional uses only. This latter addition would only be used in the most precise work for the purpose of testing the equality of the collars after protracted use, and when this is done it would be returned to its box (see cut opposite).

A Metal Mirror will be furnished with the instrument, attachable to either side of the level, enabling the operator to read the bubble without stepping aside ; a convenience which will be appreciated when working on shaky ground.

Adjustment. The adjustment of the telescope and the level must be made precisely as in an ordinary wye level. (See adjustment of the wye level, pages 51 and 54 of this hand-book.) The spirit level will be in thorough adjustment when the telescope with its *sunshade attached is focussed for a distance of about 400 feet,* when the telescope is in perfect balance and the equality of the collars is assured thereby ; for shorter distances, however, there is a small error due to the unbalancing of the telescope caused by the object slide being thrown out. Small as this error may be it can be entirely eliminated by simply bringing the bubble to the center by the use of the micrometer screw.

Explanation. The foregoing has been written at some length to give a clear understanding of the principal features of this instrument. Naturally, the question may now present itself, why not use a striding level alone, in place of the fixed or reversion level, as is done in some of the best types of instruments, particularly as the pivot arms, extending from the middle of the cross-bars, must necessarily be spread quite a distance apart, to readily permit the revolution of the telescope with the fixed level in the wyes. To this we may say, that a fixed level placed below the telescope, where it is guarded against breakage and, in a measure, from the action of the sun, is better adapted to the wants of the *Civil Engineer* in running quick and accurate levels in cities, towns, etc., than a striding level with its more cumbersome features and manipulations would be, particularly if the work was to be of the most precise character.

It is only when the collars of a telescope are badly worn or imperfectly made that the striding level has any advantage over a fixed one. As a rule a fixed level keeps in better adjustment, is simpler to manipulate than the striding level, and is free from the errors due to the uncertainty of contact of the collars and the wyes. Moreover, the construction of the new instrument is such that it has a greater stability than those of previous make. We therefore believe that the fixed level has as legitimate a standing as the striding one. (In this connection read articles on the fixed and striding levels, pages 94 to 98 of this manual.)

For the above and similar reasons the American Engineers have and will give preference to the instrument which has the level fixed to the telescope; and this has led us to the adoption of this feature in our new instruments. This idea is also prevalent among the best instrument makers and engineers in Europe, as may be seen by examining Prof. Nagel's published description of a similar instrument.

Instrument, Finish, Packing, Weight, etc. The telescope is cloth-finished, while some of the more bulky parts of the instrument are simply treated either with cloth finish or japan, in order to lessen the cost. No attempt will be made to give an elaborate finish at the expense of accuracy and utility; altogether, as all the other parts will be bronzed and lacquered in a manner customary with us, it will present a handsome appearance. This instrument is packed erect in one box in the same manner as we pack the regular engineer's wye level. It is secured to the tripod in the same manner as are all of our instruments with three leveling screws. (See page 134 for description)

The mahogany box contains a sunshade, wrench, screw-driver and adjusting pin.

Weight of instrument, $11\frac{1}{2}$ pounds ; weight of tripod, $10\frac{1}{2}$ pounds ; weight of mahogany box, $10\frac{1}{4}$ pounds ; gross weight of instrument complete, securely packed in two boxes for shipment, 60 lbs.

Price of this instrument, inverting telescope, cloth finished, metal mirror, fixed stadia wires, and a *single reading fixed spirit-level,* . **$200.00**

Extras to Engineers' Precise Level

Instrument provided with a reversion level (see Reversion Level, page 136), .	$20.00
Instrument provided with $7\frac{1}{2}$ inch auxiliary striding level, in addition to regular fixed level, ...	32.00
Center of instrument made of steel and hardened, ...	10.00
Sunshade with smaller aperture, for use with the telescope when the sun rays are too bright for accurate work, ...	1.00
Gossamer bag, to protect instrument, ...	1.00
Bottle of fine watch oil for lubricating the centers, etc., ...	0.25

† If geodesic work is to be done. a higher sensibility might be permissable, but our customary fluid would be sluggish in such a level. and the bubble tube would have to be filled with pure ether, in order to make it quick acting (see pages 7, 18, 38). An air chamber would be necessary to allow for adjustment of the bubble, which in this case changes its length rapidly for slight changes in temperature. By adding a chamber, a feature is introduced which is liable to affect the reliability of the spirit level and entail extra expense.

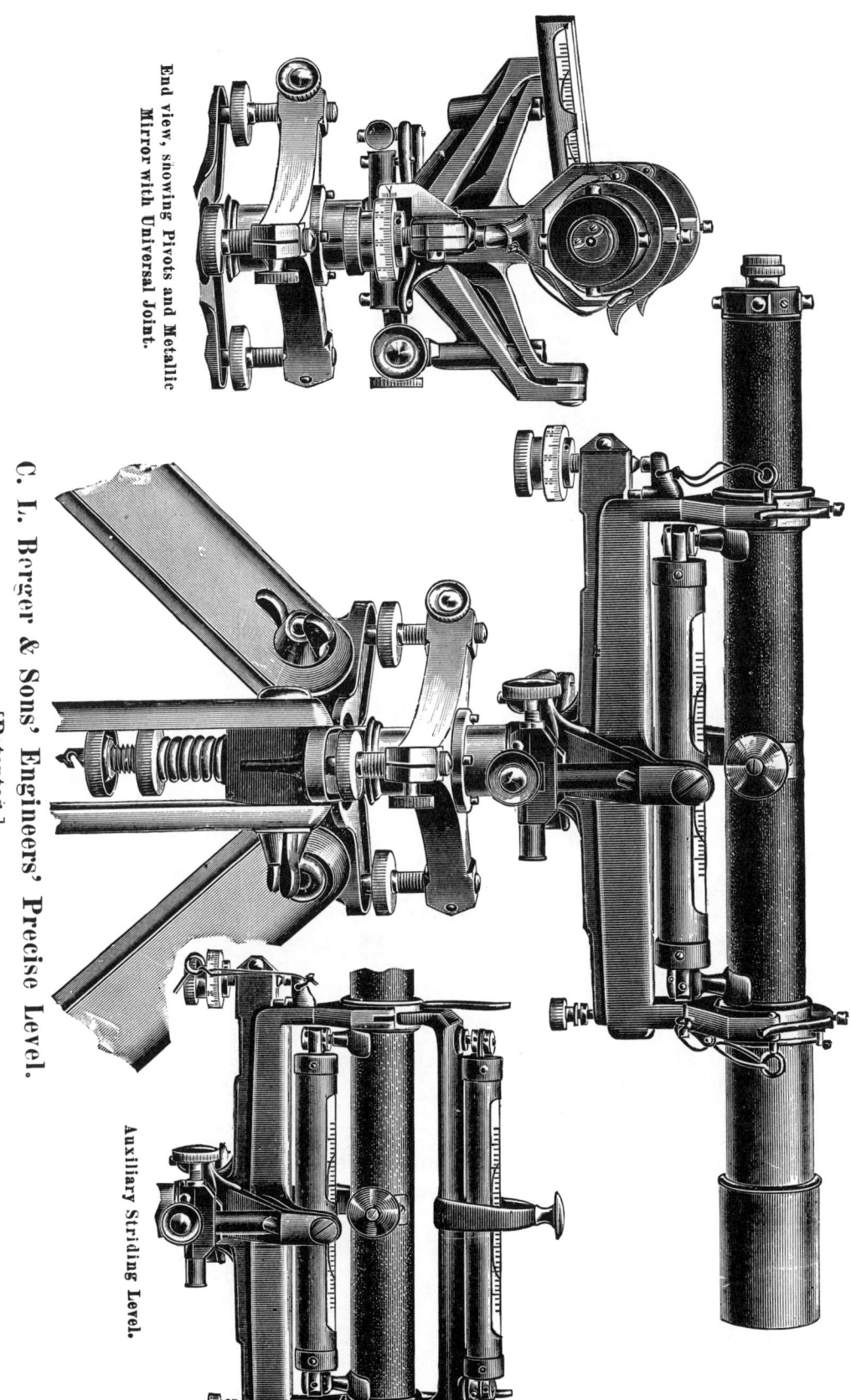

End view, showing Pivots and Metallic Mirror with Universal Joint.

C. L. Berger & Sons' Engineers' Precise Level.

[Patented.]

Auxiliary Striding Level.

Plane Table.

This instrument, made by us now in one size only, is designed to fill a want where a *high class of work in Topography* is required.

In order to obtain great rigidity and strength the bearing surface of the lower motion has a diameter of eight inches, the board rests on arms lying in a circle twelve inches in diameter and the tripod is of corresponding size. To be still portable all the essential parts are built on the skeleton plan to be light and stiff as shown in cut. To avoid a loosening of the leveling screw-fastenings, so often experienced where they are fitted into tripod heads made of wood, we make this head of composition brass, same as the other parts. To reduce this weight by the use of aluminum, see pages 122, 123. The board is 24 x 28 inches. The alidade is 22 inches long and is provided with a 16 inch inverting telescope having an aperture $1\frac{3}{8}$ inches and a power of 35 diameters. It has stadia wires, a vertical arc reading to minutes, graduated on solid silver, and a sensitive striding level. Two fixed levels are mounted on the ruler. The latter is so arranged that lines can be ruled in the vertical plane of the line of collimation, if desired.

For adjusting the line of collimation the telescope can be revolved 180° on its longitudinal axis. Materials and workmanship are of the best. The alidade is bronzed and lacquered in our customary manner, but the tripod head and lower motion are neatly japanned.

Weight of Alidade with brass ruler,		6 lbs.
" " Lower motion,		11 "
" " Tripod,	about	18 "
" " Board,	"	8 "
" " Alidade and accessories in mahogany box,	"	15 "
" " Lower motion in mahogany box,	"	18 "
" " Board in pine box,	"	23 "
Gross weight of instrument packed in four boxes ready for shipment,	"	110 "

Price of Plane Table, complete as above, including board, detached compass, screwdriver, clamps, reading glass, plumb bob, etc., in three boxes, **$300.00**

NOTE. — We have sometimes been asked to furnish a larger base to increase the steadiness of larger boards. We do not desire to make this, since we are fully confident that our standard base, as above described, is amply stiff. We are prepared, however, to make a base with arms, lying in a circle, 16 inches in diameter, should a much larger board be required. The increase in weight will be about 10 lbs. **Price, extra, $25 00.** If two tangent screws are desired for this larger lower motion, the extra cost of instrument will be **40.00.** For use with such a large board we are also prepared to furnish an alidade, 28 inches long. **Price, extra, $10.00.**

Price of Alidade, complete as above, including striding level, detachable compass, clamps, etc., in mahogany box, but *without board, lower motion and tripod,* **$200.00**

To meet a want where greater portability and lightness are thought to be more advantageous than greater rigidity and consequent accuracy, we are prepared to furnish in place of the above described lower motion of the Plane Table (shown in the accompanying cut) one of the Johnson type and character. A description of this may be omitted here, since it is described in any of the modern text-books on Plane Table Work. Suffice it to say that this motion is operated in a manner similar to that described under our Quick Leveling Attachment (see pages 39 and 126) of which it is an inverted adaptation, but is of greater size, range and steadiness.

This movement, with legs all complete, weighs only from 9 to 10 lbs.

Price of Johnson's Improved Plane Table Movement, *mounted on large tripod,* **$45.00**

Price of Plane Table Drawing Board, *in pine wood box,* . . **$15.00**

Plane Table.

As made by C. L. Berger & Sons.

Engineers' and Surveyors' Transit.

This instrument is designed for engineering work of a high class, such as is required in bridge building, water works, and for city and land surveying. The size of the circle is such that it may be graduated to read to 30″ or 20″ without fatigue to the eye. The telescope is of the best definition, and has a large aperture with perfectly flat field. The eye-piece is achromatic, and gives a large field with plenty of light. We advise our customers to order *solid silver* graduations for this instrument, also ground glass shades over the verniers for reasons given on page 6.

Transits No. 1– No. 1 c.— Horizontal circle 6¼ in. (edge of graduation), two double verniers reading to minutes; *two rows of figures* in opposite directions from 0° to 360°; *figures on limb* and *verniers* are *inclined* in the direction they should be read; verniers and graduations are protected with fine plate glass; graduations are silvered; magnetic needle 4½ inches; *adjustment for vertical plane* of telescope; improved *spring tangent screw*; improved *lower tangent screw*; *shifting center* to set the instrument exactly over a given point; improved telescope 11½ inches long; objects erect; aperture 1¼ inches; power of the telescope 24 dia., which qualifies it especially for telemeter work; eye-piece is provided with an improved screw arrangement for the accurate focussing of cross-wires; telescope is *perfectly balanced* and reverses at both ends; spirit-levels *ground* and extra sensitive; line of collimation *correct* on *all distances* without adjustable object-slide; **protection** to object-slide; *long compound centers* with heavy flanges; *improved split-leg tripod provided with thumb-nuts.*

The mahogany case has a leather strap, hooks, etc. It contains a sun-shade, a wrench, a screw driver, an adjustable plumb-bob, a magnifying glass, and several adjusting pins, and weighs from 9½ to 10 lbs.

Weight of Plain Transit, (No. 1), . . . 13½ lbs. ⎱ Weight of tripod
  "   " Transit with Level Attachment, (No. 1 a) 14   " ⎰ from 7 to 7½ lbs.
  "   " Complete Transit, (No. 1 b and No. 1 c) 14½   "

Gross weight of instrument, complete, packed securely for shipment in 2 boxes, about 50 lbs.

Extras to Transits No. 1 — No. 1c inclusive.

Verniers provided with *ground glass shades* (see page 6)	$3.00
Graduation of horizontal circle, *on solid silver*	10.00
" " reading to 30″	10.00
" " " 20″	20.00
Graduation of vertical arc or vertical circle, *on solid silver* . . .	5.00
Gradienter attachment (see page 39)	5.00
Stadia Wires, fixed	3.00
Short Focus Lens (pages 118, 169). One pair	16.00
Richards Prism, attachable to our *complete* Transits or Levels (see page 61 of Manual)	30.00
Arrangement for *offsetting at right angles*	5.00
Aluminum guard for the full vertical circle (see page 146) . . .	4.00
Variation plate	10.00
Gossamer water-proof bag, to cover transit in case of rain or dust . .	1.00
Bottle of fine watch-oil to lubricate the centers, etc., of transit . . .	0.25

NOTE.—Sometimes we are asked by those not intimately acquainted with the principles governing a telescope to place a higher power than is customary with the best makers upon a telescope of the size described above. In answer we wish to say that with the power mentioned above very good results in stadia measurement can be obtained, and that while the power could be easily increased, the light and definition of the telescope would become so diminished that it would render the instrument less efficient in more than one respect. In this connection we refer to the various articles written on the telescope in part I. of catalogue. In some cases, however, where the instrument is principally intended for use in stadia measurements, we can increase the aperture of our *inverting telescope* for Transits No. 1 from 1¼ to 1⅜ inches diameter. This increase in aperture will permit of a higher power. Thus two eye-pieces, magnifying respectively 27 and 33 diameters, can be supplied with such a telescope; but the danger of the wires getting broken, or dust blowing into the telescope, etc., in changing the eye-pieces, is so great, that in instruments of the above class the use of two eye-pieces should be as little resorted to as possible. To increase the aperture of the object-glass to 1⅜ inches adds $10.00 to the cost of the instrument, and where both eye-pieces are ordered an extra charge of $5.00 will be made. This change in aperture will add about 10 oz. to the weight of the instrument. No extra charge if telescope is to be of the inverting kind, but the instrument must be made specially.

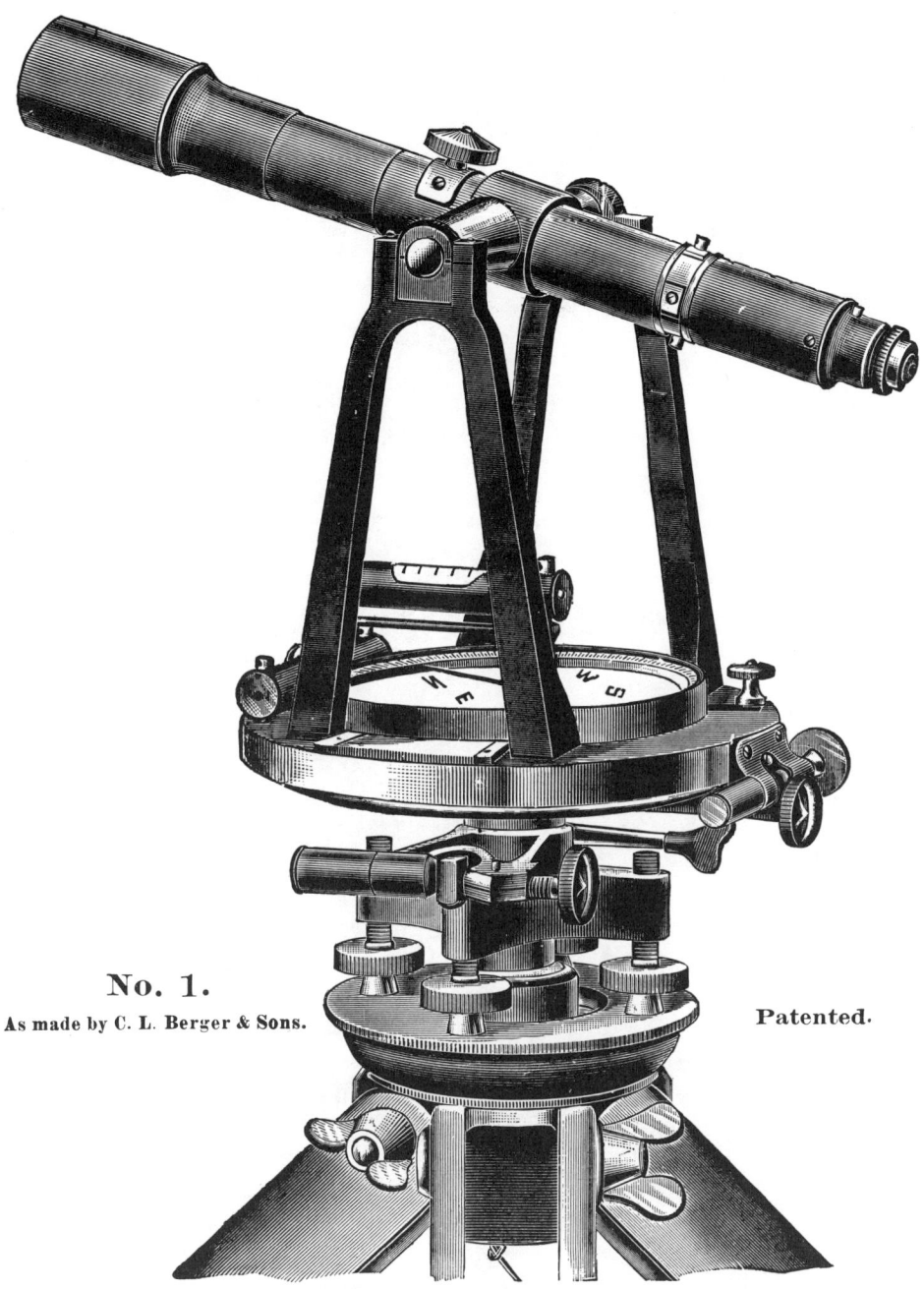

No. 1.

As made by C. L. Berger & Sons.

Patented.

Plain Transit.

Price, as above, with standards cloth finished **\$180.00**

Standards finished like instrument (*no cloth finish*) . . . extra \$5.00

For size and description of this instrument, as well as for *Extras*, see page 142.

The verniers of this instrument will be placed at an angle of 30° to line of sight unless ordered to be as in above cut.

All our transits are provided with a *fine punch mark on top of the telescope* to enable to center instrument from a point above as well as from below.

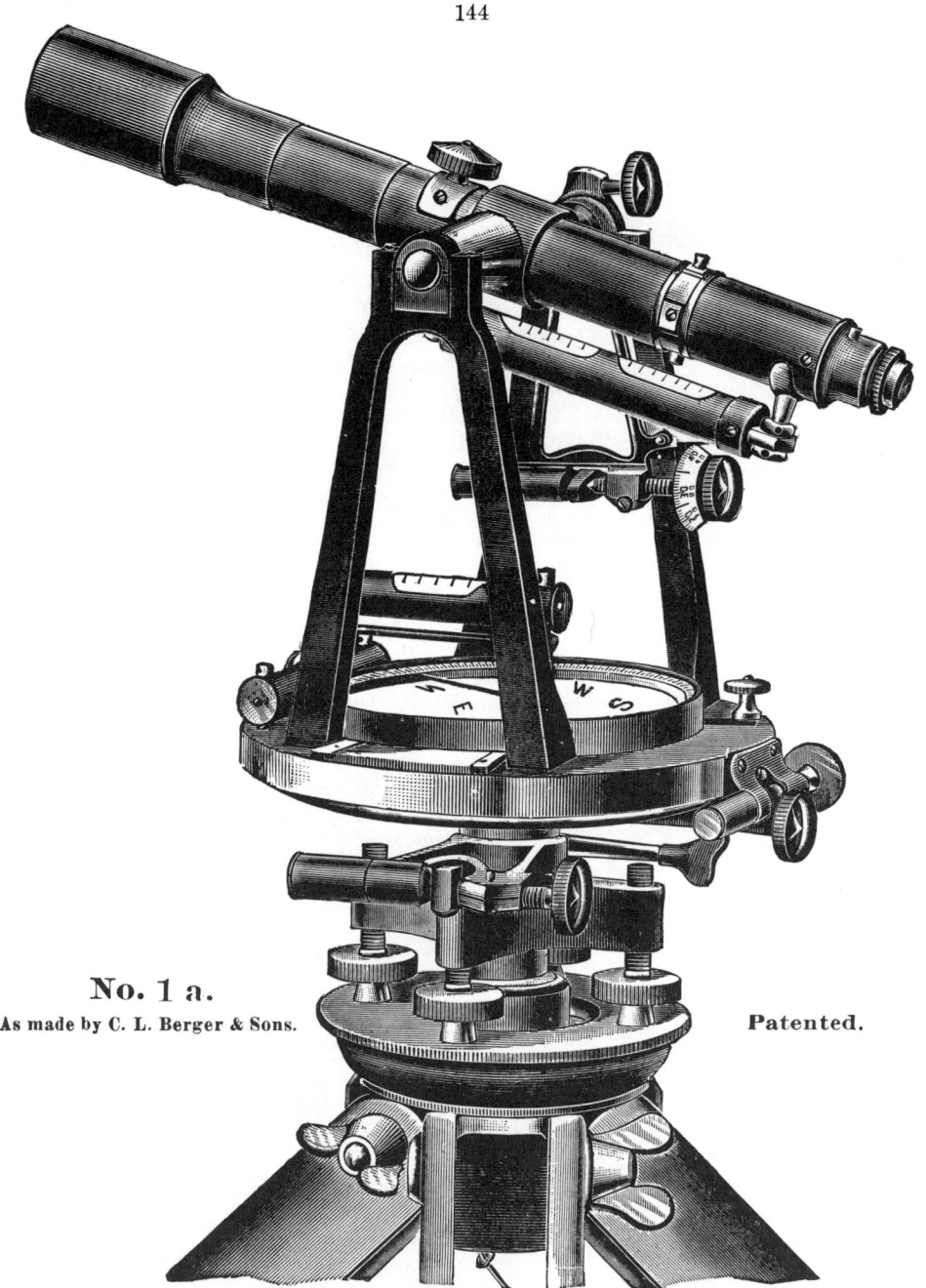

No. 1 a.
As made by C. L. Berger & Sons.   **Patented.**

Transit with Level Attachment to Telescope.

Price, as above, with cloth finish standards *without gradienter* . **$210.00**
Standards finished like instrument (*no cloth finish*) . . . extra **$5.00**

For size and particulars of this instrument, as well as for *Extras*, see page 142.

NOTE. — The character of this level attachment combined with the features of the transit, is that of a pivot-level. Its manipulation and use is similar to that described under our Hydrographic-level. The adjustment of the level to the telescope, however, must be made in the manner described on pages 48 and 54, or by means of a collimator. — With a level attachment of the above kind, good leveling can be done, as the power of the telescope and the sensitiveness of the spirit-level are equal to that of most Wye-levels.

The verniers of this instrument will be placed at angle of 30° to line of sight unless ordered to be as in above cut.

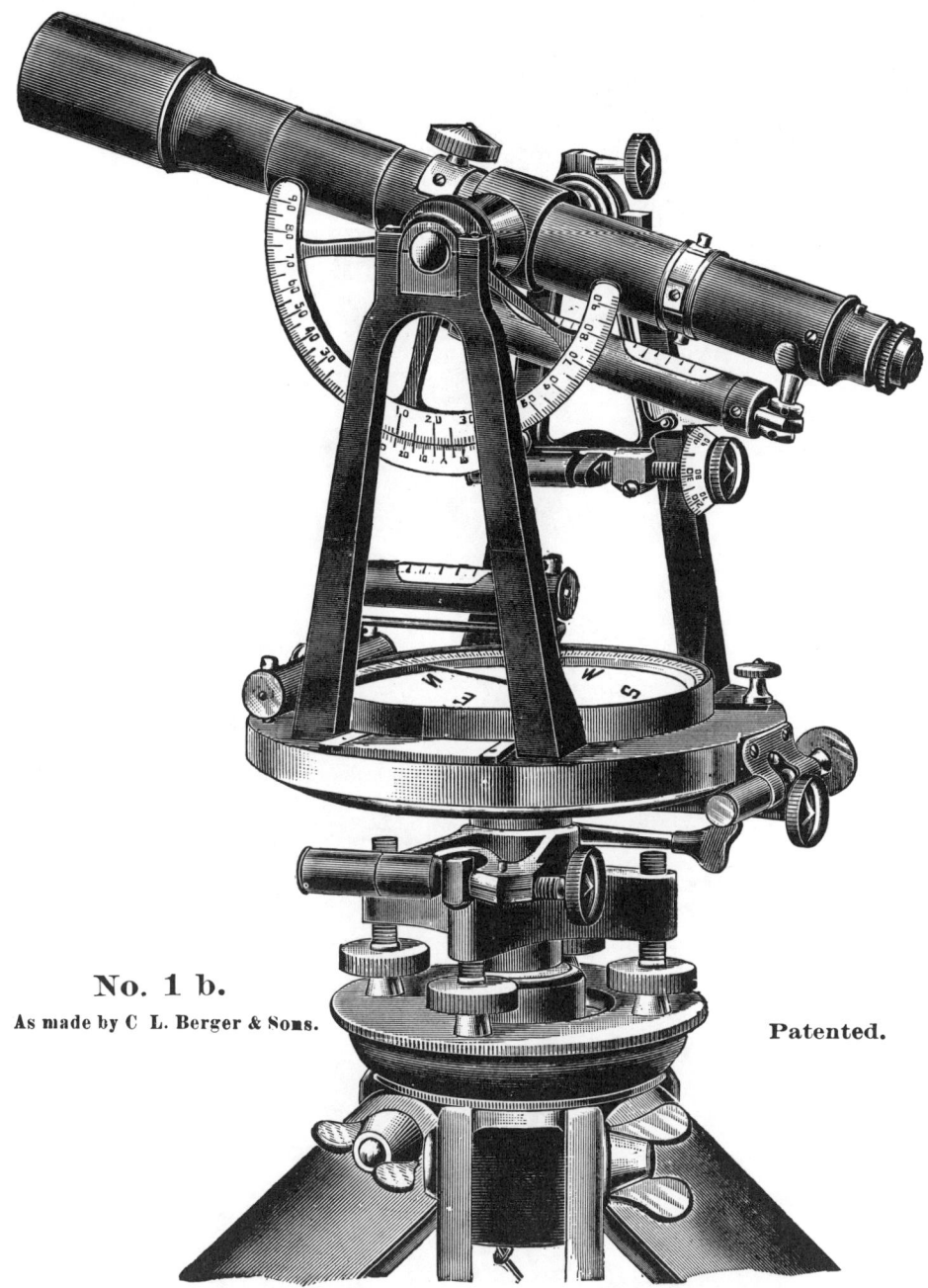

No. 1 b.

As made by C L. Berger & Sons.

Patented.

Complete Engineers' and Surveyors' Transit.

The 5-inch vertical arc is provided with double verniers reading to minutes.

Price, as above, with cloth finished standards and without gradienter, **$225.00**
Standards finished like instrument (*no cloth finish*) . . . extra $5.00

For size and particulars of this instrument, as well as for *Extras*, see page 142.

NOTE. — When stadia wires are added, this instrument becomes a *Tachymeter*. Unless ordered as is shown in the above cut, the horizontal verniers will be at 30° to line of sight.

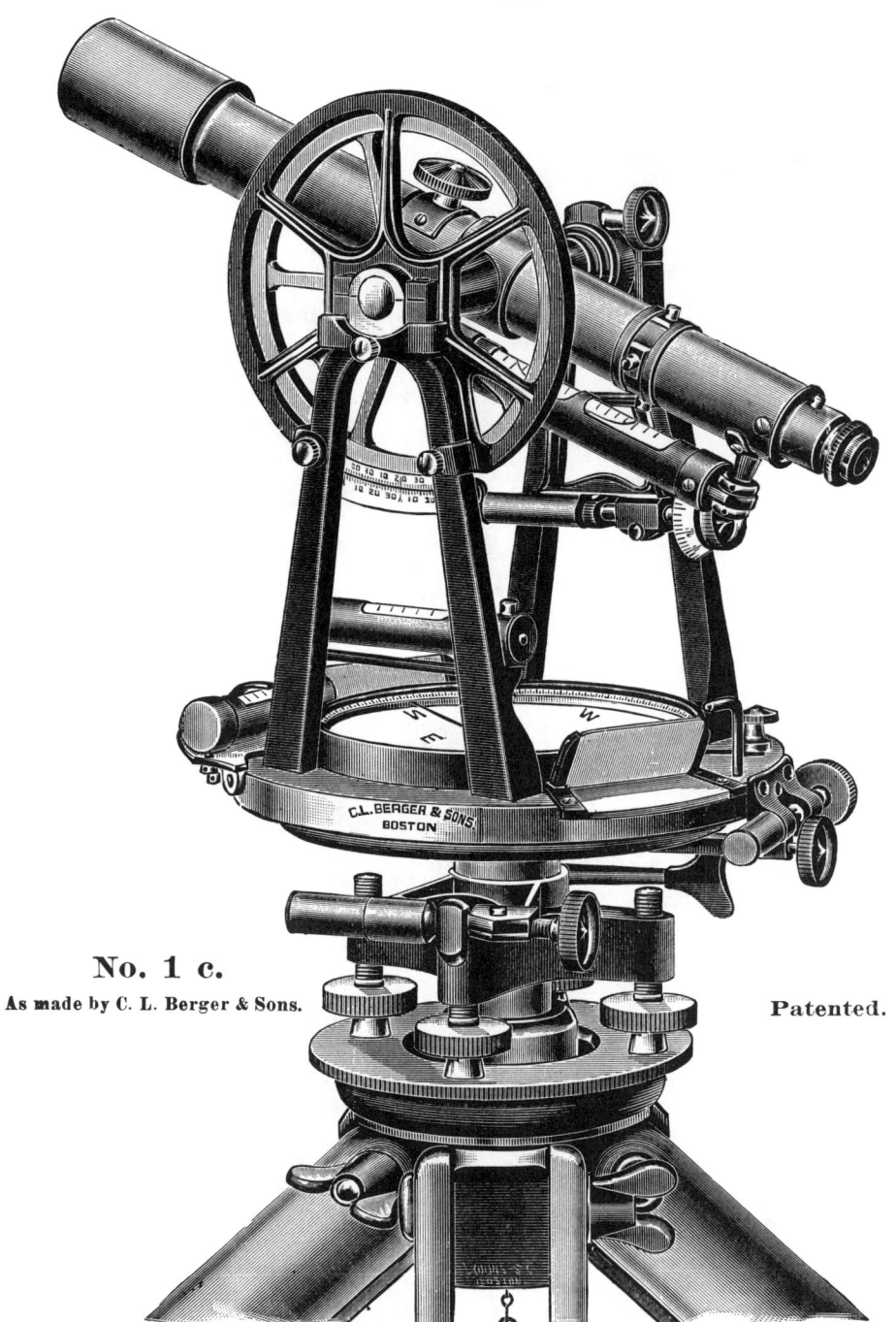

No. 1 c.

As made by C. L. Berger & Sons.

Patented.

Complete Engineers' and Surveyors' Transit.

The 5-inch vertical circle is provided with double verniers reading to minutes.

Price, as above, standards cloth finished, *with vernier shades, aluminum guard for vertical circle, but without gradienter* . . . **$237.00**

Standards finished like instrument (*no cloth finish.*) . . . extra $5.00

For size and particulars of this instrument, as well as for *Extras*, see page 142.

NOTE. — When stadia wires are added, this instrument becomes a **Tachymeter.**

No. 1 c, Style o.

C. L. Berger & Sons' Double Opposite Vernier Attachment to transits provided with a 5-inch full vertical circle.

Price, extra, $20.00

NOTE. — For most work with the Engineer's transit it is not important to read the vertical angles closer than minutes, and by estimation to 30″, and the ordinary vertical arcs and circles of our construction, as illustrated in the preceding instruments, give the fullest satisfaction in this respect. There are, however, some few cases where it may be desirable to eliminate errors and excentricities in the graduation and verniers of the vertical circle in the same manner as in the horizontal graduation by reading two opposite verniers, and the construction illustrated above has been designed to meet this want. As the vertical circle cannot be turned independently on its axis, as in repeating circles, the telescope must be reversed, when a repetition of the angle is desired. The mean of the two readings is then accepted as the true result.

In the above illustration the frame in front of the vertical circle carries two opposite verniers reading to minutes. The verniers are double, so that angles of elevation and depression can be read with ease and dispatch. For ordinary work the vertical angles may be read only from one vernier. The frame is fitted to the horizontal axis of revolution, and is circular in order to protect the graduation of the vertical circle from injury.

Two opposing capstan-headed screws, working against a projecting stud on the standard, are provided, to adjust the zero-points of the verniers to coincide with those of the vertical circle, after the instrument has been leveled up and the telescope placed in a truly horizontal position, and when adjusted so that there is no looseness between the stud and the capstan-headed screws the vernier frame maintains a fixed position, while the telescope and circle are moved in altitude. This device can be attached to transits provided with a 5-inch full vertical circle in new instruments only when so ordered.

The Tachymeter, or Universal Surveying Instrument.

On the following pages, descriptive of the Complete Engineer's and Surveyor's Transits, No. 1 c Style o, to transit No. 1 h inclusive, the name Tachymeter has been used. The want of a specific name for the complete form of the engineer's and surveyor's transit has long been felt. The term "*transit*," originally borrowed from astronomy to designate an instrument whose telescope can traverse the vertical plane, is not sufficiently comprehensive to describe an instrument in which the vertical motion of the telescope is no longer its most important characteristic. An instrument having a level on its telescope, a vertical arc or circle, and stadia wires, is adapted to the rapid location of points in a survey, since it is capable of measuring the three co-ordinates of a point in space, namely, the angular co-ordinates of azimuth and altitude, and the radius vector, or distance. The name *Tachymeter*, or rapid measurer, has been applied for many years, in Europe, to instruments of this description. The characteristic of *tachymetry* is, that all the data required for the location of points are rapidly determined by the instrument, by means of horizontal and vertical angles, and stadia measurements of distance. The compass and gradienter are auxiliaries in the measurement of angles, and an instrument having them, in addition to the essential features mentioned above, is more perfectly adapted for *tachymetric* work. We feel that we need make no apology for introducing these brief but expressive terms into our catalogue, and we venture to hope that they may come into as general use in this country as they have in Europe, and replace the inconvenient phrases now employed to describe these instruments and methods.

No. 1 c, Style p.

As made by C. L. Berger & Sons.

Tachymeter.*

Horizontal Verniers in line with Telescope.

No. 1 c, Style p. Graduations of horizontal and vertical circles on solid silver, reading to minutes; 5-inch full vertical circle with two double opposite verniers reading to minutes; glass shades over verniers; $3\frac{1}{2}$ inch striding level; gradienter attachment; fixed stadia wires; etc. Standards cloth finished.

Price, as above, $306.00.

For size and particulars, as well as for extras see pp. 142-145.

This instrument without a striding level, ... less, $20.00
" " " double opposite verniers for vertical circle, " $20.00

NOTE. — In this instrument the verniers of the horizontal circle are placed in the direction of the line of sight of the telescope, so that angles can be read from the position of the observer, without stepping aside. Unfortunately, in order to carry this out in instruments provided with a *compass upon the upper plate*, it becomes necessary either to shorten the front plate-level, which is the principal level in the Engineer's transit, and therefore should always be of standard length and character; or to raise it above the vernier, in an exposed position, where it will so shade the vernier, as to make the latter almost worthless; or, as in the cut above, to place the front plate level entirely outside of the plates, where it is also in an exposed position, and is liable to derangement though protected from breakage by a guard. A third arrangement, to place the level inside, but below the compass bottom, requires that the space between the plates be increased, thereby raising the height of the whole instrument about ½ inch, and increasing the weight; at the same time the openings in the upper plate weaken the latter and its stability to resist the rough treatment it is liable to receive in the field; and *shortening the plate levels and decreasing their sensitiveness* to such an extent as to make them but little better than the circular levels formerly used with the most ordinary instruments. None of these devices can therefore be regarded as a real improvement, except for the more ordinary instruments, where sometimes under certain conditions greater accuracy may be considered of less importance than greater convenience. In instruments without a circular compass upon the upper plate, as shown on page 179, this device can be carried out to perfection. A striding level should always accompany this instrument.

Reading glasses cannot be attached to the horizontal circle of this instrument. The glass shades on this instrument will often prove an obstacle in making solar observations when a prism attachment must be used.

* See remark on page 147.

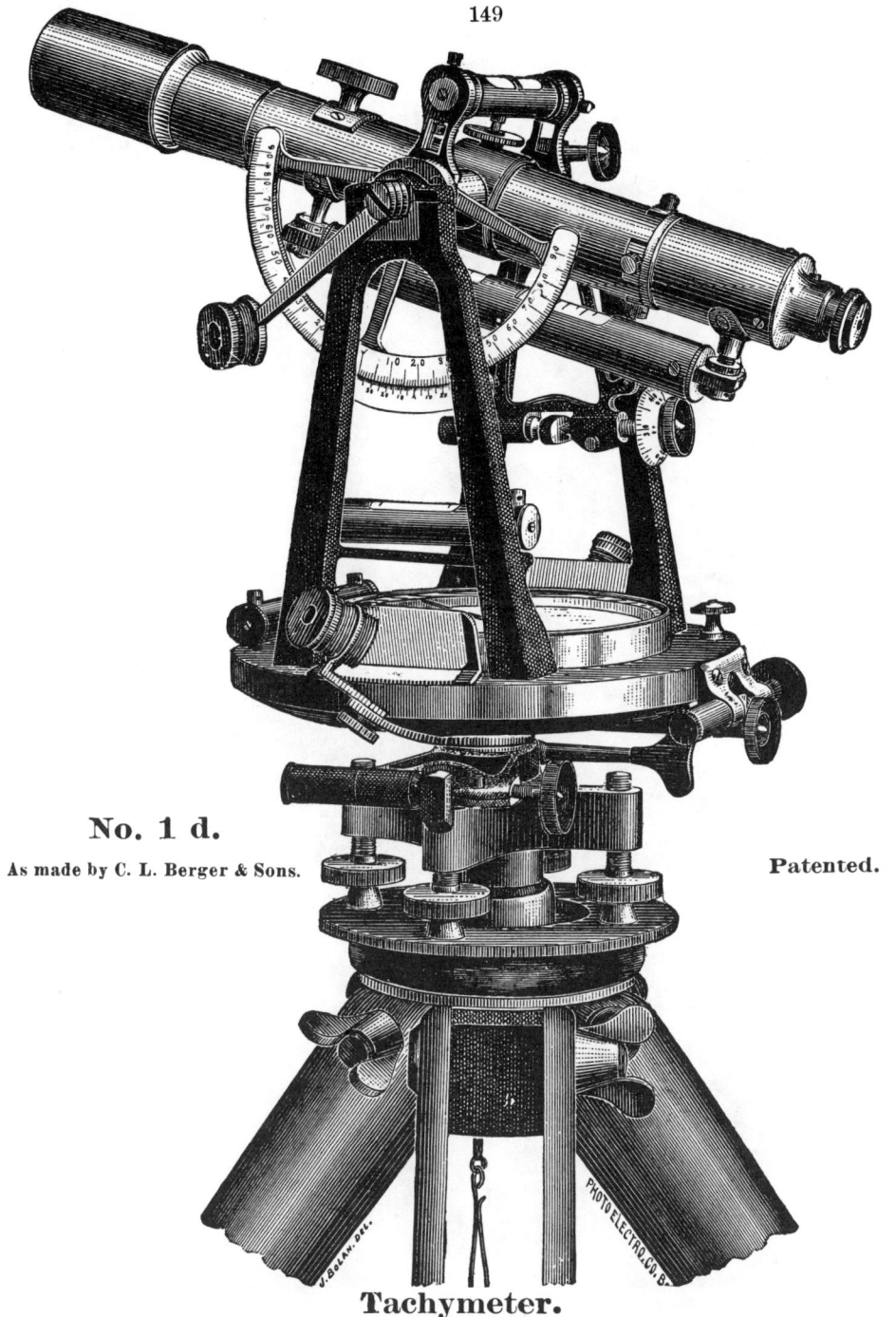

No. 1 d.

As made by C. L. Berger & Sons.

Patented.

Tachymeter.

For size and particulars of this instrument see pages 142-145.

No. 1 d, as in cut, graduation of horizontal circle on solid silver, opposite verniers reading to 20″; graduation of vertical arc on solid silver, verniers reading to minutes; glass shades over verniers; detachable reading glasses for both circles, 11½ inch telescope showing objects inverted, power 27 diameters; 3½ inch striding level; gradienter attachment; fixed stadia wires; etc. Standards cloth finished.

Price, as above, $312.00.

This instrument without a detachable reading glass to the vertical arc, less $5.00.

'   "   with a 5-inch full vertical circle in place of the arc, . extra 5.00.

NOTE. — For a description of the striding level, its use and adjustment, see page 56. The striding level and the detachable reading glasses, as shown above, can be attached only to our transits of the above description: **we cannot** attach them to instruments already made.

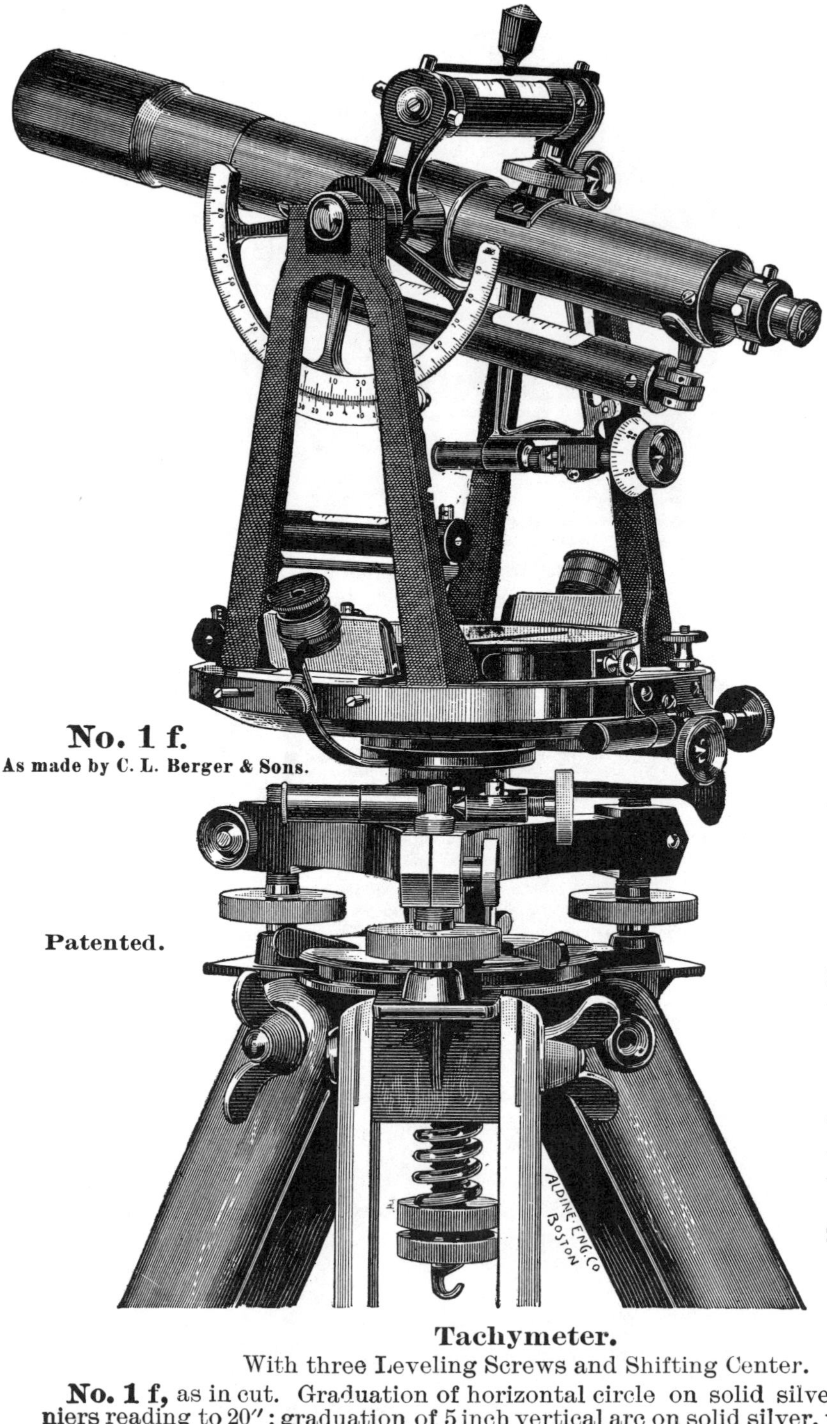

No. 1 f.

As made by C. L. Berger & Sons.

Patented.

ALDINE ENG. CO. BOSTON

NOTE.—An instrument of this size with three leveling screws requires a larger leveling base than is usual with those of the same size provided with four leveling screws.* The tripod head must also be larger in an instrument of this class, in order to make the steadiness of the tripod correspond with the increased accuracy expected from such an instrument. This increase of size not only adds about 3 lbs. to the weight of the tripod, but renders it at the same time less portable on the shoulder on account of the greater circumference of the legs at the tripod-head. The packing of an instrument with three leveling screws in its box is also more complicated. These disadvantages should be fully considered before ordering an instrument of this kind, reading as it does only to 20″ direct. We do not graduate instruments of this size and style to read to 10″; the 6¼ inch circle being too small for accurate and easy reading when divided so finely. For finer instruments see types enumerated under transit No. 11, etc.

* For more information on this point see page 38.

Tachymeter.

With three Leveling Screws and Shifting Center.

No. 1 f, as in cut. Graduation of horizontal circle on solid silver, opposite verniers reading to 20″; graduation of 5 inch vertical arc on solid silver, verniers reading to minutes; glass shades over verniers; detachable reading glasses for horizontal circle; 11½-inch telescope showing objects inverted, power 27 diameters; 6-inch spirit level parallel to telescope; 3½ inch striding level; gradienter attachment; fixed stadia wires, etc. Standards cloth finished. **Price, as above, $322.00**

For size and particulars of this instrument, as well as for **Extras** see pages 142-145.

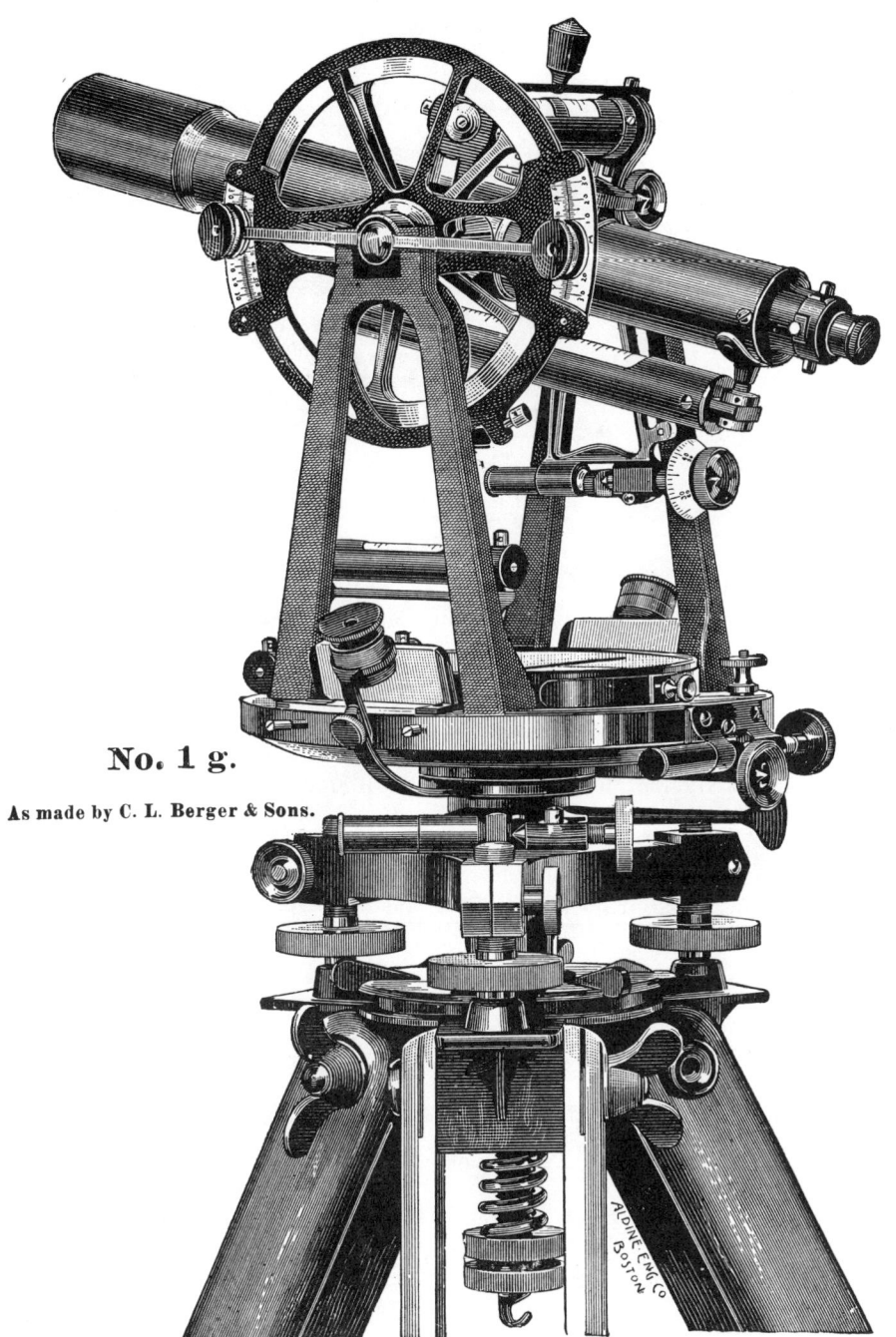

No. 1 g.

As made by C. L. Berger & Sons.

Tachymeter.

No. 1 g, *as in cut.* Same as in No. 1 f, but having a 5-inch full vertical circle with two double opposite verniers reading to minutes, and two reading-glasses to the vertical circle. **Price, as above, $352.00.**

For size and particulars of the above instruments, as well as for extras, see pp. 142–145.

Reversion Level. *

For leveling also with telescope reversed.

As made by C. L. Berger & Sons.

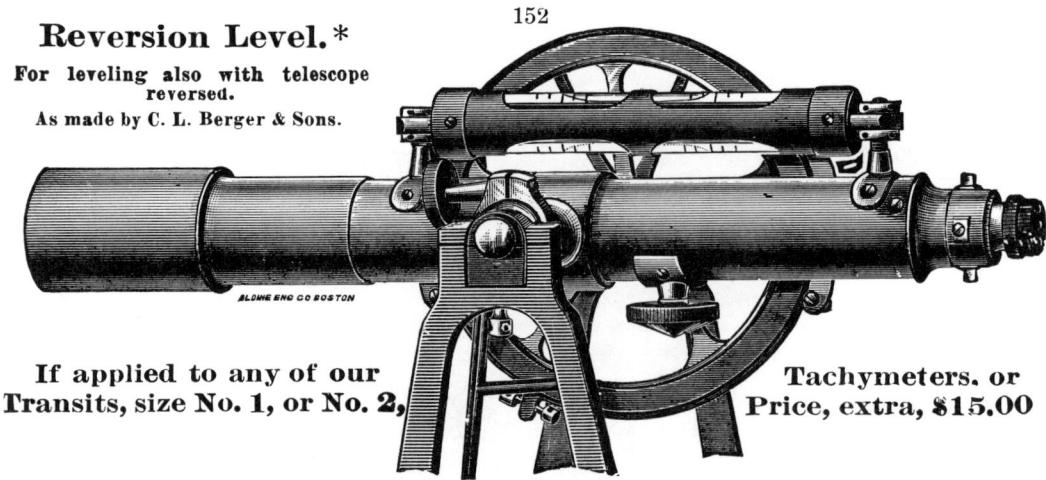

ALDINE ENG CO BOSTON

If applied to any of our Transits, size No. 1, or No. 2, **Tachymeters. or Price, extra, $15.00**

NOTE.—This level has a reversible tube which acts as a reflector, as well as a guard to protect from breakage the exposed side of the level when not in use.

* The adjustment of this level and the horizontal cross-wire has to be made in the manner described for the fixed level attached to the transit telescope, see pages 48 and 54.

Small Engineers' and Surveyors' Transit.

No. 2. Plain Transit.* The essential features of this instrument are like those *enumerated under No. 1*, with the exception of size and weight. It is designed to be used in cases *where a lighter instrument is desirable*. All the parts, the graduations, the telescope, etc., are made with as great care as in the larger instruments made by us. We can recommend it as being a very reliable and superior instrument for railroad work, for general land surveying and for mining purposes.

The dimensions are as follows:—

Horizontal limb 5 inches; magnetic needle 3¾ inches; telescope 9 inches; clear aperture 1⅛ to 1¼ inch; power 18 diameter.

The mahogany case has a leather strap, hooks, etc. It contains a sun-shade, a wrench, a screw driver, an adjustable plumb-bob, a magnifying glass, an adjusting pin, and weighs 7 lbs.

Weight of instrument 10 lbs.; weight of tripod from 6½ to 7 lbs.

Gross weight of instrument, packed securely for shipment in 2 boxes, about 40 lbs.

Price, as above, with standards cloth-finished **$180.00**
Standards finished like instrument (*no cloth finish*) . . . extra 5.00

Extras to Plain Transit.

Spirit-level 5½ inches, with clamp and tangent screw to telescope, . .	$30.00
Gradienter attachment, ...	5.00
Offsetting arrangement, ...	5.00
Graduation of horizontal circle on *solid silver*, ...	10.00
5 inch vertical arc, *double* verniers reading to minutes, . . .	15.00
" " " graduation on *solid silver*, . . .	20.00
5 " " circle *double* verniers reading to minutes, . . .	20.00
5 " " graduation on *solid silver*, ...	25.00
Glass shades over verniers (to faciliate the reading,) . . .	3.00
Stadia Wires, fixed, ...	3.00
Short Focus Lens (see pages 118, 169). One pair, . . .	16.00
Variation plate, ...	10.00
Gossamer water-proof, to protect the instrument in case of rain or dust, .	1.00
Bottle of fine watch oil, to lubricate the center, etc., of transit, . .	0.25

*A Plain Transit is one without spirit-level, clamp and arc to telescope, see No. 1, page 143.

NOTE.—If a transit is intended for very close stadia work, Transit No. 1, with its larger telescope and higher power will be best suited for that purpose. But in all cases where greater lightness and portability is a factor and where only general good results in stadia measurements, as obtained with a smaller and less powerful telescope, will be deemed satisfactory, size No. 2 should be chosen. We cannot put a telescope of the size as described in Transit No. 1 upon a Transit No. 2. It should be borne in mind that all parts of an instrument are so closely related to each other that the preponderance of any one part would simply impair the efficiency of other parts. A telescope of the size given above, but showing objects inverted, will generally give the desired result. No extra charge for such a telescope, but the instrument must be made specially. The aperture will then be 1¼ inch and the power 22 diameter.

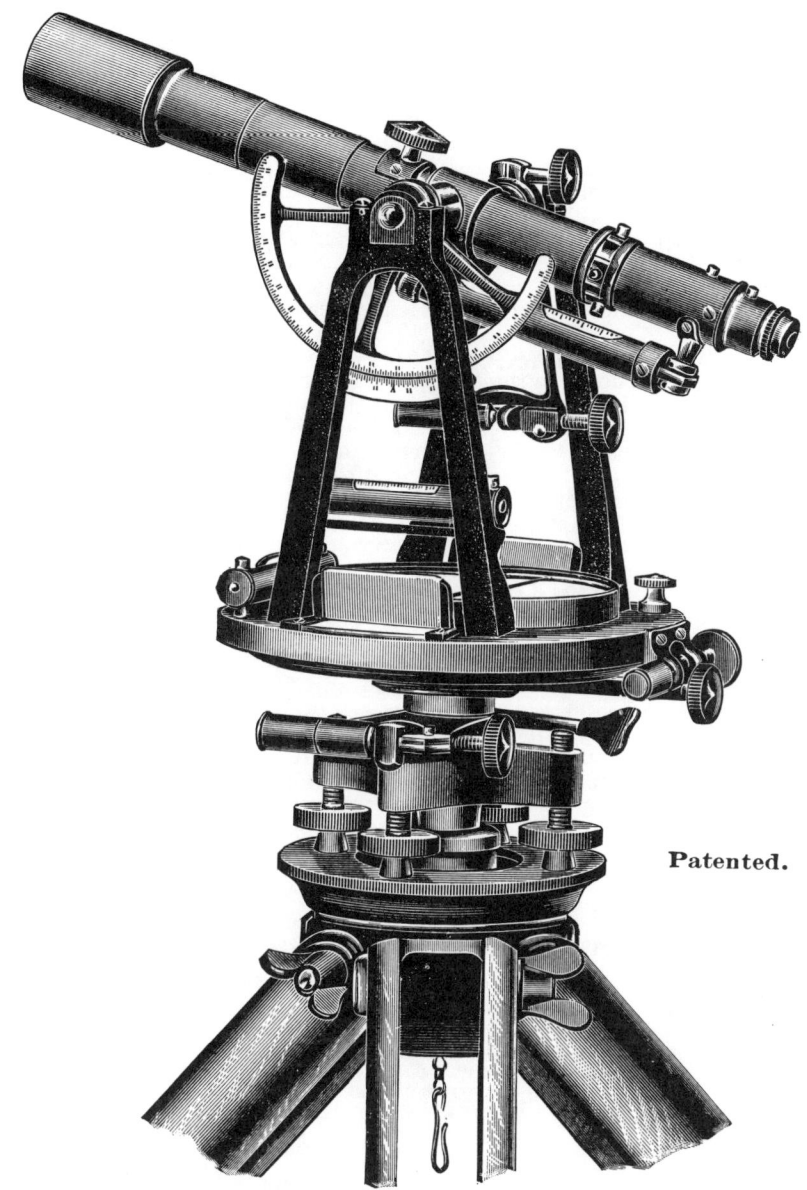

Patented.

Complete Engineers' and Surveyors' Transit
No. 2.
As made by C. L. Berger & Sons.

Unless ordered to be as in cut, the horizontal verniers will be placed at 30° to line of sight.   (See footnote to Mountain Transit No. 3.)

All our transits are provided with a fine punch mark on top of telescope to enable to center instrument from a point above as well as from below.

C. L. Berger & Sons' Solar Attachment.

Our new solar attachment or meridian finder is in principle like Pearsons' (made by us heretofore), not requiring computation, but instead of the lens-bar it has a small telescope* with $\frac{1}{2}$ inch aperture and 6 inches focal length. This consists of a single lens-objective, a diagonal eye-piece, a colored glass, and the customary cross-wires.

This solar attachment fastens by means of a screw to the end of the cross-axis of the transit telescope on the side of the clamp and tangent-screw, as shown in the cut. This solar attachment has no declination arc. The declination of the sun, and the latitude of the place of observation are both set by the vertical arc of the transit. All settings for position, viz., that of the polar axis for coincidence with the zero marks of the vertical arc and verniers, and the setting of the declination, are secured by two spirit levels. These levels are placed upon the polar axis and upon the solar telescope, as will be seen in the cut. The degree of precision and simplicity of manipulation attained thereby is commensurate with that of our engineers' transit.

To determine true meridian at any hour of a day, it is only necessary that the declination and refraction of the sun of that particular day and hour be known to the observer, and that the polar axis be raised precisely to the co-latitude of the place of observation.

By the use of our **Latitude level**** (also fastening to the cross-axis at the side of the vertical arc, as shown in the second cut), not requiring a reading of the vertical arc for every setting of the polar axis for latitude except once in a day, observations can be made repeatedly with speed and accuracy. Indeed, with the declination and refraction of the sun previously worked out for the various hours of a day, observations can be made nearly as fast as a needle can be read of the surveyor's compass.

A concise description of both attachments will be found in our manual. The accompanying illustrations represent them as applied to transits No. 1 and No. 2. Of all the different kinds in use, we believe ours to be the most efficient. Owing to their position on the transit, they can be easily manipulated. The adjustments are few and simple, and need to be verified only from time to time; besides, they can be readily proven and perfected, being similar to those in the transit.

The solar attachment can be readily attached or detached from the transit without altering its adjustments. When detached, the transit is then simply an ordinary complete engineers' transit, with vertical arc, spirit level, clamp and tangent-screw to the telescope.

The weight of the solar attachment is 1 lb., that of the latitude level about $\frac{1}{3}$ lb. Both are packed in a separate box of mahogany provided with a shoulder strap; and can also be packed in the box with the instrument.

It is not necessary to counterpoise the solar attachment in order to obtain good work, the latitude level acting in part as counterpoise. Good results will be obtained by simply watching the latitude level and the plate levels of the transit.

Price of solar attachment, as above   - - -   **$68.00.**
"   " latitude level, as above - - - - - **15.00.**

Extras to Solar Attachment.

Colored glass to apply to the telescope of the transit to observe the sun's altitude, in order to apply the correction for refraction in solar transit work, $2.00.

* The honor of first conceiving the idea of applying a small telescope in place of the lens-bar, and that of using a spirit-level for the accurate setting of the polar axis, belongs to Mr. C. L. Berger of this firm, — see on this point his article on pages 4–6 of our catalogue, published in the year 1878. The idea of setting the declination by means of the vertical arc of the transit, instead of from a separate arc, has also been anticipated by the model underlying the principle of Pearsons' solar attachment made at that time.

** This latitude level can also be used for grades and distance measurements, etc. It will be found to form a very useful acquisition to the engineers' transit, *even without the solar attachment*. In this connection we wish to say that we claim priority in attaching this device to geodetic transits for the purpose set forth, and substantially in the form as illustrated on the opposite page. An instrument of this class was furnished to Princeton College, and one to Sapporo Agricultural College, Tokio, Japan, in the year 1877. For astronomical instruments, we have furnished them in the manner described and illustrated on pages 187 and 189, of our catalogue.

Note. — The level attached to the solar telescope, for the purpose of setting off the declination and refraction of the sun, could be dispensed with. — Assuming that the normal position of the polar axis to the transit has been assured by its level, bi-sect a distant object by the main telescope, in the horizontal plane, incline the latter to the amount of declination and refraction, as explained in the manual. Pointing the solar telescope until its wires bi-sect the same object, its line of sight is at an angle with the polar axis, equal to the amount of declination and refraction. Raising the polar axis to the co-latitude of the place of observation, the solar attachment is ready for work. (It is really not necessary that the object be in a horizontal plane, any distant object, conviently found, is available. In this latter case the division mark of the vertical arc, found in coincidence with the zero mark of its vernier, becomes the zero from which to set off the declination, etc.) However, the greater convenience derived from the use of this level, reducing the operation to a mere mechanical performance, as compared with that of having to find distant objects, decided us to retain it, although it added to the weight and expense of the instrument. — Moreover, in case the polar level should become broken, the polar axis can be leveled up to its normal position on the transit by the solar telescope level alone.

C. L. Berger & Sons' Transit Solar Attachment.

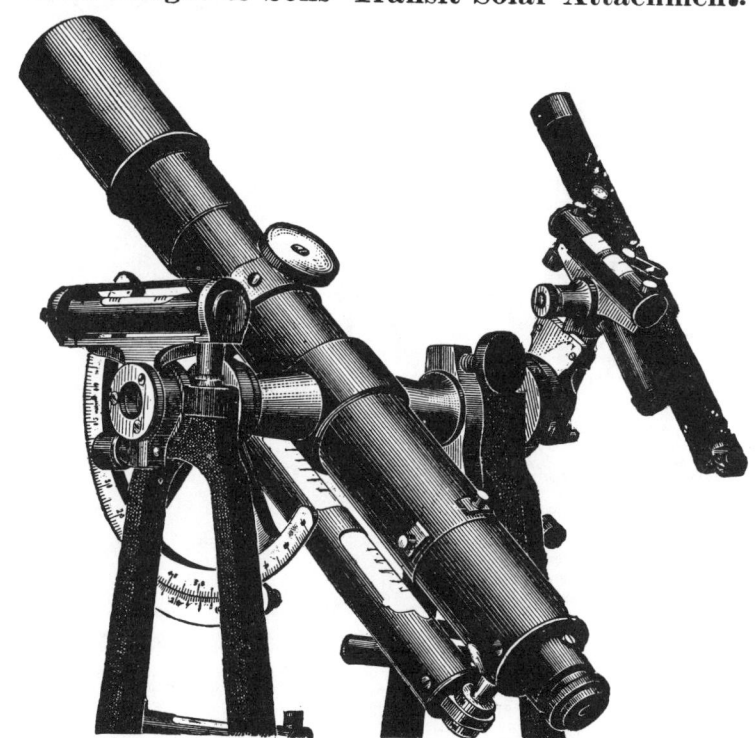

C. L. Berger & Sons' Latitude Level Attachment.

Shown as applied to a Transit with Solar Attachment.

NOTE.— Our solar attachment and our latitude level can be placed only upon Transits No. 1, No. 2, and No. 3, and then only when ordered with the instrument.

Davis' Patent Solar Attachment.

This invention is destined to supersede all other solar attachments, being by far *the most accurate, the most simple, and the cheapest* in use. The sun observations are made with the instrument's telescope direct, whereby greater range and power are secured, and limiting the adjustments to those common to the instrument proper itself. It can be attached to any engineers' and surveyors' transit which has a good vertical arc or full vertical circle. A great many have been placed on our transits (sizes Nos. **1, 2, 3** and occasionally on No. **4**), for the use of U. S. Deputy Surveyors, and others having occasion to do solar work.

However, as its manipulation involves a few mathematical calculations, differing somewhat from ordinary solar attachments, we advise our patrons to carefully read pages 73 and 75, etc., of manual, where a full description will be found.

The screen, shown in Fig. **2,** can be applied with erecting and inverting telescopes. In making an observation with an erecting telescope the full aperture of the object glass is utilized, but with an inverting telescope it must be limited to about ¼ or ⅜ inch diameter to get the wires sharply defined on the screen. To this end the telescope cap is provided with a central opening, permitting of such an adjustment, which may be closed entirely when not in use.

Attachments shown in Figs. **3** and **4** are for direct observation when the sun's altitude does not require the screen. These latter attachments are now made by us in a manner superior to those shown in these cuts on opposite page. They are mounted as in Fig. **5,** upon a frame, readily attachable to the eye-piece by means of a clamp, which can be clamped in any position most convenient for the observer. To bring the colored glasses or the prism before the peep-hole of the eye-piece, it is only necessary to revolve them, hence they can be used in rapid succession. It will be seen that these solar attachments are easy to manipulate, and therefore must insure better results than heretofore obtainable with mechanical devices of any other kind.

Price of Solar Screen as in Figs. **1** and **2,** $6.00
     "   Prism and Colored Glasses, see Figs. **3** and **4,** but of improved
          mounting, as in Fig. **5,** 12.00
     "   Solar Screen with prism and colored glasses combined, . . 18.00

If we attach the screen to instruments which are sent to us for that purpose, we must make an extra charge of $4.00. In such cases the telescope should be sent to us.

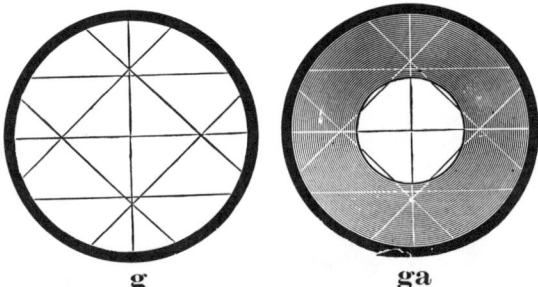

g   ga

C. L. Berger & Sons' Patent Inclined Square.

For Sun Observations with Davis' Patent Solar Attachment.

This device consists of four additional wires forming an inclined square of equal sides placed at an angle of 45° with the usual cross wires, and equi-distant from the latter's point of intersection in the Surveyor's Transit Telescope. The space contained within this square, as will be seen in the greatly enlarged Figures **g** and **ga,** is slightly smaller than the sun's disc; thus an observation of the sun for position can be made by simply setting the telescope by means of the tangent screws until the four segments, formed by the black lines against the bright disc of the sun, are of equal size. In this manner the sun's disc can be better *bisected*, as when it must be quartered by the cross lines alone — but, if desired, both methods can be applied as a check upon each other.

The arrangement of the wires in the inclined square is in *no* way confusing, as it keeps the cross and stadia wires distinctly apart for the regular work of the transit, and, in rapid work, is a help to distinguish the horizontal from the stadia wires, as shown above, which cannot be said of the erect square H — also patented, — shown on the page illustrating the various sighting wire diaphragms. Part I.

Price of Patent Inclined Square, but only, *when ordered* with the instrument, **extra** . **$4.00**
     "   also provided with Stadia Wires, as in cut,   "   .   **7.00**
     "   "   "   "   with cross and stadia wires for instruments of other make   **10.00**

Fig 1

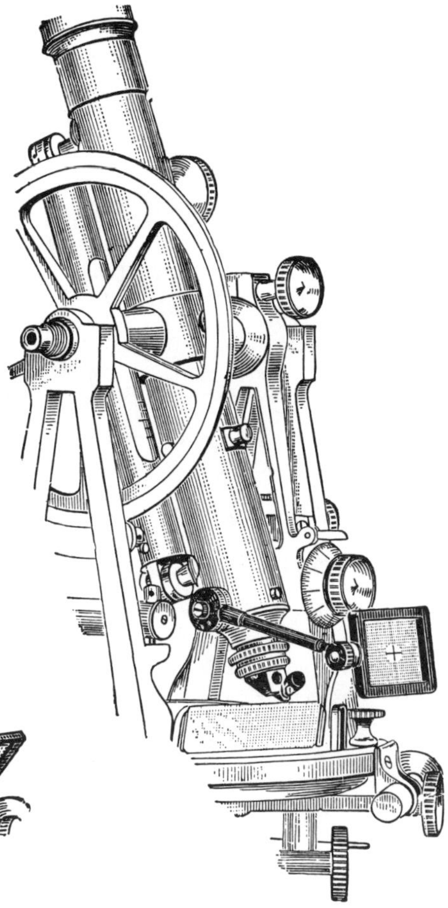

Fig 2.

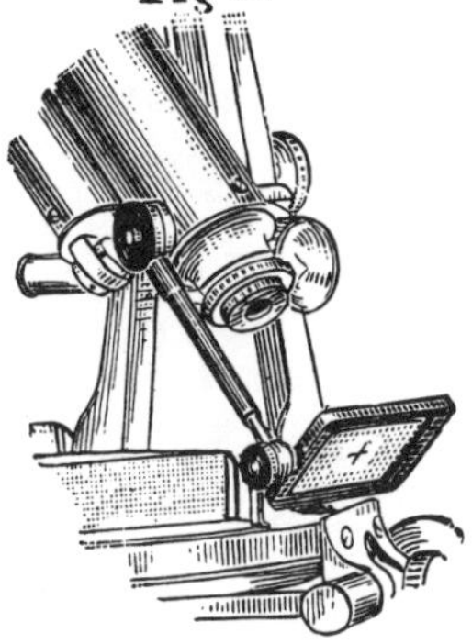

Fig. 3.

Fig. 4.

Davis' Patent Solar Attachment.

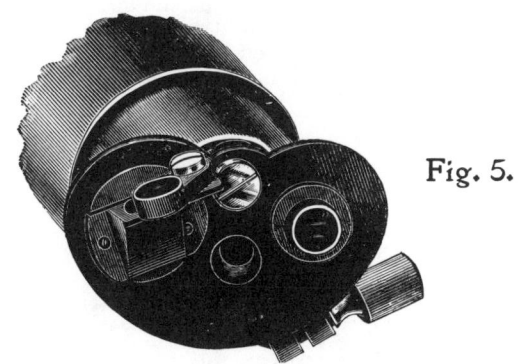

Fig. 5.

C. L. Berger & Sons' Improved Prism and Colored Glass Attachment for Solar Observations.

Mountain Transit.

No. 3. Mountain Transit.—Size as in No. 2. Provided with an extension tripod. This instrument is well adapted for use in mountainous regions, chiefly on account of its smaller size, lightness and great portability. Its work is as accurate as that of larger instruments of its class. Its weight is 10 lbs., with an ordinary tripod complete 16½ lbs., but when provided with an extension tripod three pounds are added to this weight. The graduations are on *solid silver verniers reading to minutes; ground glass shades; 5-inch vertical circle with aluminum guard; spirit-level clamp and gradienter to telescope; protection to object slide; extension tripod provided with thumb-nuts, etc.*

The mahogany case has a leather strap hooks, etc. It contains a sun-shade, a wrench, a screw driver, an adjustable plumb bob, a magnifying glass, an adjusting pin, and weighs 7 lbs.

Gross weight of instrument, packed securely for shipment in 2 boxes, about 45 lbs.

Price, complete as above, with cloth finished standards . . **$257.00**

This instrument with a 5-inch vertical arc in place of the full circle and
without aluminum guard . . . . less $9.00

Standards finished like instrument (*no cloth finish*) . . . extra $5.00

A reduction of $15.00 from the above price is made if the graduations are not on solid silver.

Extras to Mountain Transit.

Offsetting arrangement, ...	5.00
Stadia wires, fixed, ...	3.00
Variation Plate, ...	10.00
Quick leveling attachment (see manual), ...	8.00
Extra regular tripod, for use with instrument in ordinary practice, .	**16.00**
Davis' Solar Attachments, all complete, ...	**18·00**
C. L. Berger & Sons' Solar Attachment (pp. 154 and 155) . . .	68.00
" " Latitude Level "	15.00
*Prism, with colored glasses, for observing the sun's altitude, . . .	12.00
Short Focus Lens Attachment (see pages 118, 169). One pair, . .	16.00
Leather cover over case, to be strapped to the saddle of a horse. . .	11.00
Gossamer water-proof bag, to protect the instrument in case of rain or dust,	1.00
Bottle of fine watch oil, for the centers of transit, ...	0.25

NOTE.—Although the extension tripod is very slender and about 2 lbs heavier than our regular tripod, its superiority for mountain work is very apparent on account of its adaptation to sudden changes in grades. Still, for general practice, it is desirable to have the regular tripod, insuring, as it does, greater steadiness, and consequently giving increased accuracy. The surveyor will therefore find it to his advantage to order both kinds.

—It will be observed that in the cut the verniers of the horizontal circle are placed at an angle of 35° to the line of sight as in our Mining Transits, thus adapting the instrument to the work in a mountainous country. On the other hand this change in the position of the verniers requires the level in front of the telescope to be carried beyond the limit of the plate in order to be of standard length and character, and although fully protected in its partially exposed position from injury, by an improved guard surrounding it, it is, nevertheless, subject to slight changes in adjustment, as when compared with one mounted as shown in Transit No. 2, where verniers are placed at 90° to the telescope. In all cases where this change in the position of the verniers is not deemed of sufficient importance, we advise to order our Transit No. 2. A small striding level, illustrated in Transit No. 1d, can also be placed upon the telescope axis at an extra cost of $15.00. No extra charge if the telescope is ordered to be of the inverting kind when the whole instrument will have to be made specially. The aperture will then be 1¼ inch, the power 22 diameter.

* In a mountainous country, it frequently happens that a transit must be set up in places where it is extremely difficult to get standing room to take both back and fore-sights. With the aid of a prism attachable to the eye-piece, all this can be done from the side of the instrument.

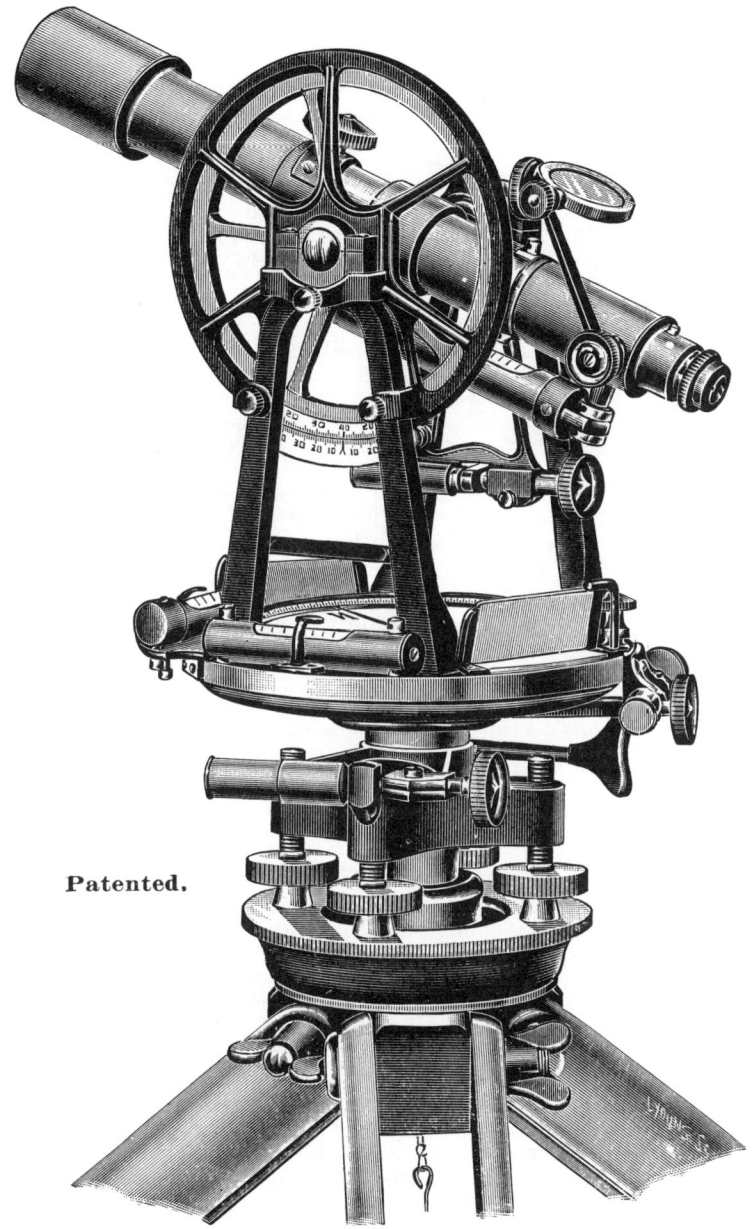

Patented.

No. 5. Mountain Transit.
Shown with Patent Solar Screen Attachment.*
As made by C. L. Berger & Sons.

For illustrations and full description of this Solar Attachment, see pages 73 and 74 of the Manual. For Price, etc., see page 156.

All our transits are provided with a fine punch mark on top of the telescope to enable to center instrument from a point above as well as from below.

Mountain, Mining and Reconnoissance Transit.

On the opposite page cut No. **4** represents a complete transit of this class. It is in every respect similar to our Engineers' and Surveyors' large transit, except in size and weight. The verniers of the horizontal circle will be placed at right angles to line of sight, only; so that both can be read without stepping aside, as could not be the case with verniers placed at 35° to line of sight. This also permits the use of a longer plate spirit level in front, an important feature in so small an instrument. The instrument is as carefully made as the larger ones, and, we believe, with careful use capable of very accurate results. For preliminary work of all kinds, as well as to fill in details, it is especially adapted. Owing to the smaller size of the telescope and its high power, we supply the inverting kind.

We are also prepared to make this size of instrument for travelers, explorers, etc., of style shown in No. **4b** when it is a transit-theodolite of the class shown on page 181, but of style shown in No. **11b**, page 182. The standard frame is cast in a single U-shaped piece to gain as much lateral stiffness as possible. The telescope can be reversed over the bearings by turning the upper covers aside, and also in the usual way through the standards. In this instrument the graduation of the horizontal circle is covered too, and the verniers are protected by glass, as in No. **4.** There is no compass needle. Instrument has a 4″ full vertical circle with double opposite verniers as shown on page 147. A level is attached to the vernier arm instead of to the telescope as in No. **4.** Striding level rests at points of contact, etc., as shown in the cut. This instrument will only be made with three leveling screws.

A more enlarged cut of this style of instrument can be seen by referring to style **11b**. Its cost of manufacture, however, is so high, as compared with No. **4,** that we are compelled to raise the price considerably when ordered in this style. — However, we believe that No. **4,** having the advantage of a magnetic needle and containing all the latest improvements possessed by the Engineers' and Surveyors' Transits, will in most cases give equal satisfaction in every respect. We therefore advise our friends to order No. **4.**

Both these instruments can be supplied with leather covers over the case, to be strapped to the saddle of a horse.

A chief danger may arise from its excessive lightness, it being apt to be overturned. Consequently careful use is required. — The regular tripod supplied with it is quite stiff and secures necessary rigidity, but the extension tripod designed for this instrument, desired by many, renders it more frail and unsteady. In all cases where an extension tripod is ordered we recommend the purchase of a *regular one* in addition. A large size tripod, weighing 9 lbs., as applied to our No. **2** Transit, can be used with a No. **4,** by the use of an adapter The extra weight incurred is more than balanced by the added stiffness. No extra charge if ordered in place of the regular 5 lb. extension tripod.

No. 4 Mountain, Mining and Reconnoissance Transit. — The dimensions, etc., are as follows : —

Horizontal limb 4 inches; graduations on solid silver, double opposite verniers reading to minutes; two rows of figures from 0° to 360° ; glass-shades over verniers; needle 2½ inches; vertical arc 4 inches; telescope 7½ inches; aperture 1¼ inch ; power from 15 to 18 diameters; 4-inch spirit-level, with clamp and tangent screw to telescope; vertical adjustment for the telescope axis; shifting tripod ; double centers; split-leg tripod, etc.

The mahogany case has a leather strap, hooks, etc. It contains a sunshade, a wrench, a screw driver, an adjustable plumb bob, a magnifying glass, an adjusting pin, and weighs 4 lbs.

Weight of instrument 5 lbs. weight of tripod 5 lbs.

Gross weight of complete instrument, packed securely for shipment in 2 boxes, 29 lbs.

Price of Mountain, Mining, etc. Transit No. 4, as above, $228.00

Price of Transit-Theodolite No. 4b as in cut, size as in **No. 4,** including additional vertical wires if desired for solar observations, . . . **$310.00**

No. 4b provided with a vertical arc or full circle, as in No. 4, level attached to telescope, . less $25.00
No. 4b without detachable reading glasses, as in No. 4, ... " 18.00
No. 4b without striding level, as in No. 4, ... " 10.00

Extras to Transits No. 4 and No. 4b.

4-inch full vertical circle (in place of arc) with aluminum guard as in Mountain Transit No. **3.**	$9.00
Fixed stadia wires, ...	$3.00
Prism and colored glasses for solar observations, (improved mounting), . .	12.00
Short Focus Lens Attachment (see pages 118, 169). One pair, . . .	16.00
Extra extension tripod, (weight 5 lbs. ; see note to No. **3** Mountain Transit), .	16.00
Leather cover over case, ...	8.50
Patent detachable side telescope with counterpoise, to transit No. 4, only, page 164, ...	$35.00
Patent interchangeable auxiliary telescope. To transit No. 4, only. Style I. p. 167 and 168, ...	45.00
Silk bag to cover transit in case of rain or dust, ...	0.80
Bottle of fine watch oil, ...	0.25

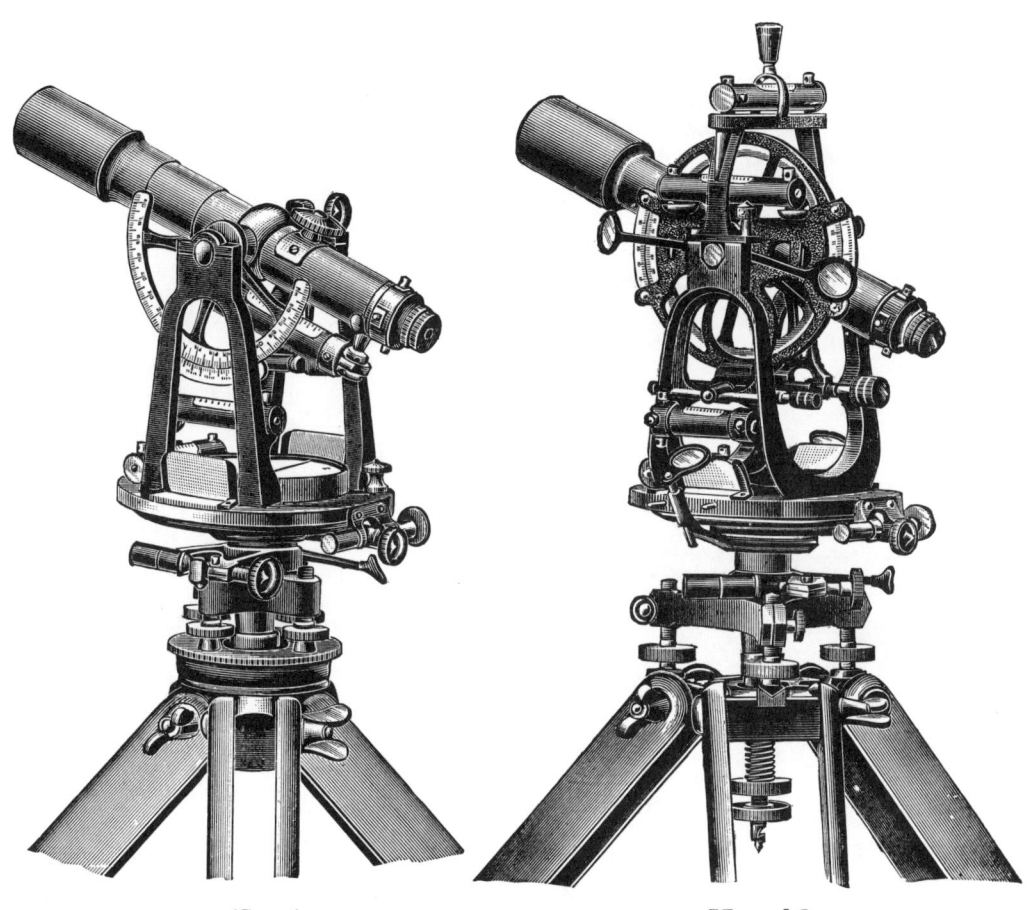

No. 4.   **No. 4 b.**

Mountain, Mining and Reconnoissance Transit.

As made by C L. Berger & Sons.

NOTE. — Not infrequently we are asked to make the telescope of this instrument to show objects erect. This cannot be done to satisfaction, since it is impossible to make so small an erecting telescope that will have the necessary power for good work and be at the same time of good quality. It seems that in all such cases the interest of our friends will be best guarded if they will order Transit No. **2** or **6**, as coming nearest to the desired end.   Having thus acquainted our friends with this fact, we would add that we are prepared to carry out their wishes in this respect, if they are willing to take the consequences upon themselves.   As will be noticed in the description of our Mining Transit, given later on, we are prepared to attach our interchangeable auxiliary telescope, Style I, to transit No. **4**, when the latter is provided with a full vertical circle.   This can be done safely since all the important parts are as stiff and unyielding as in our larger instruments.   We can also attach to the same size instrument, our new adjustable device consisting of a polar and declination axis, attachable to the central post of the telescope's cross axis as shown in Style I, by the aid of which the interchangeable auxiliary telescope is made to answer the purpose of a solar telescope and revolves parallel to the earth's equator as in our regular solar attachment, thus enabling to determine meridian lines at any hour of the day.   This must be considered a great improvement since no additional solar attachment will be necessary.   This auxiliary telescope has a higher power and adds but one pound additional in weight.

Mining Transits.

All of the foregoing instruments, *particularly No. 2 and No. 4* we recommend for general underground work. (The latter instruments chiefly on account of their greater portability and lightness.) The telescope may be either inverting or erect, and may have a prism attachable to the ordinary erecting or inverting eye-piece to facilitate sighting in inclined shafts. We frequently attach to the cross-axis a side telescope, swinging free of the plates, of nearly the same length and power, which permits of vertical sighting up or down a shaft (see cut p. 164, also pp. 104 to 106). We also furnish attached above and parallel with the line of sight of the main telescope of Nos. **4, 5** and **6,** another one of shorter length and lesser power mounted on a central pillar firmly secured by means of a threaded stud to the cross-axis of the main telescope, as in Style II; or cast on, as in Style I. (See interchangeable auxiliary telescope, pp. 165, 166, also pp. 106, 107.)

This latter telescope, which of necessity is of lesser power as compared with our regular side telescope, will in most cases be deemed sufficient, as sights in mines are short. Its power varies from 8 to 15 diameters, according as the telescope is erecting or inverting. We are, however, prepared to mount one of same length as our regular side telescope in cases where a complete revolution of cross-axis is not required. In our improved style No. II, the top telescope can be truly set parallel with the line of sight of the main telescope by means of adjusting screws. It will be seen that vertical sighting up or down a shaft can be done with ease and accuracy, and this latest device has nearly all the advantages of a telescope mounted on inclined standards with none of its faults, and is second only to our Universal Duplex Mining Transit, in so far as it entails the use of an auxiliary telescope of lesser power. When not in use the auxiliary telescope may be removed at will and stored in the box. In style II, the central pillar may be left on the instrument or taken off at pleasure. In the latter case only the stem for the counterpoise on the under side of the telescope remains attached. But in Style I, which we now strongly recommend, the central stem is permanent, as explained above. As will be seen by reference to the description and cuts of this device, pp. 106, 107 and 165, the auxiliary telescope can also be readily attached to the end of the cross-axis of our Mining Transits No. 4 and No. 6 when it becomes a side telescope. The ready interchangeability from top to side makes the auxiliary telescope one of the most desirable additions to a Mining Transit.

To avoid errors in reading cardinal points, the compass ring is figured from 0° to 360°, the same as the horizontal circle. Mining instruments should have large vernier openings to admit of as much light as possible, and all graduations should be on solid silver. For the illumination of the cross-wires a small reflector is sometimes placed in the centre of the cross-axis of the telescope of our larger instruments, but as in the smaller telescopes much light is being cut out by its use we prefer to attach a reflector shade in front of the object glass. The tripod is provided with three adjustable legs to permit of raising or lowering the instrument in cramped places.

No. 5. Mining Transit. — Dimensions as in No. **1**; graduations on *solid silver,* verniers reading to minutes are provided with ground glass shades; 5-inch full vertical circle; spirit-level, clamp and tangent screws to telescope; extension tripod, etc.
Price $253.00.

No. 6. Mining Transit. — Dimensions as in No. **2**; graduations on *solid silver,* verniers reading to minutes are provided with ground glass shades; 5-inch full vertical circle; spirit-level, clamp and tangent screw to telescope; extension tripod, etc.
Price, $253.00.

A reduction of $15.00 will be made if the graduations are not on solid silver.

Extras to Mining Transits Nos. 5, 6 and 7.

Striding level (for description and illustration see pp. 56, 148, and 178)	$20.00
Five-inch vertical circle provided with double opposite verniers, see page 147 . . .	20.00
Horizontal verniers for No. 6, placed at 35° to line of sight, as shown on page 164 (not applicable to Transit No. 7, on account of the exposed position of the level) no extra charge.	
Patent short focus lens . . . . . No. 1, $8.00; No. 2, $8.00. Nos. 1 and 2,	15.00
Reflector for illuminating the cross-wires	4.00
Aluminum Guard to vertical circle	4.00
Prism, attachable to eye-piece, ...	8.00
Detachable side telescope with counterpoise, page 164	35.00
Patent interchangeable auxiliary telescope. Style I, described on page 107 and shown on page 167, of superior construction, strongly advise it in place of Style II	45.00
Patent adjustable and interchangeable auxiliary telescope. Style II, p. 165 . . .	70.00
Patent equatorial adapter for solar observations, striding level, prism with colored glass, and diaphragm with square of coarse wires, page 168.	50.00
Arrangement for offsetting at right angles to telescope	5.00
Quick leveling attachment	8.00
Half-length tripod	13.00
Extra extension tripod	16.00
" split-leg tripod (see note to No 3)	16.00
Plummet lamp (large size)	10.00
" " (small size)	8.00
Bracket in box, with strap, auger, and lever (Fig I on page 169)	15.00
Trivet having legs 2, 4, or 6 inches long (Fig. III, page 169)	3.50
Stadia wires, fixed	3.00
Short focus lens attachment (pages 118, 169, Fig. II), one pair	16.00
Gradienter attachment for Nos. **5** and **6**	5.00
Large plumb bob, weight 4 lbs., for use in shafts	5.00
Bottle of fine watch oil	0.25
Silk bag, to cover Transit	1.00

NOTE: — The object prism, as sometimes used to enable steep sighting in mines with the ordinary transit telescope, is not enumerated here for reasons stated on page 27.

Complete Mining Transit,

With Extension Tripod, * Detachable Side Telescope, Prism, etc.,
As made by C. L. Berger & Sons.

Size as in **Nos. 5, 6** and **7.** For price of instrument and attachments, see preceding page.

* See note to Mountain Transit, page 158.

All our transits are provided with a fine punch mark on top of telescope to enable to center instrument from a point above as well as from below.

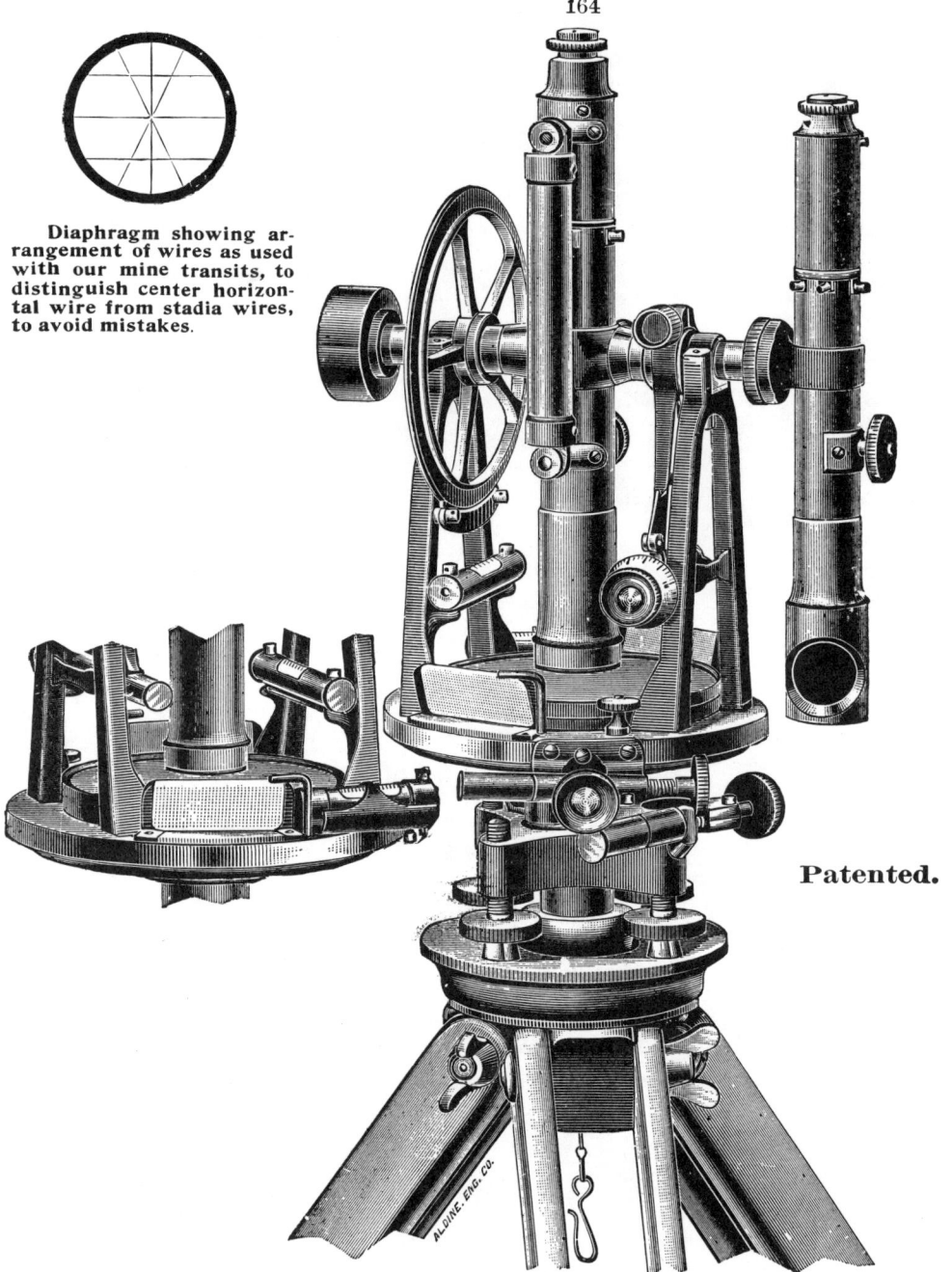

Diaphragm showing arrangement of wires as used with our mine transits, to distinguish center horizontal wire from stadia wires, to avoid mistakes.

Patented.

Complete Mining Transit,

With detachable Side Telescope and Reflector for illuminating cross wires,

As made by C. L. Berger & Sons.

Size as in **Nos. 5, 6** and **7.** For price and attachments see page 162.

NOTE.—The side telescope is the more customary attachment to mining transits. It can be readily attached or detached from the instrument proper. It can be of the same length as the main telescope when so ordered, but it is usual with us to make it an inch or so shorter, to keep it as light as possible, and to save weight in the counterpoise. When the side telescope and its counterpoise are attached to the transverse axis, they balance each other, and therefore the adjustment of the line of collimation of the main telescope is not disturbed in the vertical plane. The weight of the side telescope and its counterpoise is about 1 lb. for each.

It will be seen that in the mining transit illustrated above the horizontal verniers are placed at an angle of 35° to line of sight of telescope, to enable one to read the verniers in cramped places without stepping aside. The front plate level in this instrument is therefore of standard length and character, and in form is much improved on former styles. This will be appreciated when it is remembered that this is the main level on the transit which governs the movement of the line of collimation in a vertical plane. To place this level outside of the plates and in front of the telescope, as shown on page 148, is not permissible in mining transits, as it would shorten the range of the main telescope for steep sighting.

An instrument having its front level placed as shown above might be accompanied by a striding level for best results in steep sighting if style of instrument permits. If not permissible we can place an auxiliary level of standard length on the main telescope, near the eye end, as shown in cut on opposite page, instead.

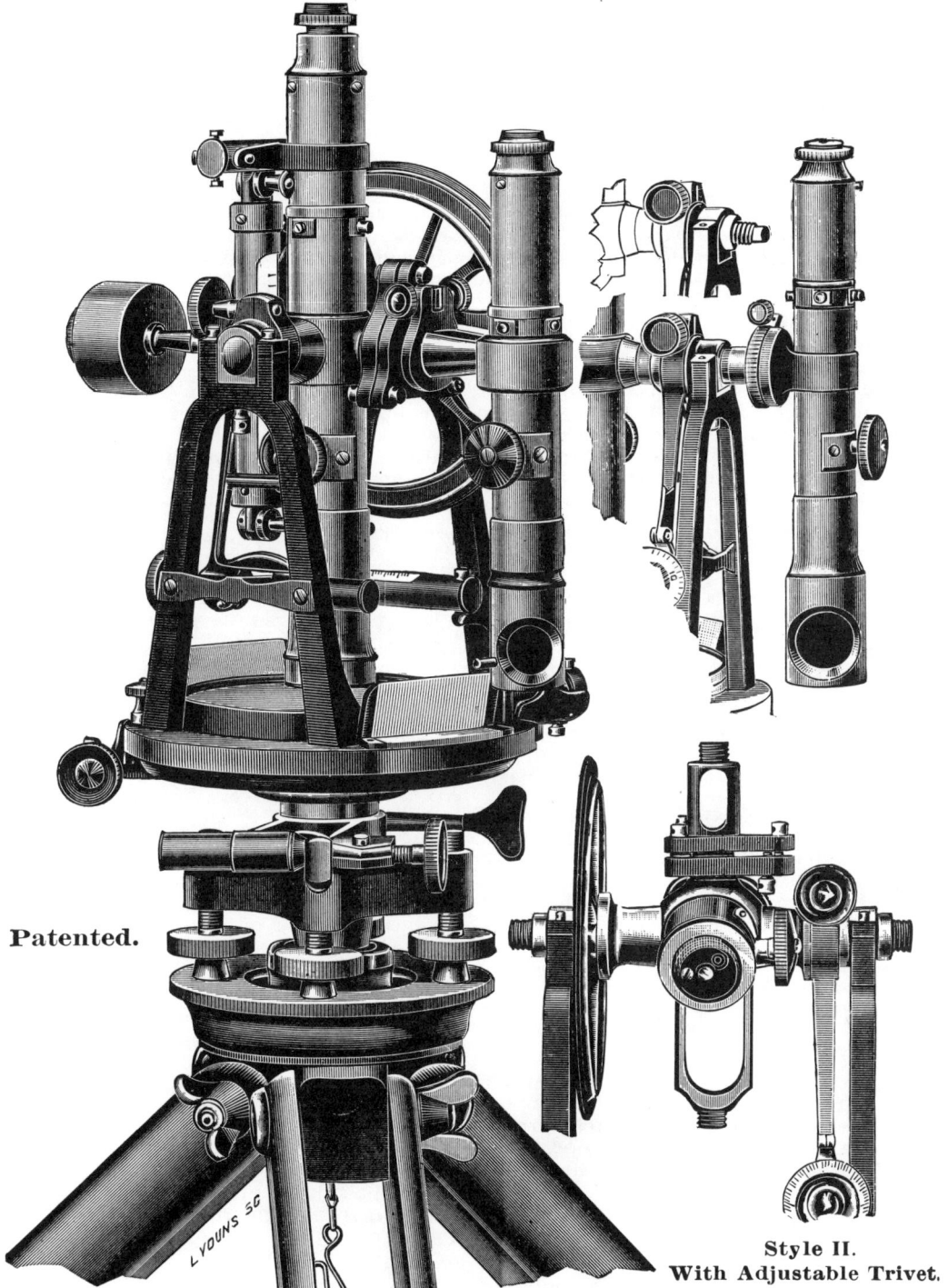

Patented.

**Style II.
With Adjustable Trivet.**

Complete Mining Transit,

with C. L. Berger & Sons' Patent Adjustable and Interchangeable Auxiliary Telescope (Style II).

(For use as top and ordinary side telescope).

Size as in **Nos. 5, 6** and **7.** For general description, price and attachments see page 162.

For full description and adjustments see pages 106 and 107.

NOTE. — The fact that this auxiliary telescope, though intended chiefly for the measurement of horizontal angles in steep mines, can also be used as a side telescope for measuring vertical angles, obviating the necessity of corrections for eccentricity in either case, we feel, will be appreciated by the Mining Engineer, who under trying circumstances in the mine must have often felt the need of just such an accessory.

As shown in the cut above, the verniers are placed at an angle of 35° to the line of sight (unless ordered to be at 90°), in consequence of which the front plate level, being of standard length and form, is carried some little distance beyond plates, which at times may render it less convenient to read, and although protected by a guard, renders it more liable to slight derangements. For reasons stated on opposite page, a striding level might be ordered, but as with this style of instrument a striding level cannot be applied, we can place instead a fixed auxiliary level on the main telescope, near the eye end, which will perform the duty of a striding level, assuring the rotation of the telescope in a vertical plane when best results in very steep sighting are desired.

Its adjustment, however, is dependent on that of the telescope in the vertical plane. This latter must be verified, as explained on page 51, before this level can be adjusted by reversing, as in the case of the plate levels. For price, see striding level "Extras," p. 162.

Complete Mining Transit No. 6D, without Compass.

Shown with our Patent Interchangeable Auxiliary Telescope, Style I.
(See pages 106 and 107.)

Responding to many solicitations to make for mines containing magnetic ore, or an electric plant, a transit similar in style and accuracy to our No. 11 (see page 180), we have designed the instrument illustrated on opposite page. It is *light, portable,* and *of the same size as our* **Nos. 4, 2** *and* **6** *transits ;* but, owing to the omission of the compass, the standards are cast in a single piece, affording greater lateral stiffness, with increased capability to withstand rough treatment. It is adapted to all the complex conditions prevailing in underground work, and is very simple in style and manipulation. It possesses all the advantages, as regards accuracy of division, highest permissible telescopic power, and sensitive spirit-levels of larger instruments. With the interchangeable auxiliary telescope added for use in steep sighting, either on top or on the side of the main telescope, as required, it becomes a most capable instrument for correctly solving what would otherwise require special instruments and methods. When the auxiliary telescope is detached, it is just as applicable to the common work in the mine or on the surface as our regular engineers' and mining transits **N os. 4, 2** and **6.**

The U-shaped standard frame of the telescope is made of aluminum, covered with a fine dark Japan not affected by moisture; all other parts are finished in the same manner as in our other instruments. The plate-levels * are of our standard character and length, mounted directly upon the upper plate, where they are easily accessible for the purpose of adjustment and ready observation, and are fully protected from falling bodies. The principal plate-level is directly under the eye-end of the telescope. The two opposite verniers of the horizontal circle are in line of sight with the telescope, and are protected from dripping water by cemented glass covers. The circle itself is provided with two rows of figures from 0° to 360°, in opposite directions, with double verniers to correspond to them (unless otherwise ordered). The vertical circle, with figures from 0° to 180°, both ways from zero, has a double vernier, to enable the observer to read angles of elevation or depression with equal facility, and is provided with an aluminum protection guard, which carries the vernier and also serves to readily adjust the latter to zero. Double opposite verniers can also be placed on the vertical circle, when the figures will run from 0° to 90° each way and back to zero. The transit has inverting telescopes (*unless otherwise ordered*). A new and important feature of the instrument, which greatly increases its value, is this: the line of collimation of the main telescope is adjusted for distant, very near, and intermediate distances, by means of our recently patented device, to a nicety never before attained; and no readjustment for near distances is necessary except after a severe accident.

The interchangeable auxiliary telescope, described on pages 106 and 107, is furnished in two styles.

Style I, illustrated on opposite page, has no trivet, and is therefore non-adjustable, but it has been so much improved that the line of collimation of its principal wire, which is the vertical one when used as top telescope, and becomes the horizontal wire when used as a side telescope, lies so nearly parallel to that of the main telescope as to be practically correct in most cases.

Style II, with the trivets, can be adjusted so that the line of collimation of *both* of its wires will be truly parallel to those of the main telescope. All of the chief features of this transit are protected by letters patent issued to us.

Weight of Mining Transit **No. 6D** . . . . 11 lbs.
 " auxiliary telescope and counterpoise, each 12 oz. .   .   .   .   .   1½ "
 " extension tripod . . . . about   9 "
 " instrument in mahogany box, with plumb-bob, sun-shade, reading-glass, etc., etc. "   22 "
 Gross weight of instrument complete, packed securely for shipment in 2 boxes   .   "   50 "

No. 6D. Mining Transit without Compass, as in cut, with Style I. Interchangeable Auxiliary Telescope.
Horizontal and vertical circles, 5 inches; solid silver graduations reading to minutes; ground glass shades; 5-inch level to telescope; 2 plate-levels; inverting telescope, 9½ inches long by 1¼-inch aperture; (if erecting, 9⅛ by 1¼ inches) ; powers, 20 diameters ; inverting auxiliary telescope, 6½ by 1 inch aperture; (if erecting, 7 by $\frac{3}{16}$ inch) ; fixed stadia wires; gradienter; 2 illuminator shades; extension tripod, etc. **Price, complete as above, $335.00**

Price, with double opposite verniers to vertical circle,   **extra,** $ 5.00
 " " prism attachment to eye-piece   "   8.00
 " " quick-leveling attachment   "   8.00
 " " style II. auxiliary telescope in place of style I.   "   25.00
     without style I. auxiliary telescope   **less,**   45.00
 " with one illuminator shade only for main telescope   "   4.00

---

* To place these levels below the upper plate and within the space formed by it and the graduated circle, as is the practice in some instruments of minor grade and in some Surveyors' compasses, has the great disadvantage of partly obscuring them from view, and that such levels are too small in diameter and length to be reliable and are apt to loose their adjustment, and that they cannot be reached in case of required treatment. But the most serious objections are that the two large openings required in the upper plate of a transit to afford a full view, weaken the latter to such an extent as to *jeopardize the stability of the superstructure upon which the permanency of the adjustments depends in case of rough treatment.*

Our Interchangeable auxiliary telescope, being of the most substantial construction and character, may also be used for finding meridian and latitude when direct observations cannot be made with the main telescope. See page 108.

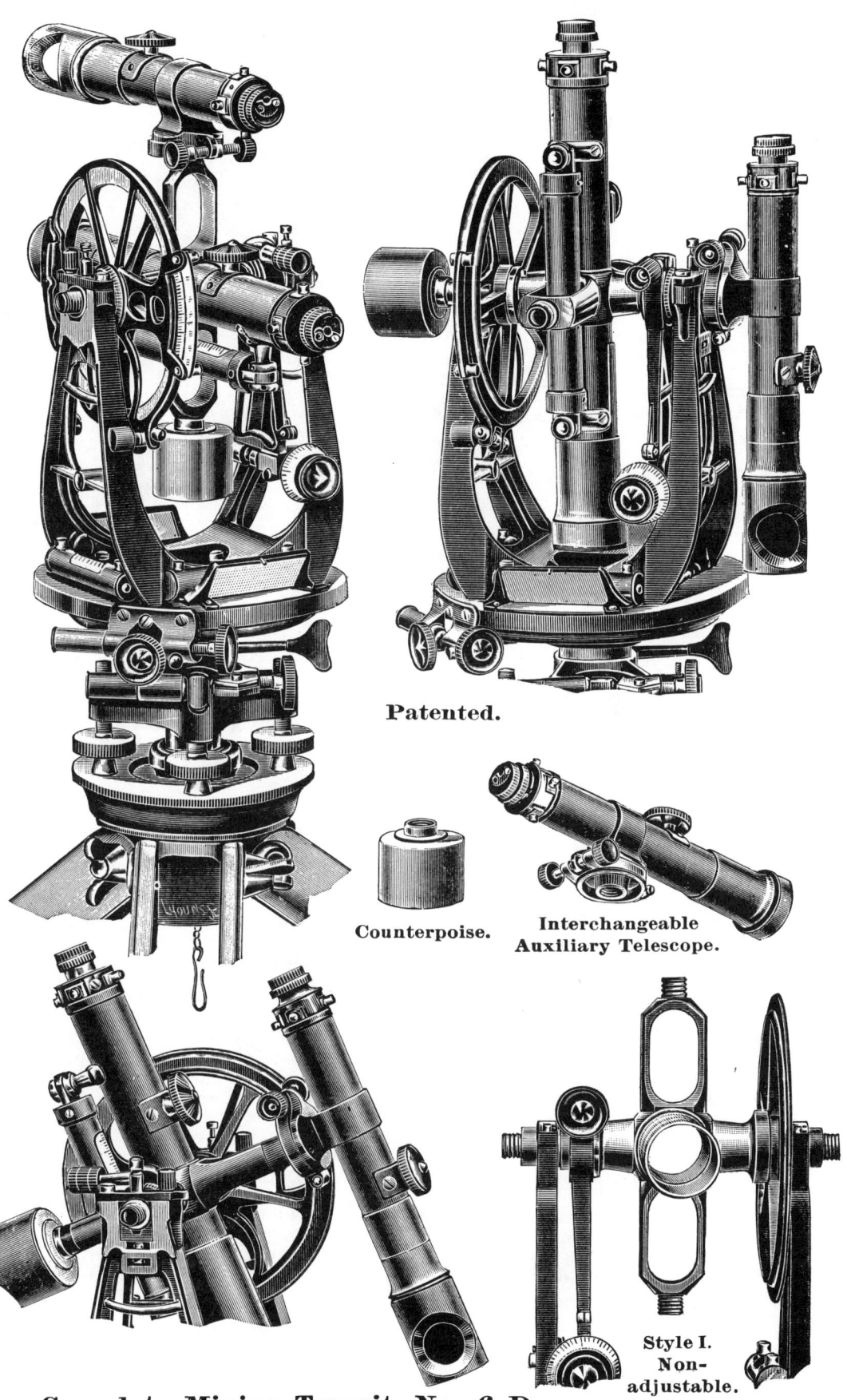

Patented.

Counterpoise.

**Interchangeable
Auxiliary Telescope.**

**Style I.
Non-
adjustable.**

Complete Mining Transit, No. 6 D,
With Patent Interchangeable Auxiliary Telescope (Style I.)
As made by C. L. Berger & Sons.

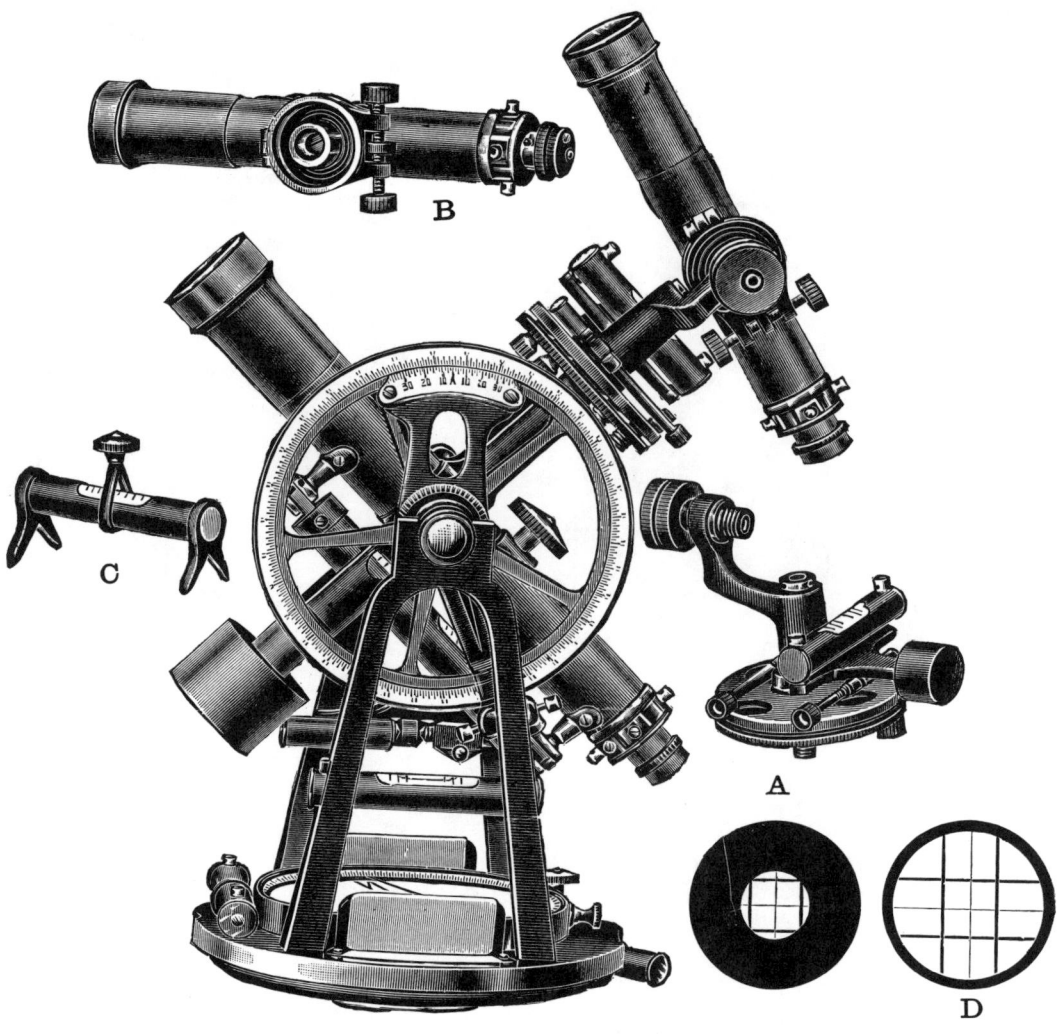

C. L. Berger & Sons' Patented Equatorial Adapter
for Finding Meridian.

The above cut illustrates our new solar attachment, more fully described on page 64a of our manual, attached to the upright post of our Style I interchangeable auxiliary telescope. It will be seen from these cuts that by introducing the equatorial adapter **A**, and the striding level **C**, the patented interchangeable auxiliary telescope used with our mining transit can also be readily converted into a solar telescope, wherewith, by attaching it to the equatorial adapter, the problem of finding the meridian at any hour of the day by the sun can be solved readily and correctly. The results obtained are more accurate than by any other solar attachment of a similar kind, as the telescope is larger and more powerful than those heretofore used. Its position on the instrument, being considerably raised above the main telescope, gives it a free motion in any direction without incumbrance, and frequently permits the sun to be observed without the use of the prism. Most parts of the adapter are of aluminum to decrease the weight. When not in use the adapter is not attached to the transit. It is so substantially constructed that the adjustments need only an occasional verification. The same prism and colored glass attachment accompanying the mining transit may be used. The patented striding level can also be used with the auxiliary telescope, when used as a side telescope, to level it. The patented diaphragm **D**, used with our auxiliary telescope, when arranged for solar observation, is provided with four coarse cross-wires equi-distant from the vertical and horizontal cross-wires as shown in cut.

We can furnish the Equatorial Adapter to any of our mining transits * Nos. **4** and **6,** new or old (**if latter is provided with style I interchangeable auxiliary Telescope**), but if to be attached to an old instrument it will be necessary to send to us the auxiliary telescope, or the whole instrument.

Price, Equatorial Adapter, striding level, prism,† fitting both telescopes with colored glass and special single colored‡ glass, for direct observations **$50.00**

* Also applicable to the Surveyor's transit when *ordered* with the instrument by providing it with an auxiliary telescope and with the vertical post, both as shown in **Style I,** at an extra expense of **$45.00**

† See figure 3, page 157.   ‡ See figure 4, page 157.

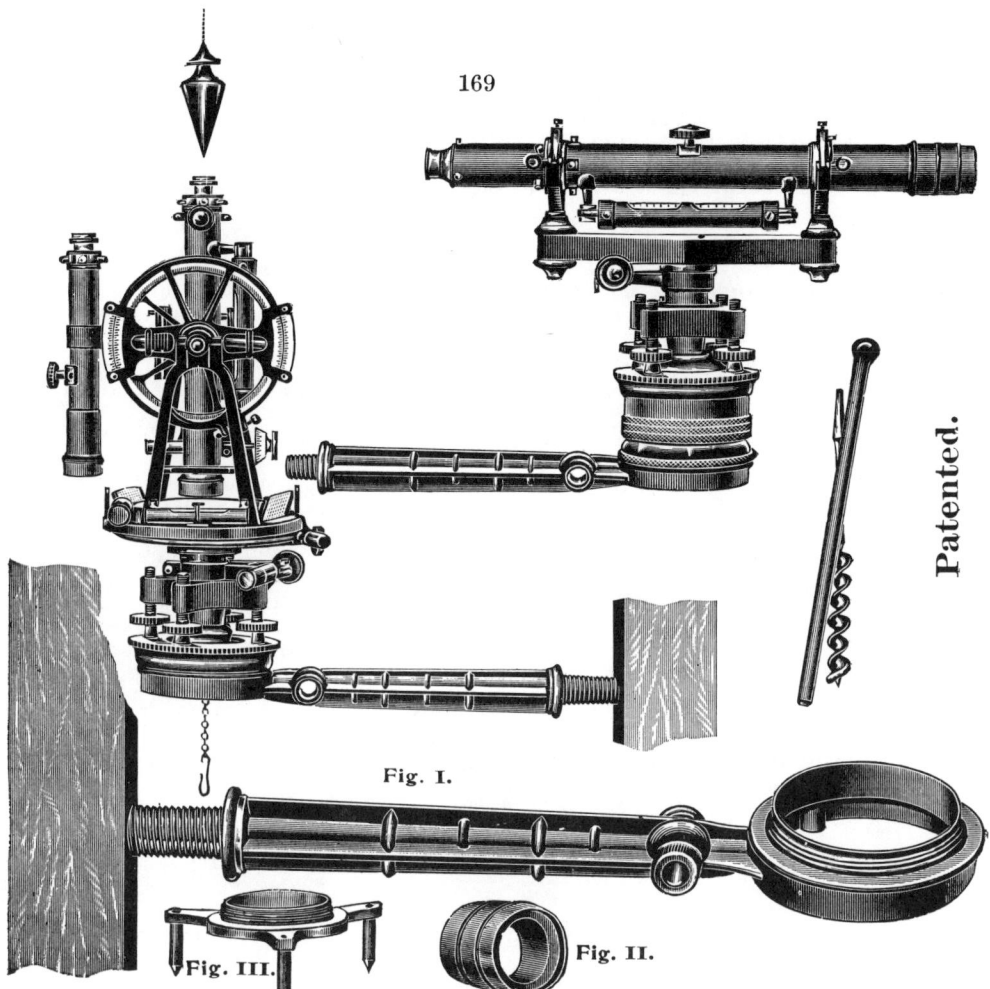

Patented.

Fig. I.

Fig. III.

Fig. II.

C. L. Berger & Sons' Bracket for Transit or Level.

This bracket is designed for supporting the instrument under conditions when the use of even our extension tripod is inadmissible, and will be found a valuable auxiliary for mining work. The instrument can be screwed upon the bracket, as on a tripod, and the transit can be centred above or below a given point. The bracket is made of brass, so fashioned as to offer the greatest rigidity, and is furnished with an auger and a lever.

Price, One bracket made for Transit No. **4,** with box, auger and lever . $14.00
" " " " " " **6,** " " " " " . 15.00
Every additional bracket extra, for either size 9.00
Trivet of brass, to mount transit on, see Fig. III 3.50

Short Focus Lens Attachment.*

The above cut of our wye level and Fig. II illustrate our patented Focus Lens Attachment, attachable to the object end of the main telescope, which permits the focusing of objects nearer than the range of the main telescope will permit. As a rule the main telescope can be made only to focus objects five to six feet distant from instrument. These lenses are generally furnished in pairs. Lens **No. 1** will permit focusing of objects about 4 feet from instrument. Lens **No. 2** will permit focusing of objects about $2\frac{1}{2}$ feet from instrument ; used together they permit focusing of objects about two feet from instrument.

The lenses are adjustable to the line of collimation of the main telescope and permit of a high degree of accuracy. They will often prove of great convenience as an auxiliary to view objects that are too near for observing without them. Attachable to transits Nos. **1, 2, 3, 4, 5, 6,** and to our wye and dumpy level.

Price, Lens No. **1** $8.50
" " " **2** 8.50
" Lenses " **1** and No. 2 . . . . 16.00

* See page 118

Mining Transit.

Interchangeable with Lamp Targets * above Leveling Screws.

The leading features possessed by this instrument, as made by us, are, that it can be interchanged with the lamp targets *above the leveling screws* (see opposite page) so that after they have been set up their relative position as to height and location will remain exactly the same upon being transferred from one tripod to another. It is an instrument especially adapted to the work of an engineer in mines, and as such involves a great deal of apparatus not possessed by instruments intended for surface work. Viewed in its entirety, as a surveying instrument, it is more complicated to handle than the more customary styles, and with the attachment of a side or top telescope it grows still more so, and is then difficult to manage in cramped places. The advantages derived from its use may, however, compensate for the more cumbersome features of this instrument. Two lamp targets are commonly supplied with an instrument of this kind, but frequently one only is ordered, according to the character of the work for which it is intended. The vertical centers of the lamp targets have to be well fitted into the sockets of the leveling heads, and in consequence they cannot be furnished separately. The engineer will therefore decide whether one or two lamp targets are necessary for this work. The lamp targets are of the same height as the transit measured from the base above the leveling screws to the line of sight, and each is provided with two spirit levels. The targets can be revolved in the vertical and horizontal planes. By means of a tubular sight, situated on top of the target, the latter can be readily set in the direction of the instrument so that its face will lie at right angles to the line of sight. The lamps can be raised or lowered at will, and can be detached whenever necessary. It is of the bull's-eye pattern; and the best lard-oil only should be burned in it. A disc of milk glass placed between it and the target furnishes an illuminated background, against which the intersection and outlines of the target are seen sharply defined.

Instrument with one Lamp.
Weight of Mining Transit No. 7, about . . . . 11 lbs.
" one lamp, " . . . . 6½ "
" Mahogany box containing instrument and one lamp, target etc., about 32 "
" one Extension Tripod, about 9½ lbs.; two Tripods, . . " 19 "
Gross weight of this instrument, complete, packed securely for shipment in two boxes, about 70 lbs.

Instrum't with two Lamps.
Weight of Mahogany box, containing instrument only, about . . . . . 22 lbs.
" " " 2 lamps, targets, etc.. " . . . . 26 "
" three Tripods, " . . . . 28½ "
Gross weight of instrument, complete, packed securely for shipment in 3 boxes, . . about 120 lbs.

No. 7. Mining Transit, size as in No. 6,‡ with one lamp target as shown on opposite page; graduations on solid silver, verniers reading to minutes, are provided with ground glass shades; 5 inch full vertical circle; spirit level, clamp and gradienter screw to telescope; illuminator shade; striding level and fixed stadia wires, 2 extension tripods, two plumb-bobs, etc. Lamp target packed in instrument-box.

Price, $380.00

No. 7a. Mining Transit, as above, but with 2 lamp targets, both packed in separate box, three extension tripods, 3 plumb-bobs, etc.

Price, $460.00

Instruments **No. 7** and **7a,** without Striding Level, less $20.00
" " " " Gradienter, " 5.00
" " " " Fixed Stadia Wires, less . . . 3.00

For price of extra attachments, see *Extras to Mining Transits, page 162.*

NOTE. — To interchange the instrument and the lamp target proceed as follows: — First withdraw the spring bolt of the lower clamp for the outer center by means of the small milled-headed nut at the end of the clamp opposite the tangent-screw. Then loosen the clamp screw immediately above the leveling head. Now upon pulling back the spring bolt, situated at the side of the clamp, the instrument or the lamp can be detached by lifting it out of the socket in the leveling head. These sockets as well as the clamps, which serve to fasten the instrument or lamp target to the leveling head, should be kept free from dirt or grit. After clamping the instrument to the leveling head, to prevent any motion in its socket, and then releasing the small milled-headed nut from its fastening on the spring bolt of the lower tangent-screw, the transit is ready for work and can be manipulated the same as other instruments of our construction.

*The difference existing between the lamp target and the lamp formerly furnished with our Mining Transit No. 7, consists in, that in the latter the flame of a paraffine candle, kept at the same height as the instrument, was to be sighted at, which of course was a source of inaccuracy, particularly for short sights. By using the lamp target shown on the opposite page, either long or short sights can be taken with rapidity and accuracy.

‡ Customary size of instrument on account of its lightness, as compared with **No. 5,** but for exceptional uses we are prepared to furnish this style of instrument of size No. 5 whenever desired, but its size and greater weight — about 4 lbs. more for the instrument and about 1 lb. more for each lamp target and tripod, etc. — makes it, however, very unwieldy in cramped places.

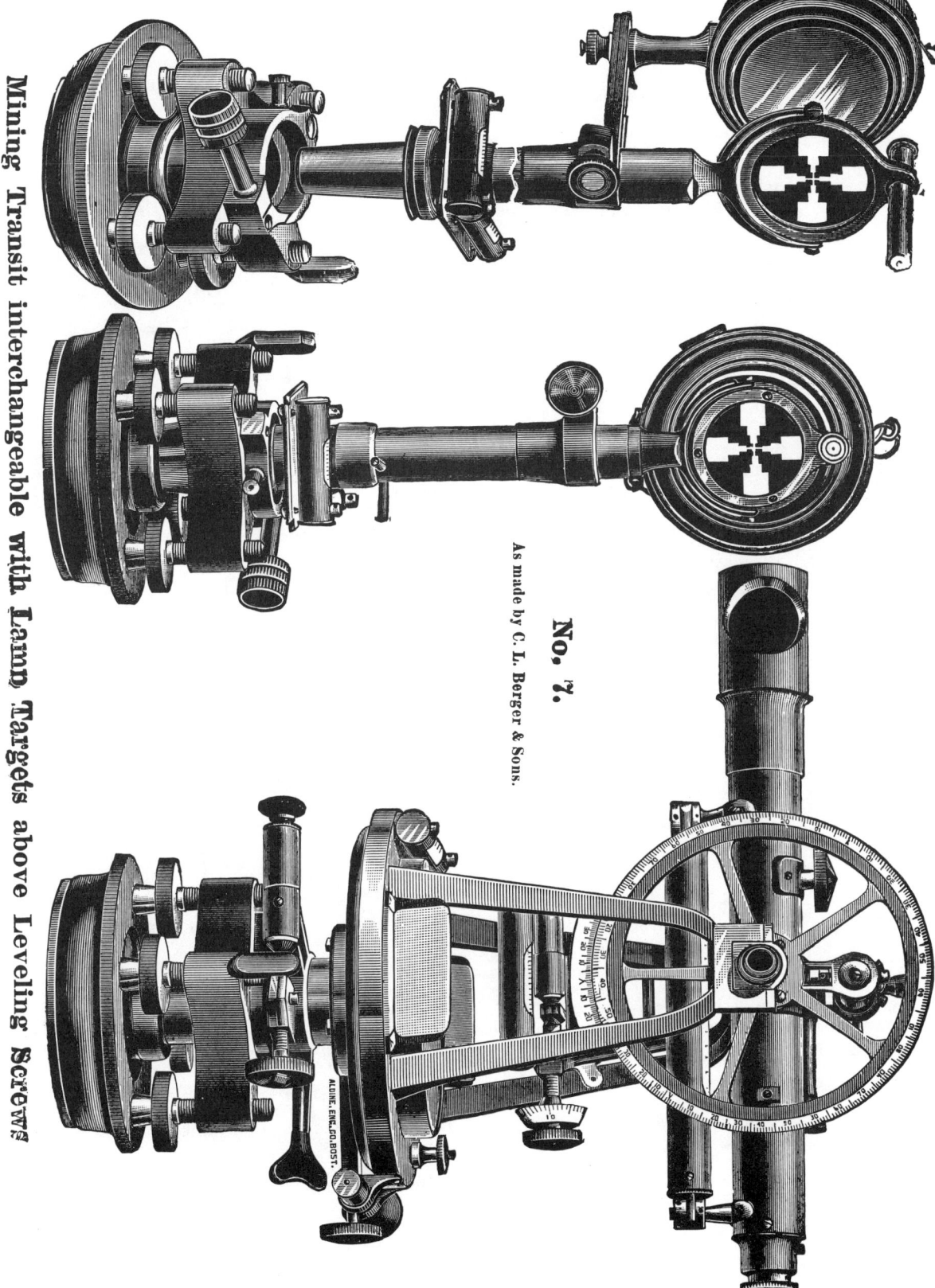

Mining Transit interchangeable with Lamp Targets above Leveling Screws

No. 7.

As made by C. L. Berger & Sons.

C. L. Berger & Sons' Universal Mining Transit with Duplex Telescope Bearings.

This instrument, represented by the annexed cuts, was designed by us June 10, 1889, in response to an urgent demand upon us to construct an instrument which could be depended upon to give the closest results under the most trying circumstances. It was designed to do accurate work in measuring horizontal angles between points, one of which may be depressed as much as eighty or ninety degrees below the horizon, while the other may be as much elevated above the horizon; and also to measure with equal accuracy angles of elevation or depression above or below the horizon. It was planned to meet the requirements of the Mining Engineer, who must have the exact location of every shaft and tunnel in a mine, the length of which may aggregate a number of miles, and necessarily the instrument must be so constructed, that it may be portable, as light in weight as consistent with the end to be accomplished, adjustable by the engineer at the bottom of a mine, and its delicate parts so protected from the dripping of water from roof of tunnel or from the shafts, that they may not be injured. The instrument having but one telescope, is as simple in construction as one with a universal adaptation can be made. It is necessarily a little crowded in order to be as strong, as compact, and as light as possible. To produce the last result, *lightness*, not only the frame of the standards, the vertical circle, its verniers and vernier frame, etc., but also all the detached parts are made of aluminum (see article "Aluminum for Instruments of Precision," page 27). No attempt at elaborate finish of the exterior surfaces of these parts has been made, as it would require a different design at the expense of simplicity, lightness, strength and general efficiency. For the most part they are treated with our cloth finish, or they will be bronzed in black or green. The form of standard is of a unique design: two arms reach out from its base and from the usual bearings, and offer an excentric bearing for the support of the telescope when it may become necessary for the engineer to direct the line of sight down or up a shaft, thus affording two bearings for the horizontal axis of revolution. One of these may be called the normal, the other the excentric bearings. To accomplish the best results in stiffness and solidity these standards are cast in one piece. There are protection clasps over the bearings of both the normal and excentric standard. Those over the excentric bearings may be omitted if so desired. A counterpoise is to be used when the telescope is in the excentric bearings. The striding level furnished with this instrument is of a most sensitive character. In cases, however, where this instrument is intended for ordinary good work only, such as could be accomplished with Mining Transits Nos. 5 or 6 when provided with a side telescope, a striding level as described on page 56, resting on special collars between the standards, can be supplied.

In order to afford increased steadiness, the extension tripod furnished with this instrument is larger than usual for its size, and the vertical centers are of the same length and diameter as those in our transits No. 1. The lamp targets (if any are ordered) are of the pattern shown in the cut. They are interchangeable with the transit on the *tripods*. The telescope should be inverting for best results.

Weight of instrument, standard frame of aluminum, about 11 lbs.
" counterpoises . . . . " 4 "
" compass, frame of aluminum, " $1\frac{1}{2}$ "
" striding level, " " " " $1\frac{1}{4}$ "
" one lamp target . . . . " $\frac{1}{2}$ "
" one tripod about 12 lbs.; two tripods " 24 "
" mahogany box, containing instrument and its attachments, etc., . about 30 lbs.
" " " " one lamp target and one plumb . . " 8 "
Gross weight of instrument, complete, packed securely for shipment in three boxes " 100 "

No. 8. Universal Mining Transit, as in cuts.—Horizontal and vertical circles, 5 inches; graduations on solid silver, double opposite verniers reading to minutes, etc.; 5-inch level to telescope; 3-inch level to vertical circle; 5-inch striding level; two plate levels. Telescope 9 inches, aperture $1\frac{3}{8}$ inches if erecting, and $1\frac{1}{4}$ inches if inverting; power 18 diameters; fixed stadia wires; illuminator shade; prism. Magnetic needle $3\frac{3}{4}$ inches. One lamp target, packed in separate box. Two extension tripods, two plumb bobs, etc.

Price, complete as above, $620.00

Price of this instrument, without lamp target, tripod, and plumb-bob, $90.00 less.
" " " " detachable compass, . . . 40.00 "
" " " " protection clasps to excentric bearings, 10.00 "
" " " with striding level resting on special collars between
the standards, see page 148, . . . 20.00 "

No. 8a. Universal Mining Transit as in **No. 8**, but with 2 lamp targets (packed in one box), 3 extension tripods, and 3 plumb bobs. **Price $710.00**

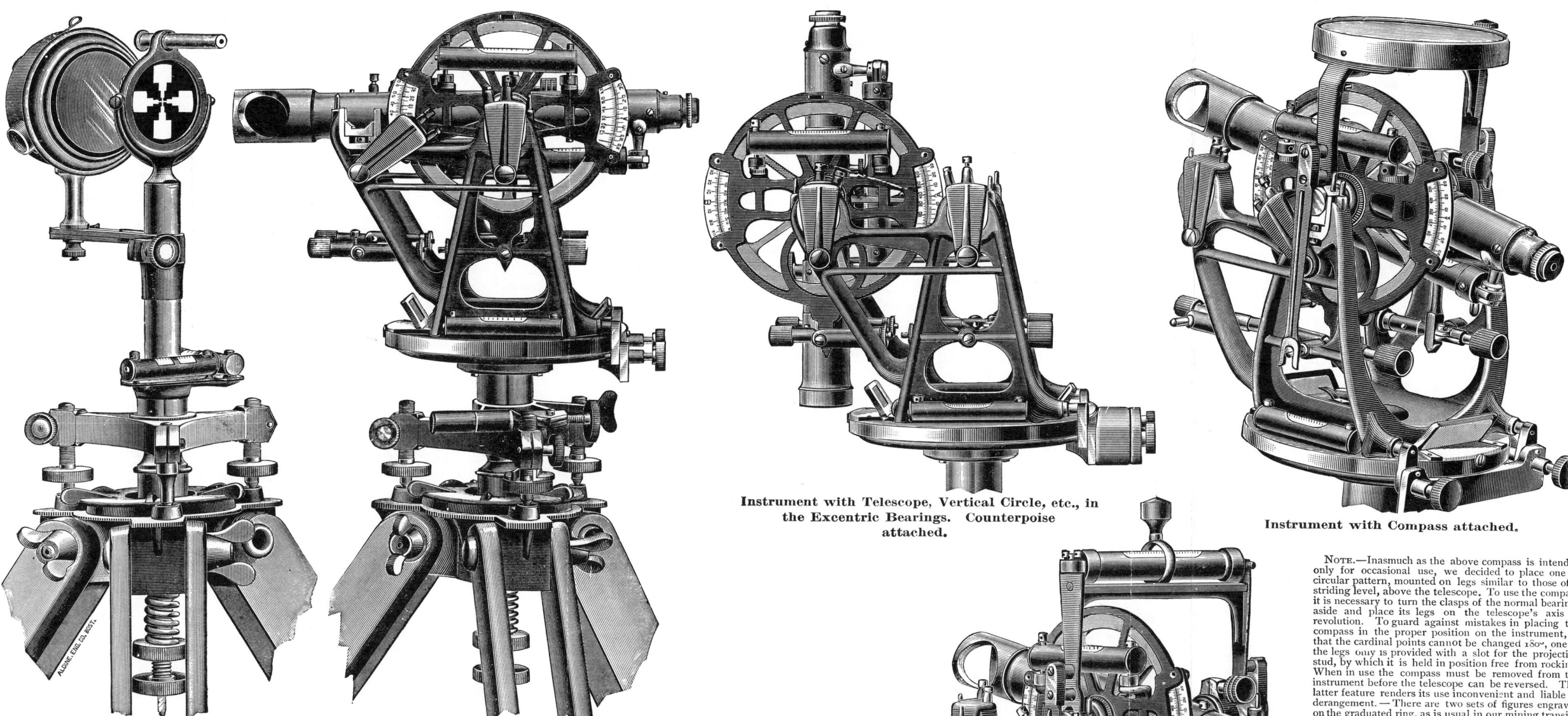

Lamp Target.

Interchangeable with Transit above Tripod Heads.

As made by C. L. Berger & Sons.

Instrument with Telescope in Normal Bearings.

Instrument with Telescope, Vertical Circle, etc., in the Excentric Bearings. Counterpoise attached.

Instrument with Compass attached.

Instrument with Striding Level attached.

[Patented.]

NOTE.—Inasmuch as the above compass is intended only for occasional use, we decided to place one of circular pattern, mounted on legs similar to those of a striding level, above the telescope. To use the compass it is necessary to turn the clasps of the normal bearings aside and place its legs on the telescope's axis of revolution. To guard against mistakes in placing the compass in the proper position on the instrument, so that the cardinal points cannot be changed 180°, one of the legs only is provided with a slot for the projecting stud, by which it is held in position free from rocking. When in use the compass must be removed from the instrument before the telescope can be reversed. This latter feature renders its use inconvenient and liable to derangement.— There are two sets of figures engraved on the graduated ring, as is usual in our mining transits. One set running from 0 to 90° and back, the other set from 0 to 360°.—To place an oblong compass directly above the telescope's axis of revolution and parallel with the telescope (the only place where one could be put) would increase the weight of the telescope and its counterpoise a great deal, besides being objectionable on account of the constant striking of the needle against the sides, before the instrument is approximately put in the meridian, thereby dulling the pivot.

The style of compass shown above will be supplied only with this instrument. It will be packed in the instrument box,— the size and weight of the latter are considerably increased thereby.

No. 8.

C. L. Berger & Sons' Universal Mining Transit with Duplex Telescope Bearings.

NOTE.—This instrument will be made in one size only and with three leveling screws.

Instructions for using our Universal Mining Transit.

Instrument and lamp target attach to their tripods in the manner described on page 26a. To secure an equal height of lamp target and instrument above tripods a slotted disk, provided with the instrument, must be placed and tightly screwed between the head of the leveling screw and the leg of the instrument's stand on the side where the lower tangent screw is situated. This is done to afford a fixed leg, so that when the instrument is leveled up, it be always of the same height. The lamp also has a fixed leg and two leveling screws and when leveled up is of the same height as the transit. When interchanging, care must be taken not to disturb the tripods, and to place the fixed legs of both instrument and lamp in the circular shaped receptacle provided for them in the sliding piece of each tripod. Instrument and lamp will then be leveled up, each with 2 leveling screws, after which their relative height and position above ground will be again the same as before they were interchanged on their tripods. Lamps and targets must be manipulated as explained under Mining Transit No. 7.

A concentric groove is provided near the bottom of the center part of the trivet of the instrument and lamp, around which a fine wire may be fastened whenever their distance apart must be measured.

All horizontal angles measure from the center of the instrument, whether the telescope is in the normal or in the excentric bearings. All vertical angles measure from the center of the telescope's axis of revolution respectively. The distance between the centers of both bearings must be determined and correction must be applied whenever, in measuring vertical angles, the telescope is in the excentric bearings. A small hole drilled in center on top of the transverse axis serves for centering the instrument under a given point by means of a plumb bob suspended from the roof when the telescope is placed horizontal.

The brass counterpoise permanently fastened to the instrument in front of the main vernier serves to balance the weight of the excentric bearings.

The large counterpoise of lead is to be fastened against the brass counterpoise by means of the milled headed screws, whenever the telescope and its attachments are used in the excentric bearings. The small counterpoise of lead is to be placed on the instrument when the striding level is to be used over the excentric bearings. Both weights are slotted to facilitate this operation.

The large counterpoise shown in the cut is not quite sufficient to balance the weight of the telescope in the excentric bearings, but as the vertical plane of the telescope, in measuring horizontal angles, is controlled by the front plate level or by the striding level, which are not affected by this deficiency in the weight of the counterpoise, and inasmuch as vertical angles are controlled by the Level attached to the vertical circle, i twas deemed best, not to add unnecessarily to the weight of the instrument. A true equipoise of the instrument can be obtained, however, by simply lengthening out the milled headed screws by which the counterpoises are secured to the instrument.

As a rule for the more ordinary purposes the plate levels alone can be depended on, but for very close work in running lines down or up a deep shaft, etc., or when horizontal angles must be measured with the telescope in the excentric bearings, the striding level should be depended on only, and then, in order to eliminate all errors of collimation and inequality, if any, in the pivots of the horizontal axis of revolution, the telescope should also be reversed over its bearings. The telescope can be reversed through the standards in both bearings, or it may be reversed over the bearings, as the case may be. Ordinarily, when in the normal bearings, it should be reversed only through the standards as being more convenient and more accurate.

Both, the normal and the excentric wye adjustment of the telescope should be made with the striding level alone. The latter being very sensitive (1 div. of level=10" of arc), there need be no uneasiness on the part of the engineer should the adjustment under ordinary circumstances be not completed within one or two divisions of its graduated tube, but, of course, as the sights are longer and approach a vertical line, it is of great importance to pay the strictest attention to the adjustments of the striding level and wye bearings of the telescope.

To lift the telescope out of its bearings, the spring bolts situated on top of the protection clasps must first be withdrawn, when the latter can be turned aside. Next withdraw about one-tenth inch the spring bolt of the telescope's clamp and also that of the vernier frame's tangent screw by means of the milled headed nut at the end of each spring box. Now lift the telescope out of its bearings and either reverse over the bearings or insert it in the excentric bearings, as the case may be, and again release to the fullest extent the spring bolts of the tangent screws under operation. This being accomplished, all settings of the telescope in the vertical plane must be made by its clamp and tangent screw alone, and no attention need be paid to the clamp of the vernier frame. It is only when vertical angles must be measured that the bubble of the level situated on the vernier frame must be brought to the center of the tube by means of its tangent screw. Good results for vertical angles may be obtained with this instrument, although the plates may not be leveled up accurately, if due regard is paid only to the fact that the bubble of this level must be placed in the center of its tube before a reading can be made.

To make the adjustment of this level proceed thus:— Place the telescope in the horizontal plane by means of its tangent screw, then move the vernier frames' tangent screw until the zero line of the double verniers, marked A, is in coincidence with the zero line of the vertical circle, and now raise or lower the adjusting screw of this level, as the case may be, until the bubble is in the center of its tube.

It is now supposed that the zero line of the double opposite verniers, marked B, are also in coincidence with that of the vertical circle. If not, the verniers marked B can be moved after releasing the capstan-headed screws, until both zero lines on that side of the vertical circle are also in coincidence. However, this is a very laborious proceeding for those uninitiated in this work, and as it cannot always be made quite exact, owing to the mode of mounting the telescope on its axis, it will be found easiest to eliminate errors of excentricity in the graduation of the vertical circle and verniers by reversing the telescope and taking the mean of the readings. The vertical circle is graduated from 0° to 90° and back, and the verniers are double, so that angles of elevation and depression can be read with ease and dispatch. For further information see page 147.

Tunnel Transit.

Under this heading we wish to say that the general form of the transit-theodolites, described under Nos. **8, 11** and **12,** is best adapted for tunnel engineering. The telescope should be inverting and be provided with a diagonal eye-piece, or a prism. A sensitive striding level should be added to rest on top of the telescope axis to establish the vertical plane correctly.

The EXTRAS in addition to those enumerated under No. **11,** are as follows : —

4½ inch sensitive striding level,	$20.00
Diagonal eye-piece,	12.00
Prism, attachable to eye-piece,	8.00
Lamp, of brass or copper, with ground lens,	7.00
Small table to attach lamp to standard of transit to illuminate wires,	3.00
Small reflector in the center of telescope, holes drilled through the telescope axis for the illumination of wires,	10.00
Plummet lamps,	$15.00, 10.00, 8.00

TUNNEL TRANSITS CONSTRUCTED WITH ECCENTRIC TELESCOPE TO ORDER.

Straight Line Instruments.

For running Straight Lines on the Surface of the Earth and in Tunnels.

No. 9 and No. 10. These Instruments are without circles and graduations. They rest either on three or four leveling screws, and are so arranged as to readily adjust exactly over a given center, after an approximate setting with tripod legs. The telescope axis is of hard bell metal, and reverses in its bearings ; the bearings are adjustable in vertical plane ; striding level is provided with a handle in the center and is highly sensitive. We make two sizes, viz. : —

No. 9. Aperture of object-glass, 1⅜ inches in diameter ; focal length 18 inches ; objects erect or inverted, etc.

Price $190.

No. 10. Aperture of object-glass, 2 inches ; focal length 18 inches ; etc.

Price $270.

With the addition of a glass or spider line micrometer, lamp and reflector for the illumination of the micrometer, both these instruments may be used for taking time.

NOTE.— If desired, an extra striding level can be placed upon two concentric rings of equal diameters parallel to line of collimation of the telescope, thus making a leveling instrument of great power and precision for long sights. This attachment will cost $35.00.

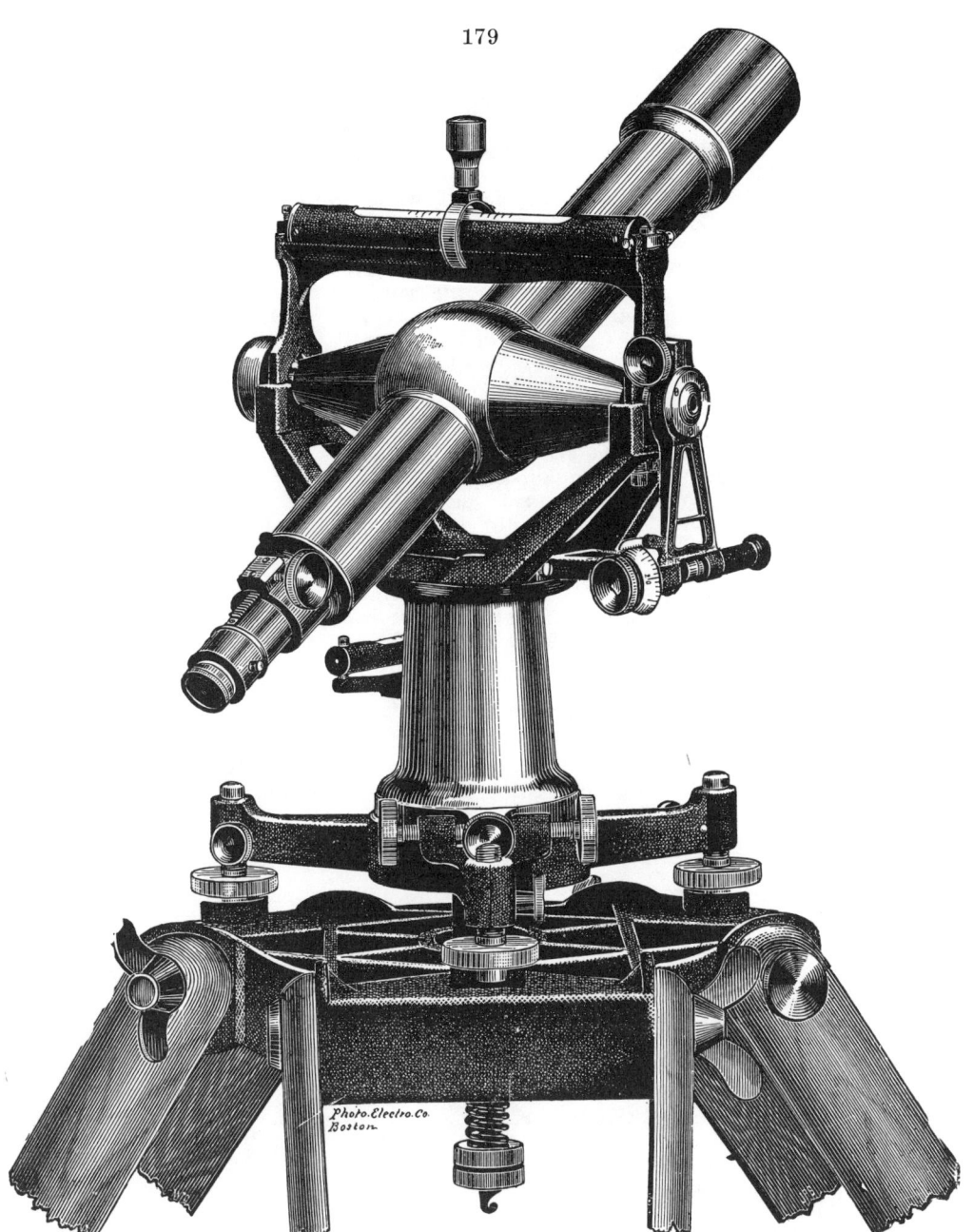

No. 10.

Straight Line Instrument.

With Gradienter Attachment.

As made by C. L. Berger & Sons.

Transit-Theodolite.

For use in cities, in tunnels, and for triangulation.

No. 11. Many years ago we found it desirable to design a transit for use when instruments of ordinary construction would fail to give satisfaction, or do not permit of rapid work where the highest degree of accuracy is required. The instrument as shown in the accompanying illustration has no compass, and, therefore, the upper frame mounting the telescope is in one piece, which is provided with ribs, and which rests directly on the top-flange of the inner center. The result of this is that great lateral strength is obtained, which permits of mounting the telescope axis by means of cylinders at each end in wyes. On top of each bearing of the telescope axis is mounted a cap provided with an adjusting screw, with which the necessary friction for the revolving telescope is obtained, and these caps are also provided with two milled-headed screws, which can be removed readily when the telescope is to be reversed for straight line measurements over the bearings. These caps are so arranged as to completely exclude dust from the axis. In this new arrangement the telescope can be reversed as in ordinary instruments through the standards, as well as over the bearings, as is usual in triangulation and for aligning straight lines; and last, but not least, the movement of the telescope in the vertical plane is the most accurate known. Ordinary transits cannot fulfil these functions owing to the peculiar form of the bearings wherein the telescope axis revolves, which bearings are so made to give lateral stiffness to the telescope, on account of the slenderness of the standards resting on the upper plate. Many of these instruments are in use in the survey and triangulation of our largest cities and have given great satisfaction. The instrument is provided with three or four leveling screws. The verniers can be placed at right angles to the line of sight, or as shown in the cut. The dimensions, etc., are:—

Horizontal plate, 6¼ inches, *graduation on solid silver* protected as in our regular engineers' instrument; two double opposite verniers, reading to 30″; two rows of figures in opposite directions; long compound centers; 11-inch telescope; (inverting or erect); 1¼ inches clear aperture; power 24 diameters; *protection to object-slide;* single spring tangent screws for the upper and lower plates; 4 leveling screws; shifting center, split-leg tripod, case, etc.

Weight of instrument with a 7-inch circle, and all complete, 14 lbs.; weight of tripod 7½ lbs.

Price, for Plain Transit-Theodolite, $240.00.

Extras to Plain Transit-Theodolite.

For Extras to Tunnel Transits, see Extras to Mining Transits.

Horizontal limb 6¼ inches in diameter, verniers reading to 20″*				. .	$10.00
" " 7 " " " 10″*				. . .	35.00
Vertical arc 5 " " " minutes,				. .	20.00
Full 5 inch vertical circle,		" " "			25.00
" " " " " double opposite " " "					50.00
Two reading glasses, with ground glass shades to verniers,				. . .	15.00
Instrument provided with 3 leveling screws, as in cut,				. . .	10.00
†Shifting center for instrument with 3 leveling screws (improved)				. .	5.00
Stadia wires, fixed,	. . . . .		. . . .		3.00
" " adjustable,	. . . .		. . . .		10.00
Striding level,**	. . . . . .		. . . .		20.00
6 inch spirit-level with reversible clamp, tangent and *gradienter* to telescope,					40.00
Oblong compass, with motion for setting off the variation. (Three-inch needle reads only a few degrees each way from zero.)				. . . .	20.00

*These graduations should always be ordered with reading glasses and ground glass shades *attached* to the instrument.

**Unless ordered to the contrary the striding level of this instrument will not rest directly over the points of contact in wyes as represented in Transit No. 12. For greater convenience it rests on the telescope axis at points between the standards, which has the advantage that the striding level can be left on the axis while the telescope is revolved in the vertical plane. A striding level should always be ordered with an instrument of the class described above.

Note.—A Plain Transit-Theodolite is without a level, clamp and arc to telescope.

—For increase of aperture in inverting telescopes see note to Transit No. 1, page 142.

†Note. — For description, use, etc., of our shifting center for a transit with three leveling screws, see manual.

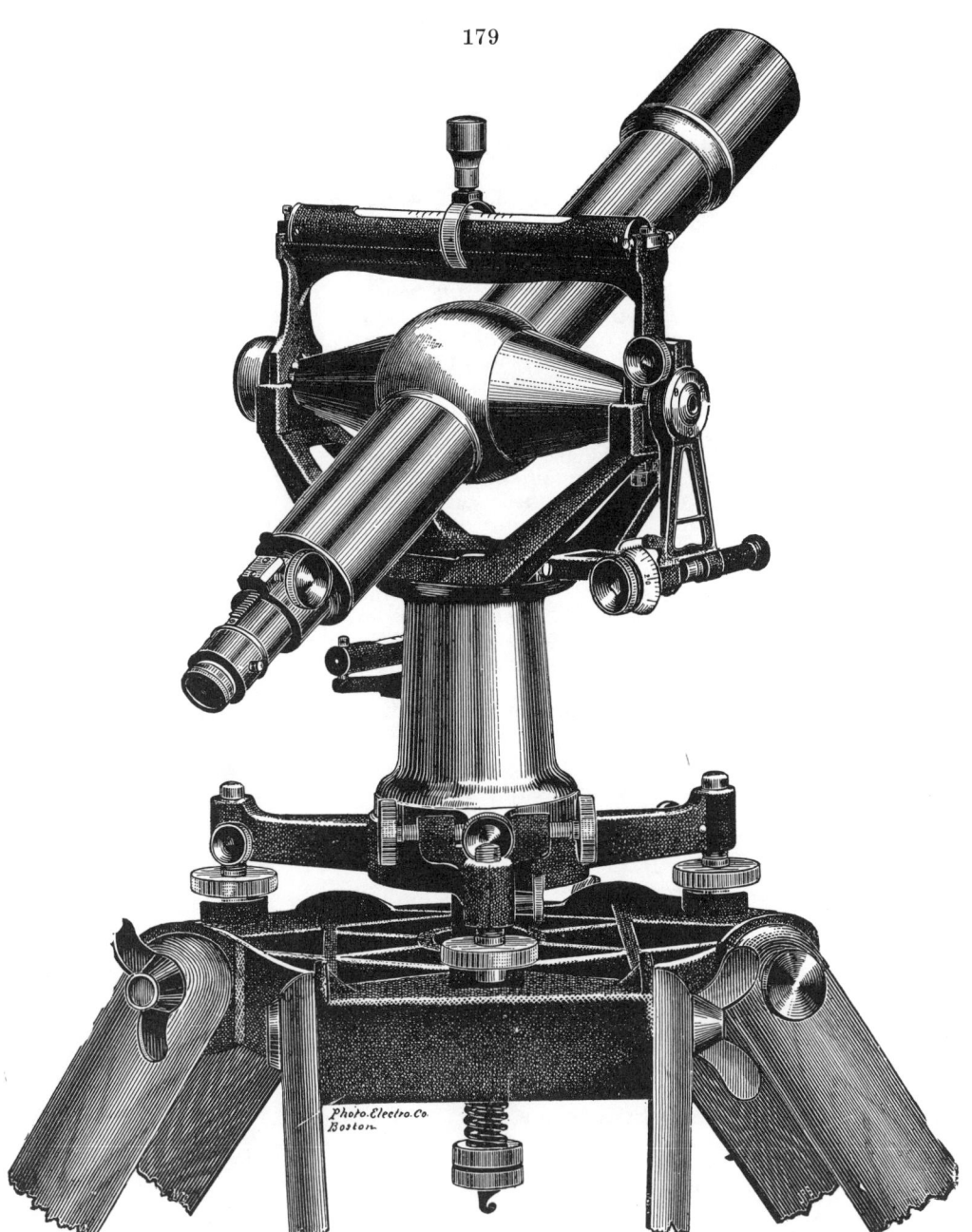

No. 10.

Straight Line Instrument.

With Gradienter Attachment.

As made by C. L. Berger & Sons.

Transit-Theodolite.

For use in cities, in tunnels, and for triangulation.

No. 11. Many years ago we found it desirable to design a transit for use when instruments of ordinary construction would fail to give satisfaction, or do not permit of rapid work where the highest degree of accuracy is required. The instrument as shown in the accompanying illustration has no compass, and, therefore, the upper frame mounting the telescope is in one piece, which is provided with ribs, and which rests directly on the top-flange of the inner center. The result of this is that great lateral strength is obtained, which permits of mounting the telescope axis by means of cylinders at each end in wyes. On top of each bearing of the telescope axis is mounted a cap provided with an adjusting screw, with which the necessary friction for the revolving telescope is obtained, and these caps are also provided with two milled-headed screws, which can be removed readily when the telescope is to be reversed for straight line measurements over the bearings. These caps are so arranged as to completely exclude dust from the axis. In this new arrangement the telescope can be reversed as in ordinary instruments through the standards, as well as over the bearings, as is usual in triangulation and for aligning straight lines; and last, but not least, the movement of the telescope in the vertical plane is the most accurate known. Ordinary transits cannot fulfil these functions owing to the peculiar form of the bearings wherein the telescope axis revolves, which bearings are so made to give lateral stiffness to the telescope, on account of the slenderness of the standards resting on the upper plate. Many of these instruments are in use in the survey and triangulation of our largest cities and have given great satisfaction. The instrument is provided with three or four leveling screws. The verniers can be placed at right angles to the line of sight, or as shown in the cut. The dimensions, etc., are: —

Horizontal plate, 6¼ inches, *graduation on solid silver* protected as in our regular engineers' instrument; two double opposite verniers, reading to 30″; two rows of figures in opposite directions; long compound centers; 11-inch telescope; (inverting or erect); 1¼ inches clear aperture; power 24 diameters; *protection to object-slide;* single spring tangent screws for the upper and lower plates; 4 leveling screws; shifting center, split-leg tripod, case, etc.

Weight of instrument with a 7-inch circle, and all complete, 14 lbs.; weight of tripod 7½ lbs.

Price, for Plain Transit-Theodolite, $240.00.

Extras to Plain Transit-Theodolite.

For Extras to Tunnel Transits, see Extras to Mining Transits.

Horizontal limb 6¼ inches in diameter, verniers reading to 20″* . . $10.00
    "   " 7   "   "   " 10″* . . . 35.00
Vertical arc   5   "   "   " minutes, . . 20.00
Full 5 inch vertical circle,   "   "   " 25.00
  " " "   " " double opposite   "   "   " 50.00
Two reading glasses, with ground glass shades to verniers, . . . 15.00
Instrument provided with 3 leveling screws, as in cut, ... 10.00
Shifting center for instrument with 3 leveling screws (improved) . . 5.00
Stadia wires, fixed, ... 3.00
  "   " adjustable, ... 10.00
Striding level,** 20.00
6 inch spirit-level with reversible clamp, tangent and *gradienter* to telescope, 40.00
Oblong compass, with motion for setting off the variation. (Three-inch
    needle reads only a few degrees each way from zero.) 20.00

*These graduations should always be ordered with reading glasses and ground glass shades *attached* to the instrument.

**Unless ordered to the contrary the striding level of this instrument will not rest directly over the points of contact in wyes as represented in Transit No. 12. For greater convenience it rests on the telescope axis at points between the standards, which has the advantage that the striding level can be left on the axis while the telescope is revolved in the vertical plane. A striding level should always be ordered with an instrument of the class described above.

Note.—A Plain Transit-Theodolite is without a level, clamp and arc to telescope.

—For increase of aperture in inverting telescopes see note to Transit No. 1, page 142.

†Note. — For description, use, etc., of our shifting center for a transit with three leveling screws, see manual.

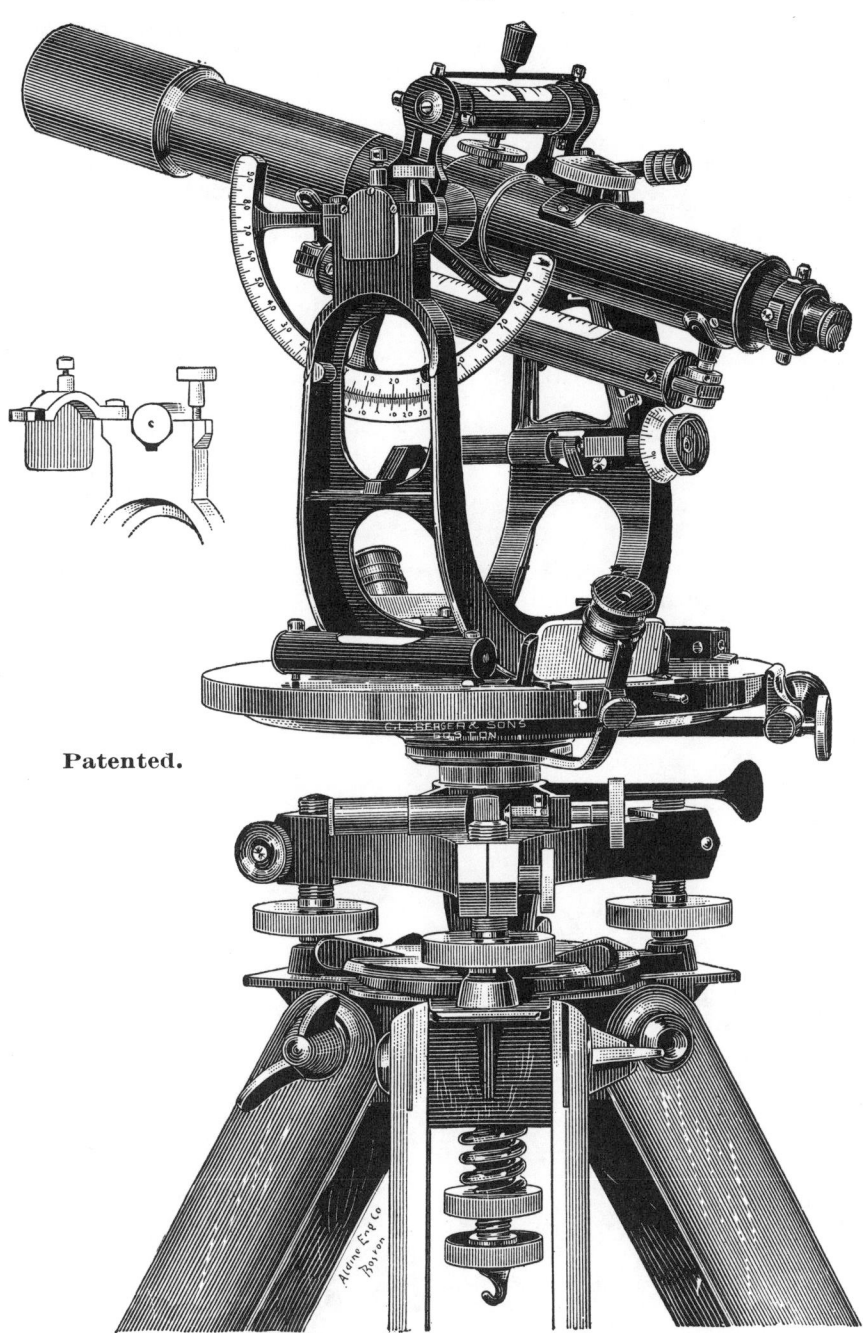

Patented.

No. 11.

Complete Transit-Theodolite.

For use in cities, in tunnels, and for triangulation.
As made by C. L. Berger & Sons.

If a fine performing instrument is desired at minimum expense, the **U** shaped standard frame carrying the telescope, instead of being finely polished and lacquered, can be **cloth finished.** Price of instrument so treated . . . . **less $15.00**

No. 11 a. Complete Transit-Theodolite, with striding level attachment and vertical circle as in cut. (Lower half of instrument as shown in cut of Complete Transit-Theodolite No. 11.)

Instrument has three levelling screws with shifting center; 6¼-inch horizontal circle, graduation on solid silver reading to 20″; two rows of figures from 0 to 360° in opposite directions; 5-inch vertical circle; double opposite verniers reading to minutes; two reading glasses and glass shades for each circle; two plate levels; 11½-inch inverting telescope, aperture 1¼ inches, power 27 diameter; 4½-inch striding level to telescope with reversible clamp and gradienter to telescope; 4½-in. transverse striding level as in No. 11 b; fixed stadia wires.

Weight of instrument about 15 lbs.; weight of tripod 10 to 12 lbs.

Price, as above, $445.00

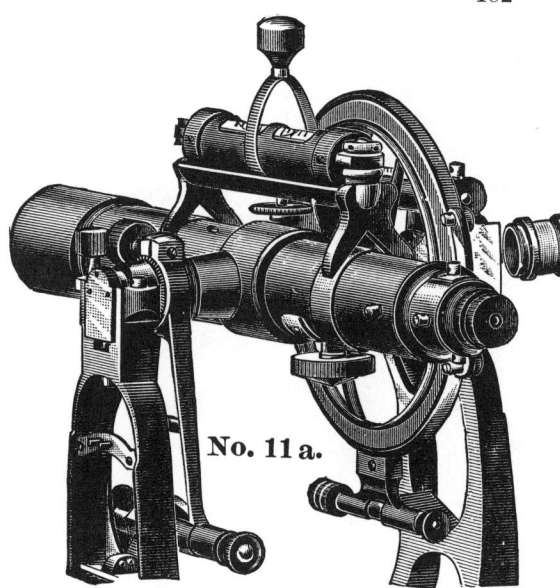

As made by C. L. Berger & Sons.

No. 11 a.

NOTE.— In place of a fixed level the above instrument has a striding level parallel to line of collimation, which can be mounted on the telescope whether it is in a direct or in a reversed position. This attachment renders the adjustment of the telescope for leveling nearly as simple in the Transit as in the Wye level. On the other hand a fixed level to the telescope commends itself for this class of instruments on account of *greater compactness, simplicity and readiness for use, and when properly adjusted it is quite as accurate and more apt to stay in adjustment than the striding level with its greater number of pieces and adjustments, not to speak of its greater liability to injury.*

No. 11 b,

As made by C. L. Berger & Sons.

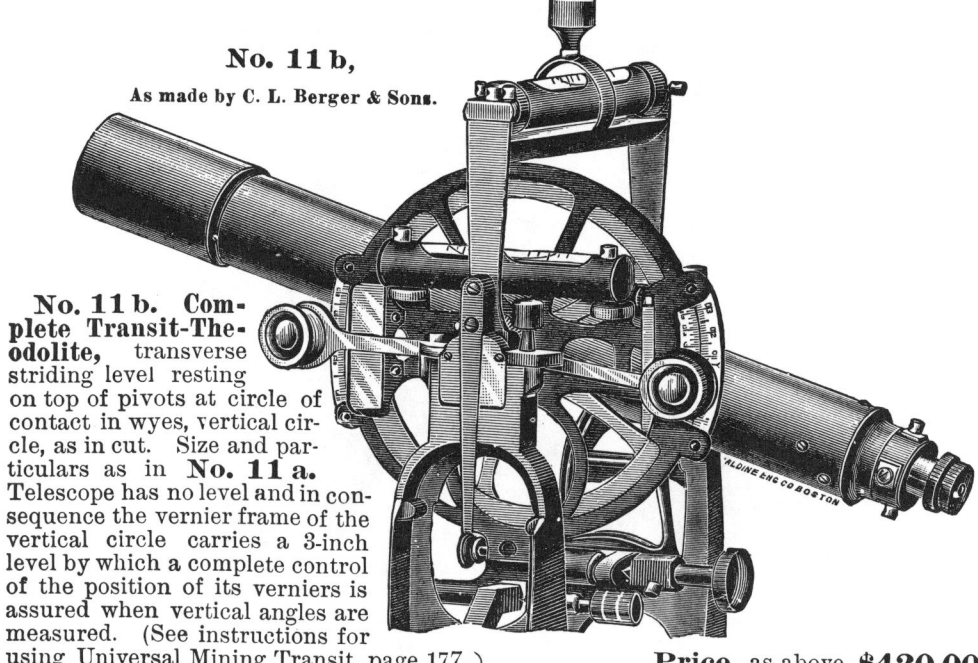

No. 11 b. Complete Transit-Theodolite, transverse striding level resting on top of pivots at circle of contact in wyes, vertical circle, as in cut. Size and particulars as in No. 11 a. Telescope has no level and in consequence the vernier frame of the vertical circle carries a 3-inch level by which a complete control of the position of its verniers is assured when vertical angles are measured. (See instructions for using Universal Mining Transit, page 177.)

Price, as above, $420.00

No. 11 a and No. 11 b, 7-inch horizontal limb graduated to read to 10″,	Price, extra,	$25.00
No. 11 b, provided with a 4½-inch striding level to telescope,	" "	40.00
No. 11 b, provided with a 6-inch fixed level to telescope,	" "	20.00
No. 11 a, 3-inch level to vertical circle, as in No. 11 b,	" "	10.00
No. 11 a and No. 11 b, without reading glasses to vertical circle,	Less	10.00
No. 11 a and No. 11 b, without a striding level for the telescope axis,	"	35.00

NOTE.— If desired, a striding level as in No. 11 a, with reversible clamp and gradienter to telescope can be attached to our **Plain Transit-Theodolite, No. 11,** in place of a fixed level. **Price, $60.00**

No. 12.

8-inch Transit for Triangulation.

As made by C. L. Berger & Sons.

No. 12. The form of frame chosen for mounting the telescope is as in the cut, which permits the reversal of the telescope through the standards, as well as over the bearings.

The inverting telescope has a clear aperture of 1½ inches, and is of 11 inches focal length, and has a power of 20, 25 or 30 diameters as ordered. It is provided with a 5 inch vertical arc, graduated to read to minutes, a 4½ inch striding level to rest on the pivots of its axis, a 6 inch spirit-level (parallel to line of collimation) together with a reversible clamp and tangent screw. The horizontal circle is 8 in. in diameter, opposite verniers reading to 10″. Its graduation is open, but it is protected by a rim which is raised above the limb. This instrument has three leveling screws which have a larger base than is usual, and the tripod-head being proportionately large the instrument has great stability.

Weight of instrument 16 lbs.; weight of tripod 12 lbs.

Price, as above, $450.

This instrument without arc and clamp to telescope, $50.00 less.

Transit - Theodolite.

As made for the U. S. Corps of Engineers.

Two of these instruments, shown in the accompanying engraving, were designed and constructed for use in the Geographical Exploration and Survey West of the one-hundredth meridian, and exhibited at the Centennial, in the United States Government Building, by Lieut. George M. Wheeler.

No. 13 and **No. 14** represent this instrument as used in the field for triangulation, and as used for astronomical observations. It is designed to combine in a portable form of construction, the efficiency for field use usually obtained with the larger classes of instruments.

Horizontal limb is 8 inches in diameter, opposite verniers reading to 10″; vertical circle 5 inches in diameter, opposite verniers reading to 20″; object-glass 1½ inches clear aperture; focus 11 inches; powers of two direct eye-pieces respectively 30 and 40 diameters; power of diagonal eye-piece 40 diameters; spirit-level attached to telescope and striding level capable of reading to seconds of arc; three leveling screws; split-leg tripod; low standards are *cast* on vernier-plate; two *extra* standards for astronomical observations; 3½ inch needle; round level for vernier-plate; spider-line micrometer; instrument is ribbed throughout; lamp, arm and adjustable plane reflector; sunshade; adjusting pins; case and strap.

The weight of whole instrument is 14½ lbs.; weight of tripod 8½ lbs.

Price $650.

Theodolite.

No. 14. Price of instrument, same in size as that above, arranged for triangulation only, without extra standards, diagonal eye-piece, etc. . **425.**

No. 14.

No. 13.

Astronomical Transit-Theodolite.

As made by C. L. Berger & Sons.

No. 15.
Alt.-Azimuth.

As made by C. L. Berger & Sons.

Repeating horizontal circle eight, non-repeating vertical circle 6 inches in diameter. The former can be provided with 2, 3 or 4 verniers reading to 10″, the latter is provided with 2 verniers reading to 20″. The telescope has a clear aperture of 1½ and a focus of 11 inches. Striding level and filar micrometer read to seconds of arc. Mahogany box, etc.

Price, all complete as in cut, $680.00.

No. 15 a.*

Alt.-Azimuth.

Alt.-Azimuth, *as in cut.* Graduations of 5⅛ inch circles on solid silver, **two** opposite micrometer-microscopes for each circle reading to 10″, and by estimation to 2″. Both circles can be shifted, so as to bring different parts of the graduation under the micrometer-microscopes. The telescope is 10 inches long, has an aperture of 1¼ inch and a power of 24 diameters. Telescope is provided with a level on top and with 3 horizontal wires for leveling and for stadia measurements, and if desired with 5 vertical wires for star observation. The telescope must be reversed in its bearings by hand. Telescope axis is of hardened steel. The striding and microscope levels read to 5″ of arc. Two ordinary small levels attached to the instrument serve to place it in an approximate horizontal position. Complete in box.

Price, as above, $580.00

---

* See Preface.

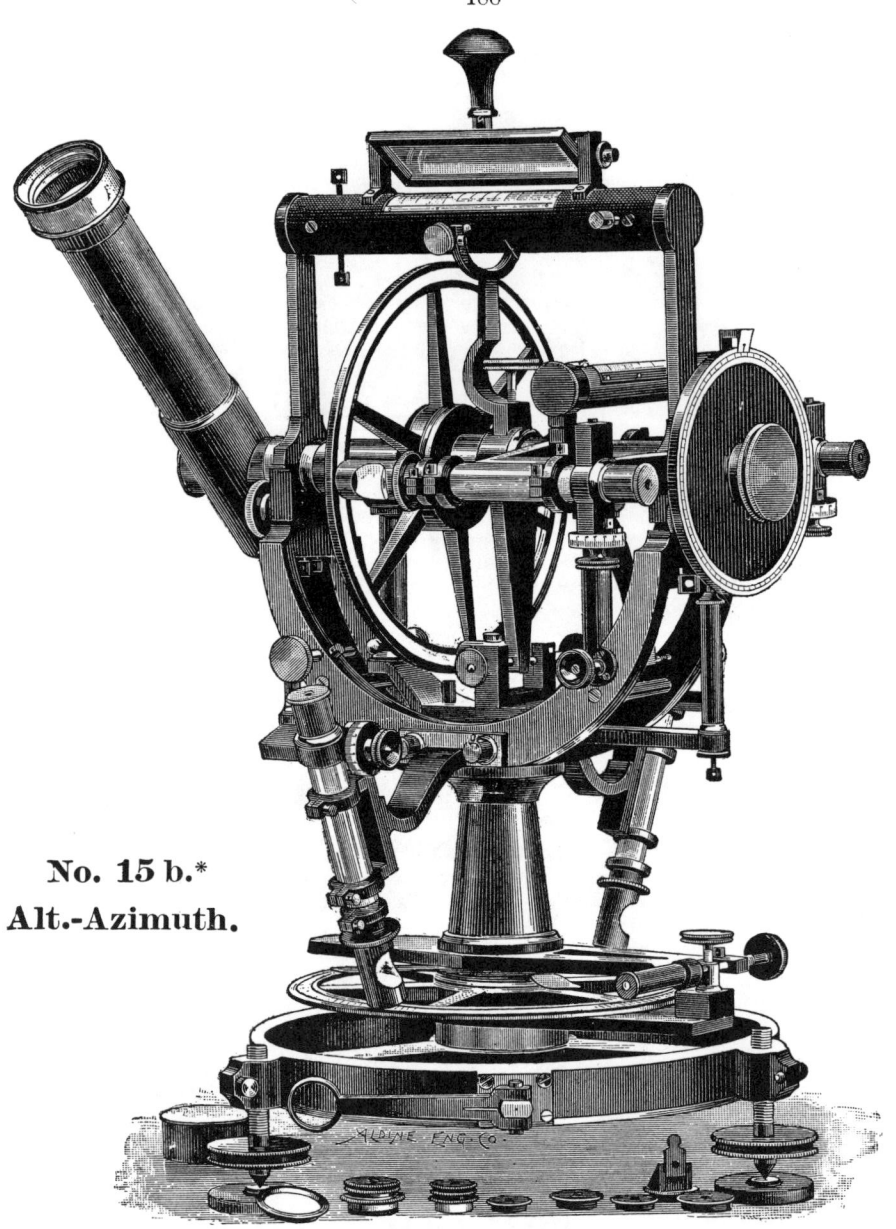

No. 15 b.*
Alt.-Azimuth.

Alt.-Azimuth, *as in cut.*   Circles 8¼ inches diameter, micrometer-microscopes reading to 5 seconds direct, and by estimation to single seconds.   Telescope, 1.6 in. aperture; focal length, 16½ inches; power, 32 and 48.   Telescope axis is of hardened steel and balanced by friction rollers.   Reversing apparatus.   Complete in one box.

Price, as above, $920.00

This instrument without reversing apparatus, ...   less, $100.00

No. 15 c,* **Alt.-Azimuth,** *as in cut above.*   Circles 10¼ inches diameter, micrometer-microscopes reading to single seconds direct.   Every single degree figured. Telescope, 1¾-inch aperture; focal length, 20¼ inches; power, 40 and 60.   Telescope axis is of hardened steel and balanced by friction rollers.   Complete in two boxes.

Price, as above, $1300.00

* See Preface.

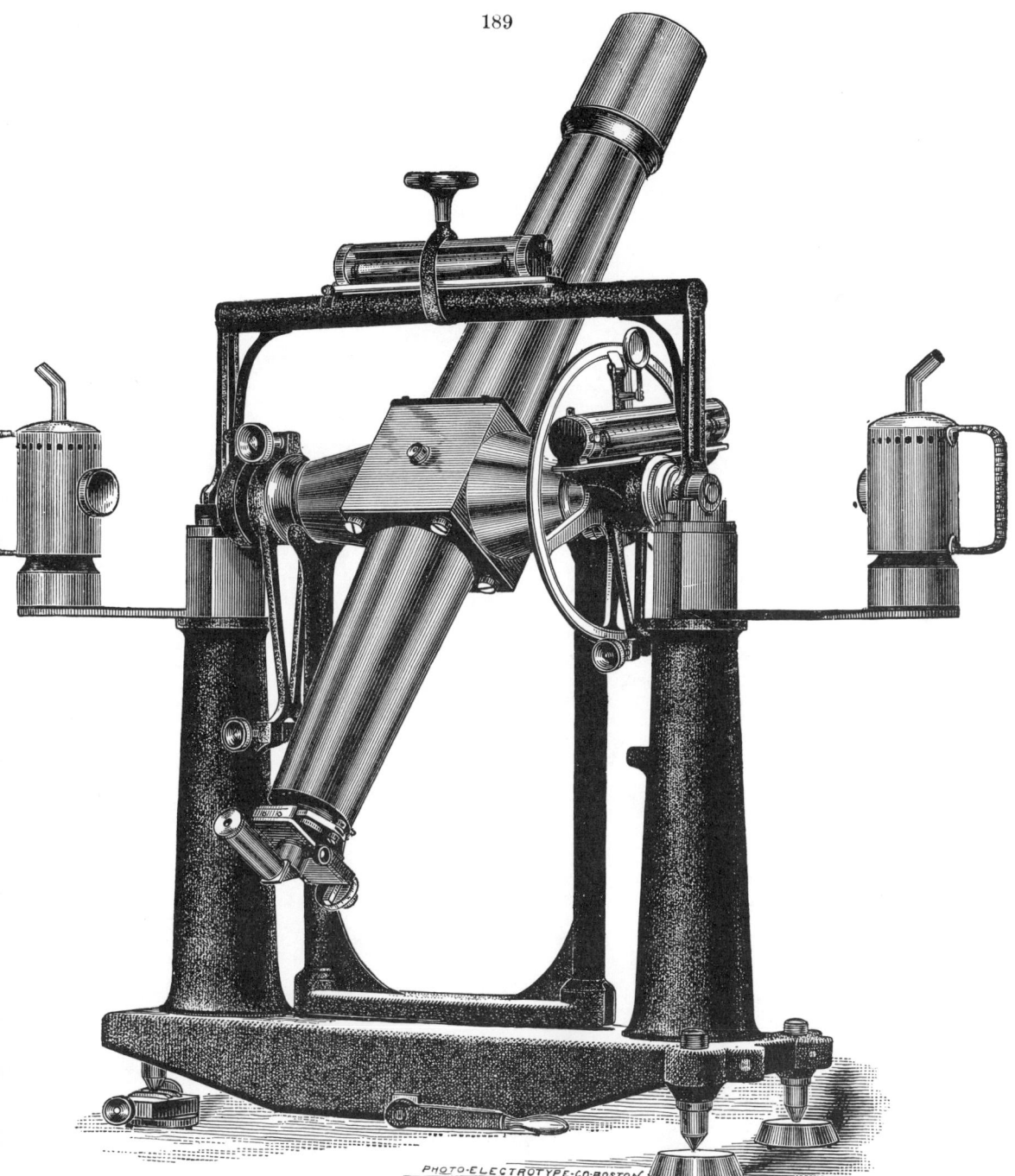

Portable Transit Instrument for Latitude Observations.

As made by C. L. Berger & Sons.

No. 16. Aperture of object-glass 3 in.; focus 28 in.; spider-line or glass micrometer; micrometer screw reads to seconds of arc; spirit-levels read to seconds of arc; diagonal eye-piece 80 dia.; Ramsden eye-piece 40 dia.; vertical circle 8 in. in dia.; bell-metal pivots, two lamps and arms, adjustable reflector; reversing apparatus; two cases, etc.

Price $980.

Portable Astronomical Transit Instrument.

No. 17.   Aperture of object-glass 3 in.; focus 39 in.; spider-line or glass micrometer; diagonal eye-piece magnifies from 90 to 120 dia.; Ramsden eye-piece magnifies 75 dia.; striding level reads to seconds of arc; adjustable mirror to read the level from below: reserve level; pivots of hardened steel; small adjustable plane reflector; two lamps and arms; reversing apparatus; two finding circles each provided with double verniers; cast-iron frame rests on three leveling screws of steel, which are provided with foot-plates — one of them is adjustable to set instrument in the meridian; two cases, etc.

Price $1300.

(Notice of this Instrument, with full description, in Johnson's New Universal Cyclopædia, under article "Transit.")

No. 17

Astronomical Transit Instrument.

As made for U. S. Lake Survey.

Equatorials.*

No. 1. Portable Equatorial Telescope, as in cut. Cast iron pillar; clock; prismatic illuminating arrangement; tangent screw motion brought down to the eye-end. The telescope rests in a cradle-piece to which it is firmly attached by two brass clasps. The telescope-tube is of brass, polished, and provided with rack and pinion adjustment to focus; finder; five astronomical eye-pieces 60, 120, 200, 300 and 400; one solar eye-piece and one terrestrial pancratic eye-piece. The declination circle is graduated on silver, two opposite verniers reading to minutes. The hour circle is graduated on silver and has two sets of graduations and verniers.

Price, as above, $1350.00

No. 2. Fixed Equatorial Telescope. Aperture, 5 inches. Cast iron pillar. Telescope is made of brass, tapering towards both ends. Rack and pinion motion to eye-end; finder; five astronomical eye-pieces 60, 120, 200, 300 and 400; first surface reflecting prism for viewing the sun; diagonal eye-piece; transit eye-piece; position micrometer at eye-end of telescope, graduated on silver and reading to minutes, with quick and slow motion clamp and tangent screws; large declination circle graduated on solid silver and read by microscope from eye-end, with coarse graduation on edge for rough setting; hour-circle graduated on silver with two sets of graduations; driving clock, which can be changed from sidereal to lunar rate, and additional slow motion in right ascension and declination by means of rods and handles brought down to the eye-end; striding level to determine the horizontal position of the declination axis in order to use the instrument as a transit; prismatic illumination arrangement for micrometer, declination and position circles. Best qualiiy.

Price, as above, $2150.00

No. 3. Fixed Equatorial Telescope, as in No. 2, best quality, but with 6-inch aperture and 6 eye-pieces 35. 85, 155, 240, 360 and 490.

Price, $3150.00

---

* See Preface.

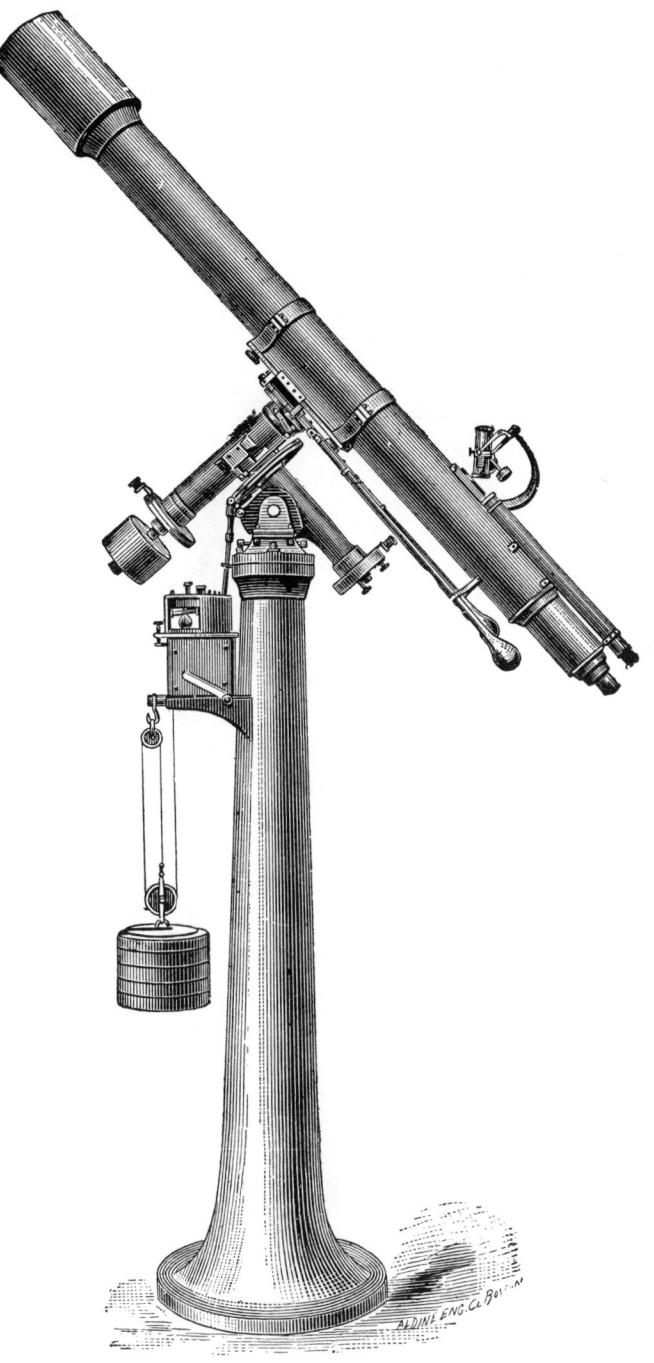

No. 1.

Portable Equatorial Telescope.

Aperture, 5 inches.

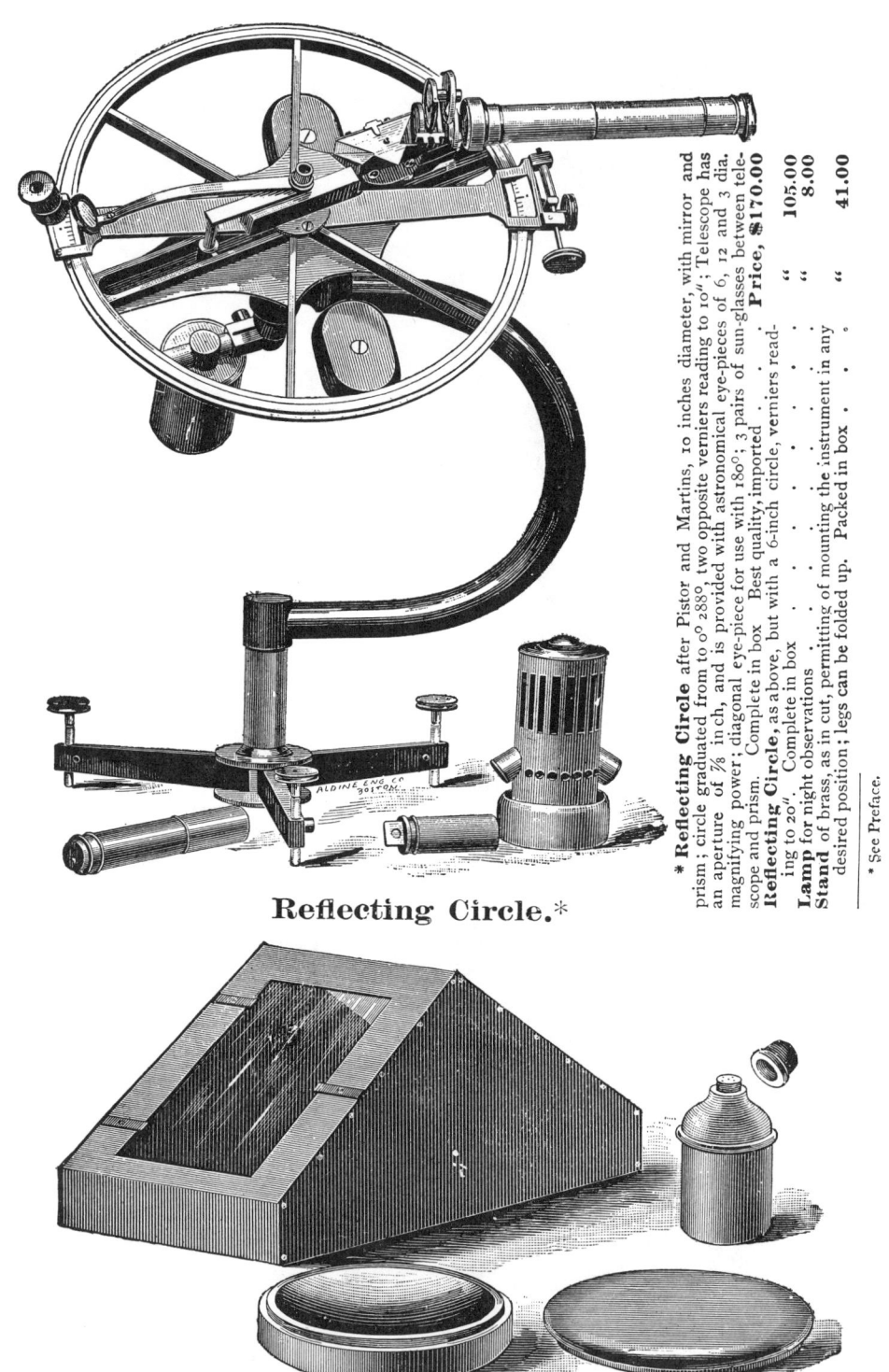

Reflecting Circle after Pistor and Martins, 10 inches diameter, with mirror and prism; circle graduated from to 0° 2880, two opposite verniers reading to 10″; Telescope has an aperture of ⅞ inch, and is provided with astronomical eye-pieces of 6, 12 and 3 dia. magnifying power; diagonal eye-piece for use with 180°; 3 pairs of sun-glasses between telescope and prism.   Complete in box   Best quality, imported . . . . **Price, $170.00**

Reflecting Circle, as above, but with a 6-inch circle, verniers reading to 20″.   Complete in box . . . . " **105.00**

Lamp for night observations . . . . " **8.00**

Stand of brass, as in cut, permitting of mounting the instrument in any desired position; legs can be folded up.   Packed in box . . . . ° **41.00**

* See Preface.

Reflecting Circle.*

Artificial Horizon.*

Mercury Horizon of boxwood, with silver-plated copper bowl; bottle of box-wood for mercury; brass rectangular roof with glass covers made of parallel glass. All complete, packed in a box.   Best quality, imported. . . . **Price, $50.00**

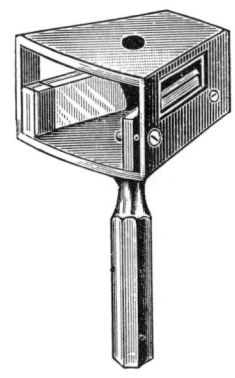
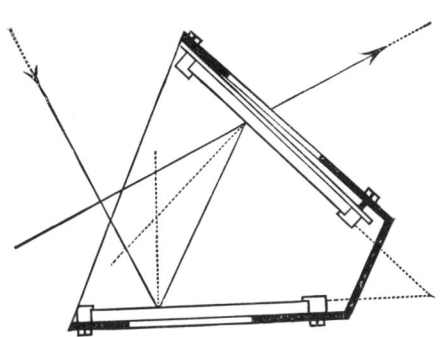

Angle Mirror or Optical Square, ... **$6.00**

See Preface.

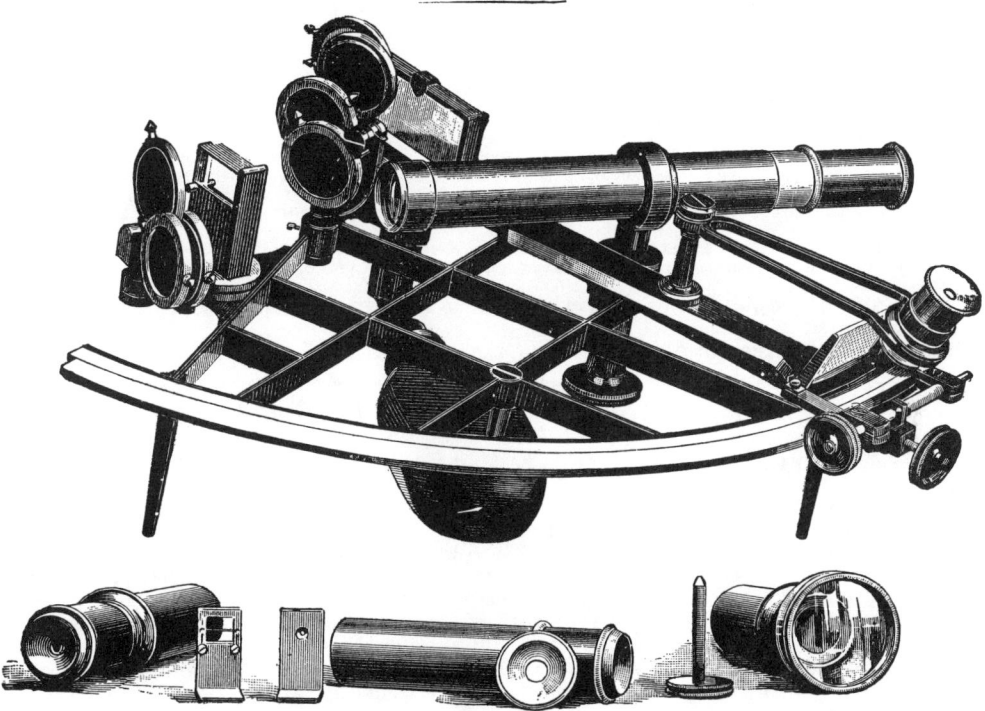

Sextant.*

Sextant. Radius, 7 inches, 145°; four sun-glasses between the large and the small reflecting mirror, and three sun-glasses behind the small reflecting mirror, all of which can be turned on their axis 180°; graduation on solid silver, reading to 10″; telescope ¾ inch aperture; two astronomical eye-pieces with powers of 6 and 10 dia. One Galilean telescope with extra large objective, power 3 dia.; one fixed reading glass; two sights for examination and correction of the large reflecting mirror. All complete in box. Best quality, imported. . . **Price, as above, $130.00**

Sextant, as above. Radius 10 inches, all complete in box. . **Price, 150.00**

Pocket Sextant, best quality . . . . " **43.00**

* See Preface

Current Meters.

The types of current meters, as shown in Figs. I, II, and III, in our former Catalogues, have been omitted, owing to the many improvements made and embodied in the Meter, as shown in Figs. IV, V, and VI, this Catalogue. We are, however, prepared to make to order Current Meter No. III, as designed by Mr. Clemens Herschel, if so desired.

Current Meter No. IV.

The electric form of meter shown in Fig IV is especially adapted for observations upon large rivers, arms of the sea, etc. It has its registering apparatus above the surface of the water, or on the bank of a river, and current measurements may be made with it at any depth, and may be continued for a week, or longer, without stopping, if desired. Half a dozen or more of these meters may be strung on one and the same vertical rod or wire, and *simultaneous* observations then taken of the velocities at different depths below the surface.

This form was used upon the gauging of the Connecticut River * by General Ellis, and was designed particularly to avoid the catching of floating substances, such as leaves and grass, upon either the vanes or the axis, and to render the record of the instrument independent of the position of its axis with respect to the line of the current, also, so as to get less friction upon the axis so as to measure low velocities accurately.

This current meter is constructed upon the principle of Robinson's Anemometer, turning by the difference of pressure upon opposite vanes of the wheel. The vanes of this meter, however, instead of being hemispherical cups with a straight stem, are made conical at the ends, and are hollow and taper to the central hub, so as to offer no obstruction to the slipping off of straws, leaves, or grass, as the wheel revolves. The central hub is made tapering, so that any object can slide off easily, and it extends over the joints at the ends of the axis, so as to enclose and protect them from floating substances. The axis runs in iridium bearings. The forward end of the frame which carries the wheel can be turned and secured in any position, so that the wheel can be horizontal, vertical, or at any desired angle.

The electrical connection is made by carrying an insulated wire from near the center of the instrument, where the insulated wire from the battery is attached to it by a binding screw when in use, out to the end of one arm of the wheel frame, where it ends in a fine platinum wire resting upon a ring in the hub of the wheel. This ring is made of alternate interchangeable sections of silver and hard rubber, secured in place by screws, so that their position can be changed to register whole or part revolutions as desired. There is also a socket and set-screw in the body of the frame near the center, for the return current, which can be carried through a plain wire slightly twisted around the insulated wire so as to form one cord. If the instrument is run upon a wire, or has a metallic connection with the surface, the return current can be made through that. A better method now in vogue is to use a "twin" insulated wire.

The universal motion at the center of the frame and the tail are of the usual construction. This meter can be used in connection with any apparatus for registering the revolutions of the wheel by the breaks in the electric circuit.

Price complete, as in Fig. **IV,** with electric register and one battery etc., packed in three cases, ... **$195.00**
Price of this instrument without electric register and battery . . . **135.00**

* For further information on this point, see Gen'l G. K. Warren's Report of Surveys and Examinations of Connecticut River.

We can have this meter, as well as Nos. V and VI, carefully rated at an additional expense of from $15.00 to $25.00. Unless ordered otherwise, the instruments will be sent unrated.

Current Meter No. V, and No. VI.†

An illustration of this instrument will be found on page 198.

This form of Current Meter is specially adapted for observations upon smaller rivers, streams, conduits, flumes, etc. It is provided with a registering apparatus. For more extended observation upon rivers, etc., an electric register and battery similar to those used with No. IV can be supplied with this instrument. The dial wheels are completely protected by a glass cover, as shown by cut of meters V and VI. The counting mechanism is operated by a string, by means of which the dial wheels are thrown in and out of gear. One short pull on the string throws them in gear, and the succeeding pull will throw them out again ; the next one in, and so on.

Price of Current Meter No. V, supplied only with the ordinary registering apparatus, as shown in the main cut on page 198, and with 12 feet of brass tubing, made in sections of four feet, and graduated in feet and tenths. Complete in two cases, ... **$160.00**

Price of Current Meter No. VI, in all respects similar to that above, but in addition to the ordinary registering apparatus this instrument is provided with an electric register, one battery and copper wire, as shown in the smaller cuts on page 198. Complete in four cases, . . **$220.00**

† For further information on this Current Meter, read "Description of some experiments on the Flow of Water, made during the Construction of Works for Conveying the Water of Sudbury River to Boston," by A. Fteley and F. P. Stearns (Transactions of the Society of Civil Engineers, Jan.–March, 1883). Also, "On the Current Meter, together with a Reason why the Maximum Velocity of Water Flowing in Open Channels is Below the Surface," by F. P. Stearns; a paper read at the Annual Convention of the American Society of Civil Engineers, St. Paul, Minn., June 21, 1883. (Transactions, etc., Vol. XII., August, 1883).

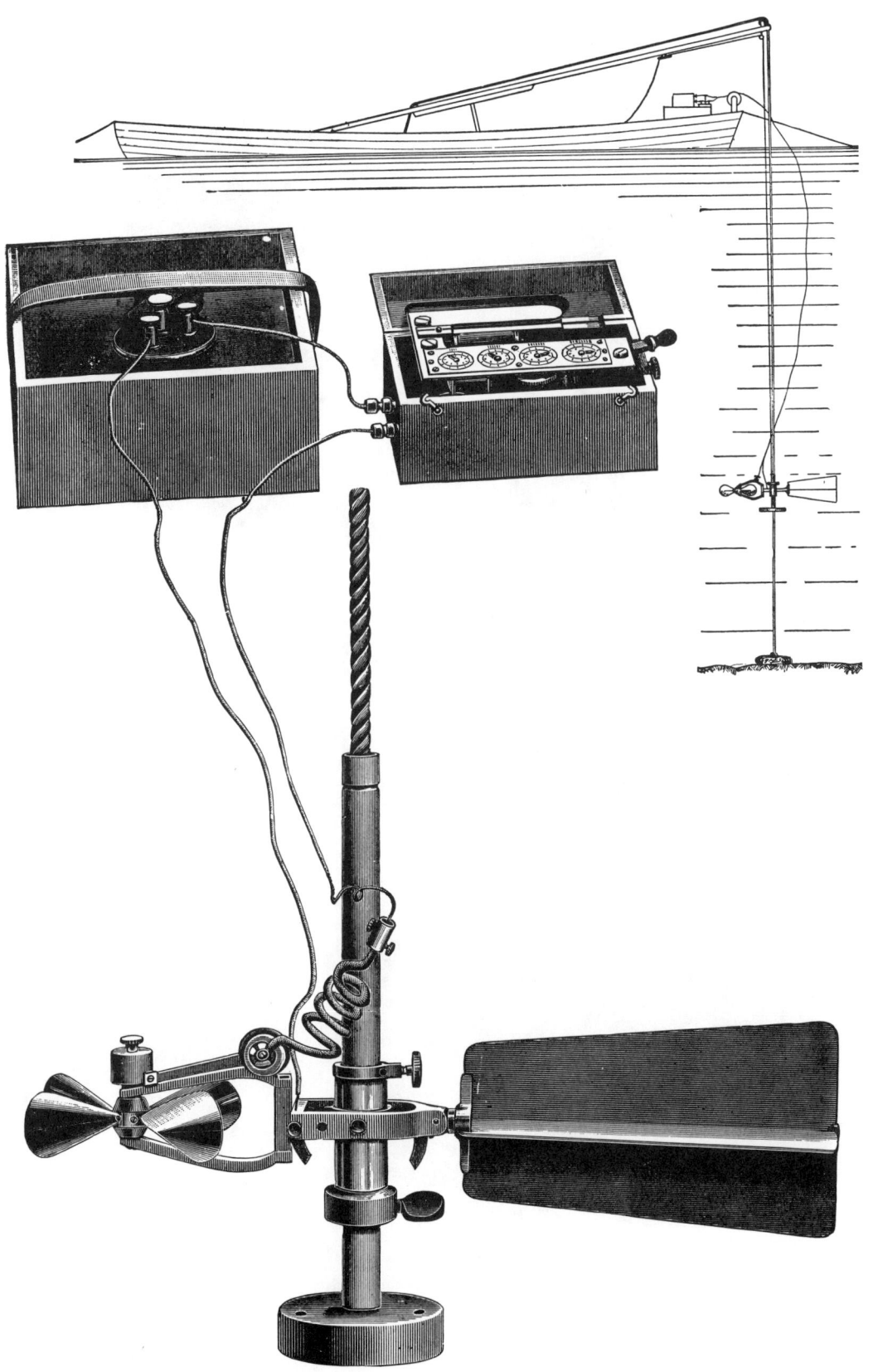

Current Meter No. IV.

As made by C. L. Berger & Sons.

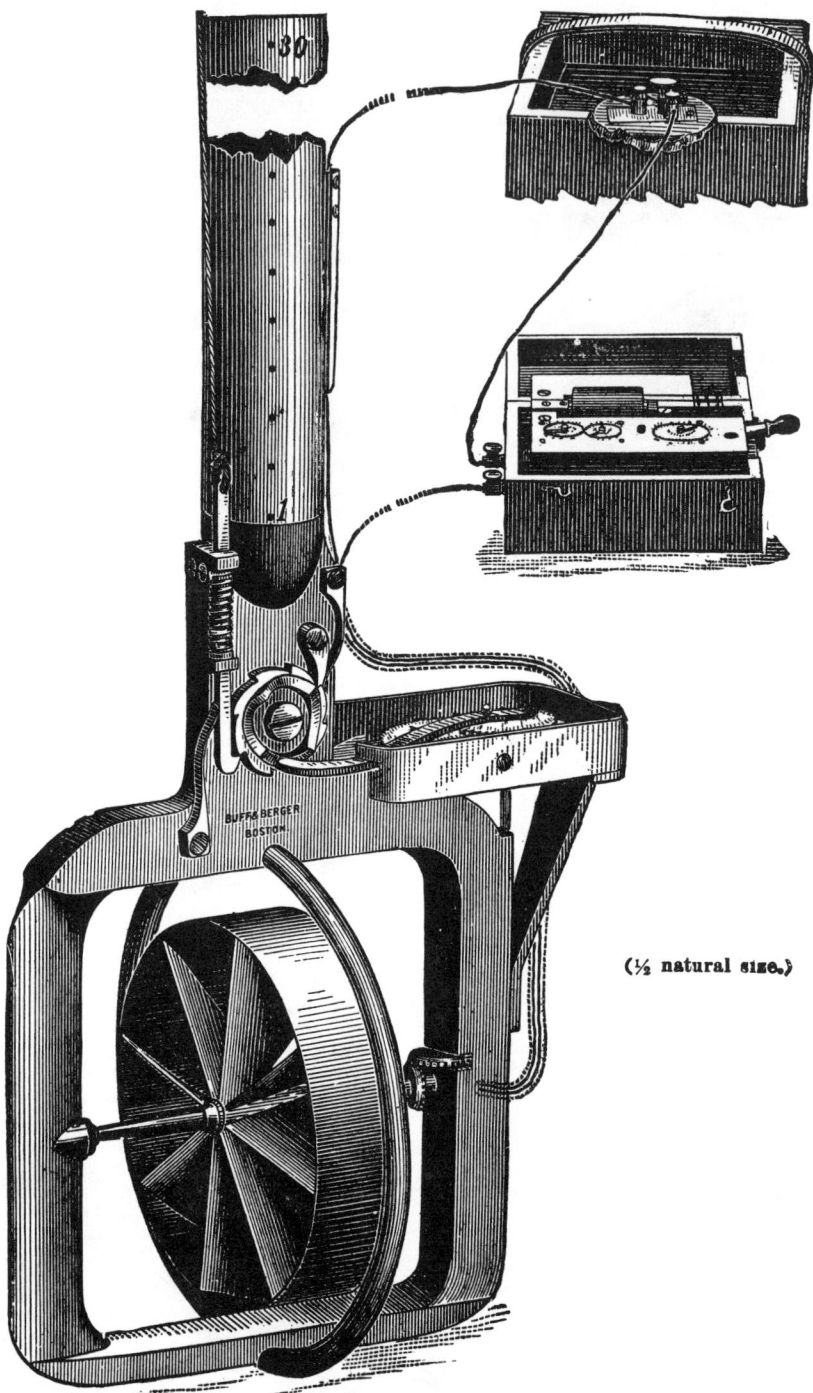

(½ natural size.)

Current Meters No. V and VI.

As made by C. L. Berger & Sons.

For price, etc., see page 196.

Precision Pantographs.

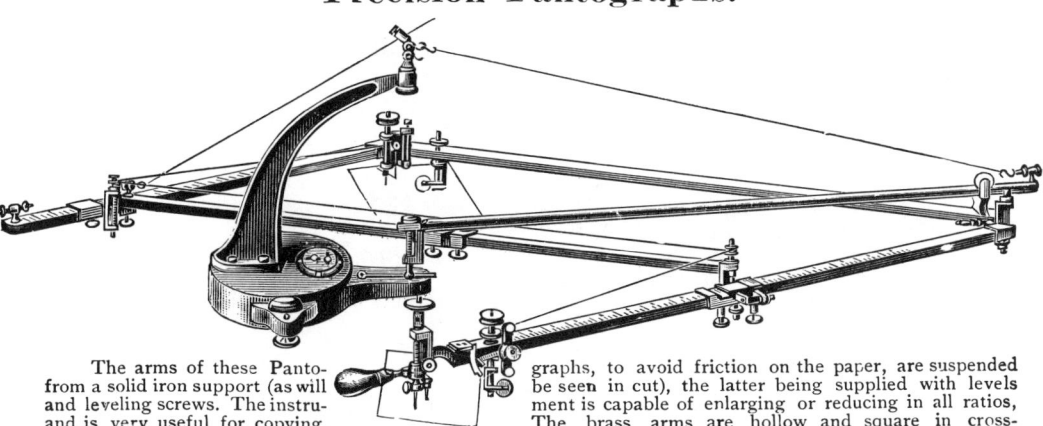

The arms of these Panto- graphs, to avoid friction on the paper, are suspended from a solid iron support (as will be seen in cut), the latter being supplied with levels and leveling screws. The instru- ment is capable of enlarging or reducing in all ratios, and is very useful for copying. The brass arms are hollow and square in cross-section, and are divided to millimeters with verniers reading to $\frac{1}{10}$ mm. For the accurate setting of the verniers slow motion screws are provided. All swivel joints turn upon center points. The disengaging mechanism is a special convenience. The ratios from $\frac{2}{3}$ to $\frac{1}{20}$ are set with pole at end, those from $\frac{2}{3}$ to $\frac{1}{1}$ to $\frac{3}{2}$ are set in the middle. The pole and pencil-holder are therefore interchangeable.

No. 99. **Suspended Pantograph,** arms about 24 inches long, in wooden case.

Price, as above, $150.00.

No. 100. **Suspended Pantograph,** arms about 38 inches long, in wooden case.

Price, as above $180.00.

Note — The Pantograph with 24-inch arms when set at $\frac{1}{2}$ can circumscribe a 19-inch square, or an oblong $15\frac{1}{2} \times 24$ inches, approximately.

The Pantograph with 38-inch arms can circumscribe a 31-inch square, or an oblong $27\frac{1}{2} \times 39$ inches, approximately.

Planimeter.

For description see elsewhere.

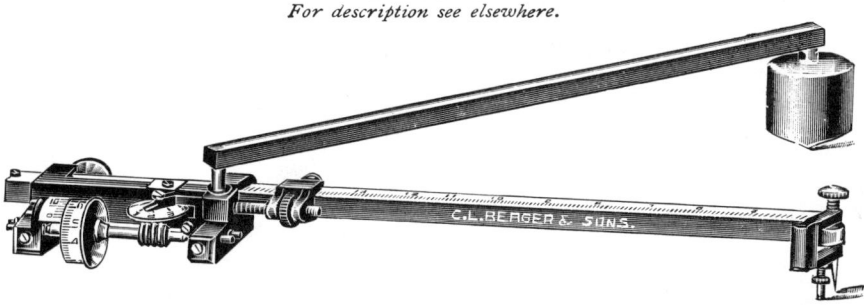

An instrument for measuring the areas of plane surfaces, by passing a pointer around their periphery. It is of great convenience to all classes of engineers, and practically applicable to a great variety of purposes. To measure the areas of figures that are bounded by irregular lines, such as :—drainage areas ; lots bounded by rivers or creeks ; contour lines of ponds, etc. ; to get the true average of observations taken at irregular intervals ; to measure indicator and other diagrams, and for many other portions of engineering work. As these instruments will not only give the area of any figure, but also any multiple of such area, and the sum of any number, or series of such multiples, at one operation, they may be used to very great advantage in the calculation of the cubical contents of solids ; as in the calculation of earth-work, etc. See on this point an article by Clemens Herschel, Esq., Civil Engineer, of Boston, in the *Journal* of the Franklin Institute for April, 1874, and the directions for use which we furnish with each rated instrument. Earth-work measurements, made in the manner indicated, do NOT *require the plotting of cross-sections.* The planimeters graduated by us are rated to read square inches of area, square centimeters of area, any multiple of these areas, and so as to give the cubic yards in any cut or fill, if used according to the directions that will accompany each rated instrument. Two consecutive measurements of the same area need never differ by more than 0.02 of a square inch ; and by repeating the measurement in the same manner that angles are repeated with a transit instrument, the error of observation may be reduced to but a small fraction of one hundreth of a square inch of area.

No. 107. **Price** of instrument, when rated as above indicated, . . **32.00.**

" 108. " " not " but with all our improvements, **28.00.**

Precision Planimeters.

These Planimeters are very much more accurate than the ordinary Polar Planimeters, as represented on preceding page. The graduated rollers do not touch the paper at all, but roll, instead, on a hard, highly polished surface of steel, thus eliminating all errors due to the irregularities of the paper surface.

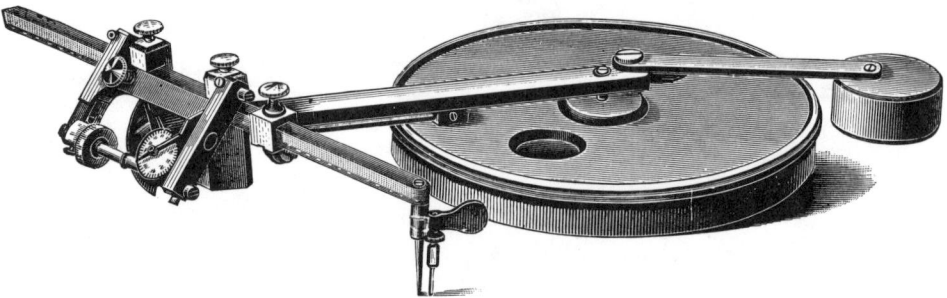

No. 109. Large Suspended Ball Planimeter.

This instrument is capable of doing very accurate work. The tracer arm is $11\frac{3}{4}$ inches long, the pole arm is $6\frac{1}{4}$ inches long, and the diameter of the toothed circle on the pole is $6\frac{1}{4}$ inches. The angular motion of the tracer arm is about 90°.

Surfaces from $2\frac{1}{2}$ x 4 inches to 7 x 10 inches can be measured without moving the pole.

Price of instrument complete packed in morocco box . . . . **$75.00.**

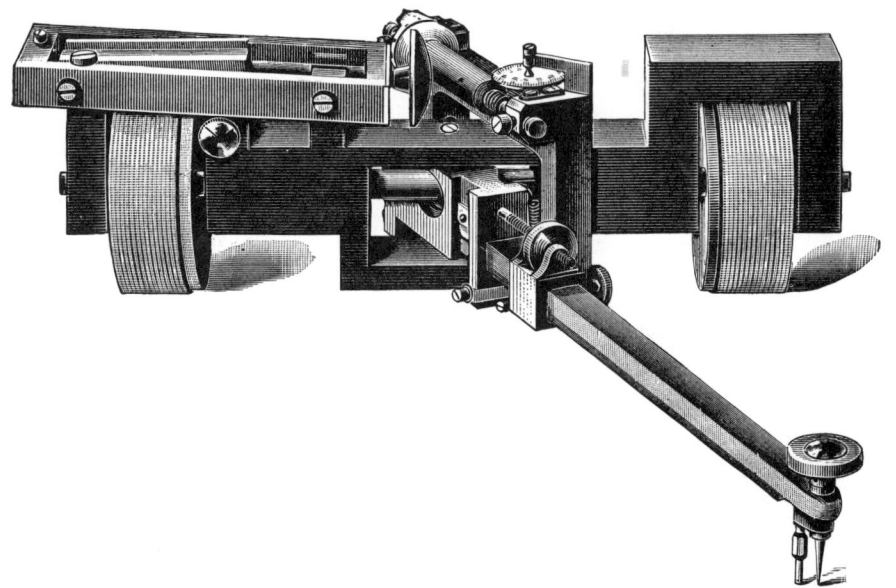

No. 110. Large Rolling Ball Planimeter.

This instrument is capable of doing work more accurately than any planimeter yet made, both on large and small surfaces. The tracer arm has a length of $11\frac{3}{4}$ inches which can be increased by a lengthener to $19\frac{3}{4}$ inches. Its angular motion is about 90°. The two rollers are made of exactly equal diameters, ensuring a motion of the instrument, as a whole, in a straight line.

A surface of any length and of a width of 20 inches can be measured with the $19\frac{3}{4}$ inch tracer arm.

Price of instrument complete in morocco box with lock . . . **$85.00.**

Surveyors' Pocket and Marine Compasses.

No. 111. Burt's Solar Compass, with adjusting socket and leveling tripod, **$220.00**

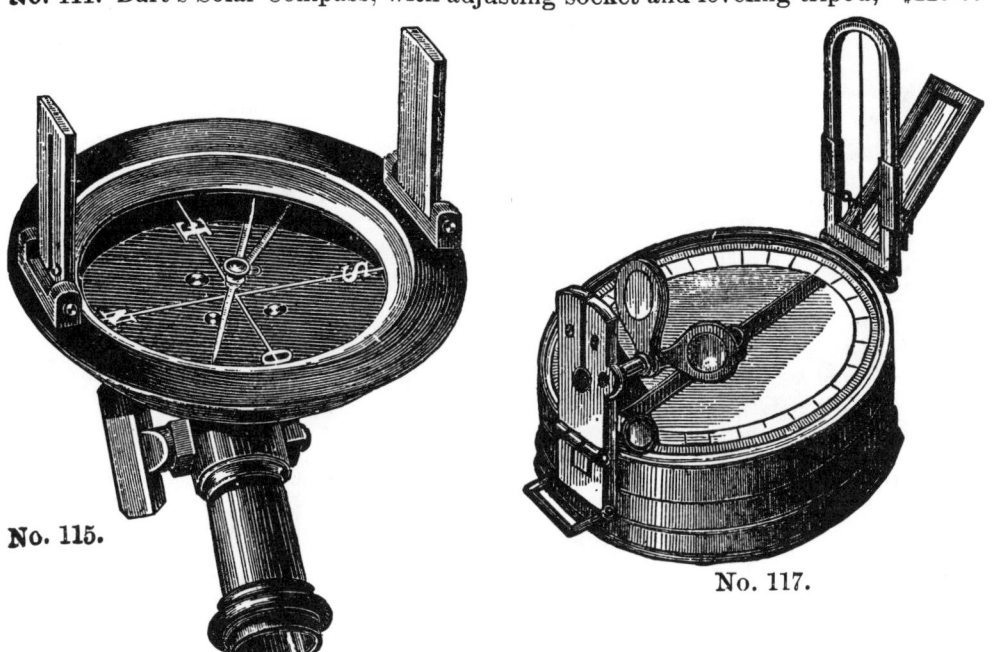

No. 115.

No. 117.

No. 112. Pocket Compass, with folding sights, 2½ inch needle, . . . $ 8.00
" 113. " " 2½ inch needle, Jacob Staff mountings, . . . 10.00
" 114. " " 3½ " " " . . . 11.00
" 115. " " with level, folding sights, 4-inch needle, with ball and
socket joint, ... 13.00
" 116. Vernier Pocket Compass, 4½ inch needle, "Tripod" and 2 levels . 23.00
" 117. Prismatic Compass, complete, with azimuth glasses, and divided aluminum ring, 3 inch Leather Sling Case. Best kind, ... **30.00**
" 117a. Hutchinson's Prismatic Compass bronzed, of improved pattern nearly
enclosed top, floating card dial, 2 inch, in morocco case . . 11.00
" 118. Pocket Compass, watch pattern, brass, 1½ inches in diameter with
hinged cover and stop to needle, ... 1.75
" 119. Pocket Compass, gilt, watch pattern, with stop, enamelled dial and
agate centre; 1 or 2 inches in diameter, ... 5.00
" 120. Ritchie's Patent Liquid Compasses, of all sizes, from $33.00 to $35..0,
$45.00 and $55.00.

Miners' Compasses.

No.125.

No. 125. Miners' Compass, provided with stop and glass covers, for tracing
iron ore, 3 inch Norwegian needle, ... **$13.0**
" **126.** Miners' Compass, provided with stop and glass covers, 4 in. Norwegian needle, ... 15.00

Leveling Rods.

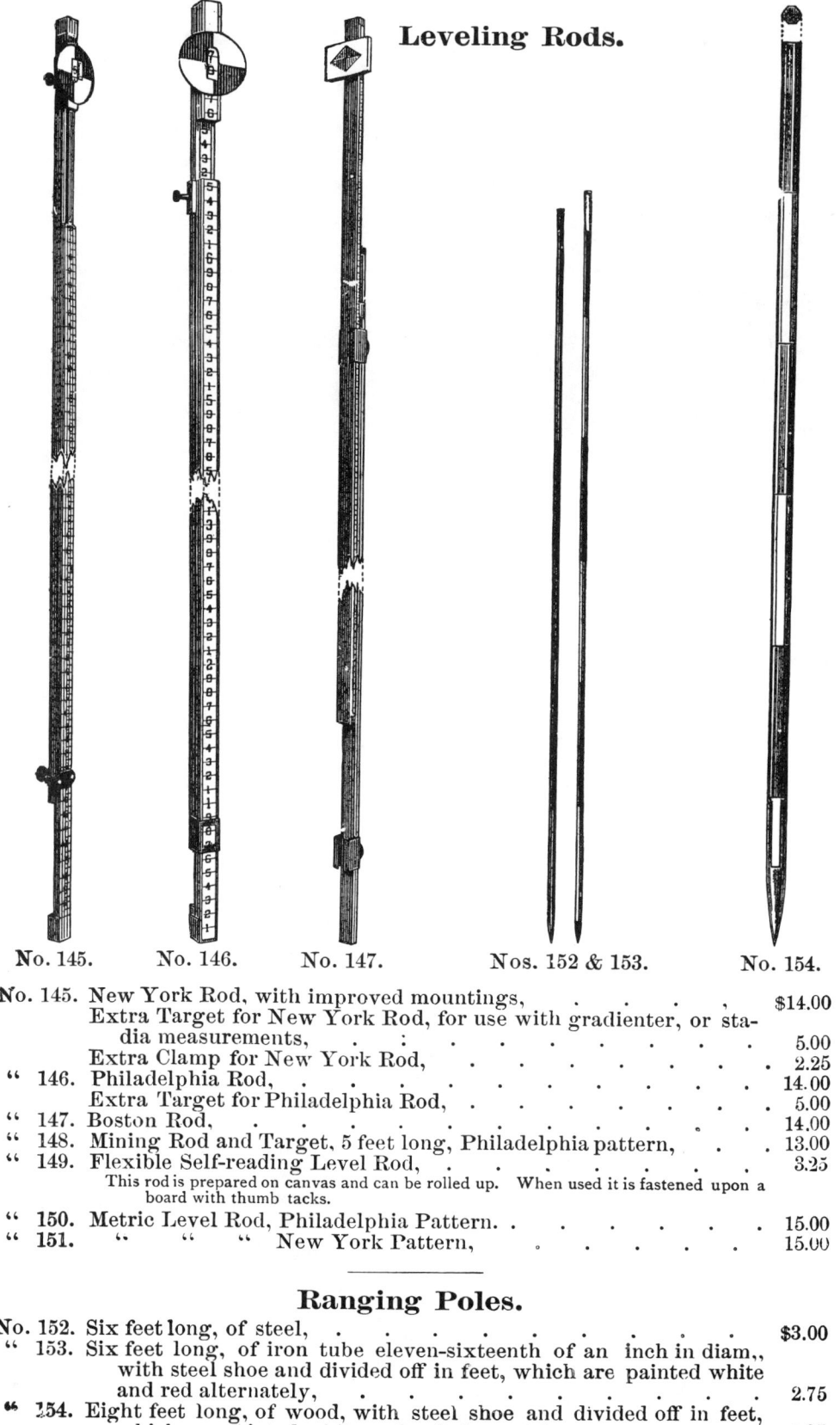

| No. 145. | No. 146. | No. 147. | Nos. 152 & 153. | No. 154. |

No. 145. New York Rod, with improved mountings, ... $14.00
  Extra Target for New York Rod, for use with gradienter, or sta-
    dia measurements, . : 5.00
  Extra Clamp for New York Rod, ... 2.25
" 146. Philadelphia Rod, ... 14.00
  Extra Target for Philadelphia Rod, ... 5.00
" 147. Boston Rod, ... 14.00
" 148. Mining Rod and Target, 5 feet long, Philadelphia pattern, . . 13.00
" 149. Flexible Self-reading Level Rod, ... 3.25
    This rod is prepared on canvas and can be rolled up. When used it is fastened upon a
    board with thumb tacks.
" 150. Metric Level Rod, Philadelphia Pattern. 15.00
" 151. " " " New York Pattern, ... 15.00

Ranging Poles.

No. 152. Six feet long, of steel, ... $3.00
" 153. Six feet long, of iron tube eleven-sixteenth of an inch in diam,,
    with steel shoe and divided off in feet, which are painted white
    and red alternately, ... 2.75
" 154. Eight feet long, of wood, with steel shoe and divided off in feet,
    which are painted white and red alternately, , . . . 2.25

Paine's Steel Tape Measures.

¼ inch wide. In Leather Cases, with flush handles.

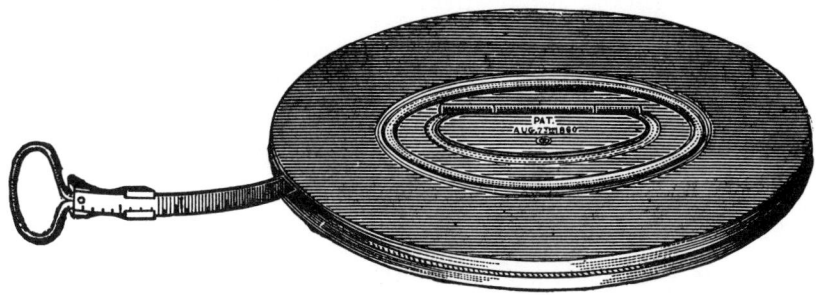

No. 160. 100 feet Paine's Steel Tape, divided in 10ths,　　•　•　•　$11.00
" 161.　50　"　　"　　"　　"　　"　　"　.　.　.　.　6.00
" 162. 100　"　　"　　"　　"　　"　　" on one side, on the other
　　　in centimeters,　.　.　.　.　.　.　.　.　.　.　15.00

Chesterman's Steel Tape Measures.

⅜ inch wide. In Leather Boxes.

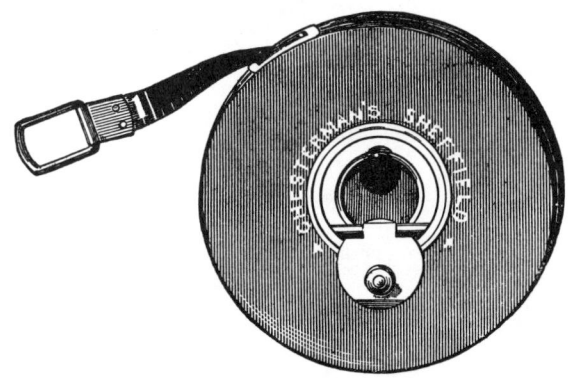

No. 163. 100 feet Chesterman's Steel Tape, divided in 10ths,　•　•　•　$11.00
" 164.　66　"　　"　　"　　"　　" .　'　•　•　8.00
" 165.　50　"　　"　　"　　"　　" .　•　•　•　6.00
" 166.　33　"　　"　　"　　"　　" .　•　◦　◦　5.00

Chesterman's Pocket Steel Tape Measures.

In German Silver Cases, with spring and stop.

No. 167. 3 feet long, divided in 10ths.　.　.　.　.　.　.　.　.　$1.50
" 168. 5　"　　"　　"　　"　.　.　.　.　.　.　.　2.00
" 169. 5　"　　"　　"　　" on one side, and in centimeters and mil-
　　　limeters on the other side,　.　.　.　.　.　.　.　2.25

Excelsior Steel Tape Measures.
½ inch wide. Patent Brass Frame with Handle.

No. 170. 100 feet Excelsior Steel Tape, divided in 10ths, ... $11.50
" 171. 50 " " " " " 6.60

Excelsior Steel Tape Measures.
½ inch wide. In Leather Boxes.

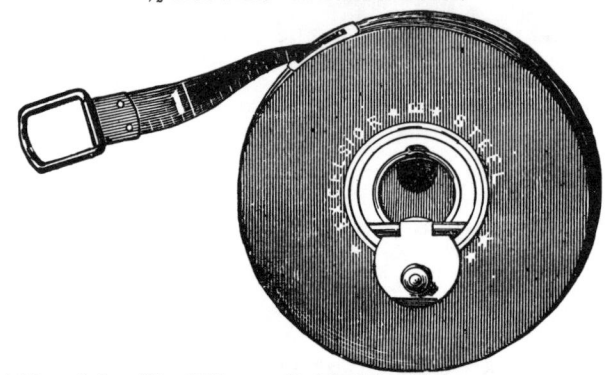

No. 172. 100 feet Excelsior Steel Tape, divided in 10ths, ... $11.50
" 173. 66 " " " " " 8.00
" 174. 50 " " " " " 6.50

Lufkin Steel Tape Measures.
⅜ inch wide. In leather case.

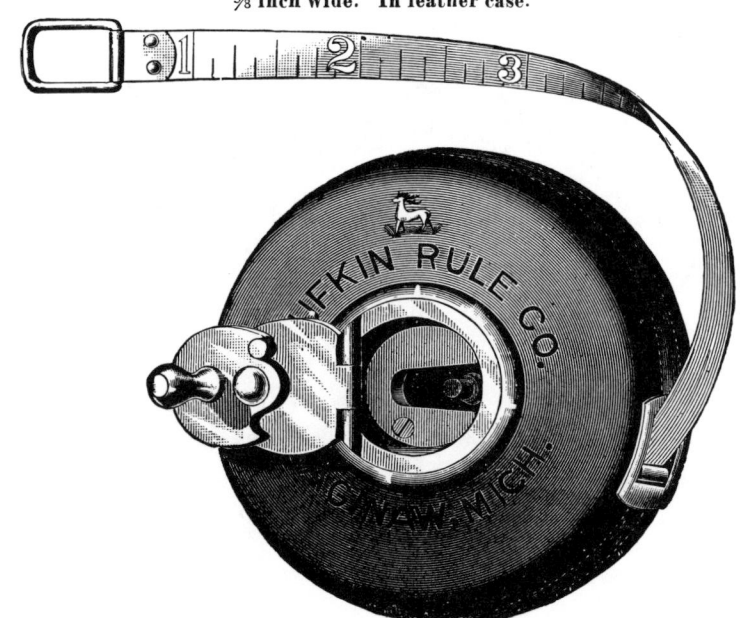

No. 206. D. 100 feet Lufkin Steel Tape, divided in 10ths, ... $11.00
No. 203. D. 50 " " " " " 6 00
No. 103. D. 50 " " " " " ¼ inch wide; 2¾
inch dia.; 5 oz. in weight; can be carried in vest pocket . . . 4.00

Lucas's Improved Steel Tapes.
¼ inch wide.

All tapes of this manufacture are made from the best quality of clock-spring steel, one-fourth of an inch wide, and of thickness best adapted to strength and flexibility, tempered straight, and graduated under tension, being drawn on steel bars made to correspond with U. S. standard, and are guaranteed to be as accurate as any tapes made in this country. The graduations are made each five feet or links, according to the style of tape, by brass or german silver bands firmly soldered to the tape, and marked each side with plain figures in such a manner as to be conveniently read from either end without liability of error in count. The intermediate points of feet or links are marked by a small brass rivet through the tape, with raised head on each side so as to be easily seen. Each end foot on Engineers' tapes is graduated to tenths of a foot. The adjustments for taking measure are so arranged that no difference is made by the use of large or small marking pins, the measurement being made and taken from the same side of the pin. Soldering of the number bands secures them from peeling up or rusting underneath. The method of numbering avoids the necessity of changing ends of tape, it works same with either end forward.

No. 178 **D.** 100 feet, Engineer's, graduated to feet, each five feet by soldered bands marked with figures, end feet to tenths of a foot $4.00

No. 178 **K.** 66 feet, Surveyor's, graduated to links, with figured bands every five links 3.50

Folding Brass Reel, $2.00 extra. Brass Snap Handles, 30 cents per pair extra.

Tension and Temperature sent with each tape when sold.

Roe's Steel Tapes on Brass Reel.
¼ inch wide.

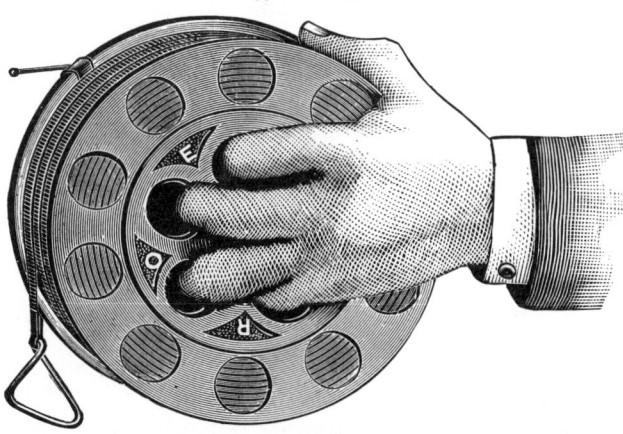

These tapes are made of superior steel, ¼ inch wide, graduated every foot by a brass rivet, end feet in tenths. Every five feet has a brass plate with the numbers, and every ten feet has a copper plate with numbers.

They are graduated from a standard tape certified to by an official of the U. S. Coast Survey Department as correct at a temperature of 62° F.

No. 179 **1A.** 100 feet long, graduated every foot, end feet in tenths, . $5.00

7A. 50 " " " " " " " " " . 4.00

Prices above include a Patent Brass Reel and pair of Patent Brass Detachable Handles.

Brass Reel, without Tape, ... $1.50

Detachable Handles, per pair, ... 0.30

Standard Steel Tape Measures.

For city and bridge engineering, in lengths from 100 to 1000 feet.

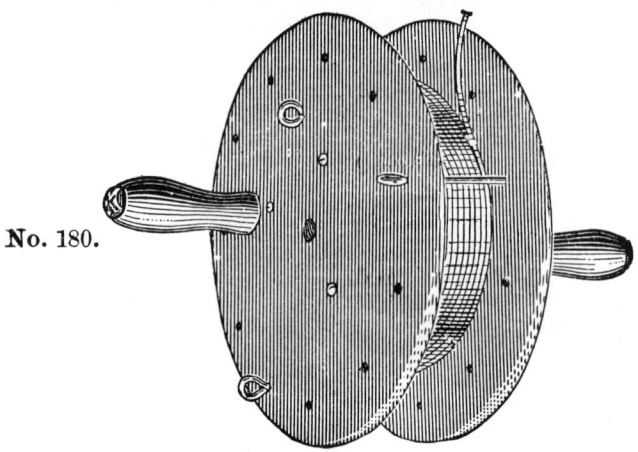

No. 180.

These tapes are of exact United States Standard and have no joints. They are generally made in lengths of 300 feet with graduations at every 10 feet, the last 10 feet graduated in single feet, and the last foot into 10ths. For railroad and under ground work we frequently furnish them in lengths of 400 and 500 feet. A clamping handle can be furnished to attach to the tape at any desired length, if shorter measures than the whole length are intended to be made. We also can furnish a small brass clamp to fasten on the tape in order to mark lengths that are used repeatedly.

Price of tape 100 feet, graduated at every 10 feet, the last 10 feet graduated in single feet, the last foot in 10ths, ... $8.00
Price of tape 200 feet, graduated as above, ... $11.50
" " " 300 " " " " 17.00
" " " 400 " " " .. " 22.50
" " " 500 " " " " " 28.00

Extras to Standard Steel Tape Measures.

Each additional graduation and figuring, ... $0.20
Reel, handle and stop to wind up tape. 3.50
2 large brass handles to unship, ... 2.50
Clamping handles, each, ... 1.50
Small brass clamp to fasten on tape, ...75

Metric Steel and Metallic Tape Measures.

In Leather Boxes.

No. 191 20 Meter Steel Tape, divided in meters and centimeters, 9 mm. wide $11.00
" 192 10 " " " " " " 9 " 6.00
" 193 20 " Metallic Tape, divided in meters and centimeters, 17 mm. wide 3.50
" 194 10 " Metallic Tape, divided in meters and centimeters, 17 mm. wide 2.75

Chains.

No. 195. Surveyors' Chain, 2 poles, 50 links, No. 12 best steel wire, brazed
links and rings, ... $5.50
" 196. Surveyors' Chain, 4 poles, 100 links, No. 12 best steel wire, brazed
links and rings, ... 10.00
" 197. Engineers' Chain, 50 feet, 50 links, No. 12 best steel wire, brazed
links and rings, ... 6.00
" 198. Engineers' Chain, 100 feet, 100 links, No. 12 best steel wire, brazed
links and rings, ... 11.00

Metric Chains.

No. 199. 20 Meter Chain, 100 links, No. 12 best steel wire, brazed links and rings, 10.00
" 200. 10 " " 50 " " " " " " " 5.50

Extras to Tapes and Chains.

No. 201. Pocket Thermometer, ... $1.50
" 202. Spring Balance and Level, ... 5.00

Marking Pins.

No. 203. Set of Marking Pins, eleven in a set, steel wire, No. 6, . . $1.80

Odometer.

No. 204. An instrument for measuring distances traveled by carriage, . $15.00

Pedometer.

No. 205. An instrument for measuring distances walked, in german **silver**
case, of the size of a watch, ... $5.00

Plummet Lamp.

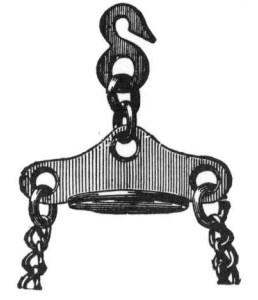

Lamp for Mining Engineering

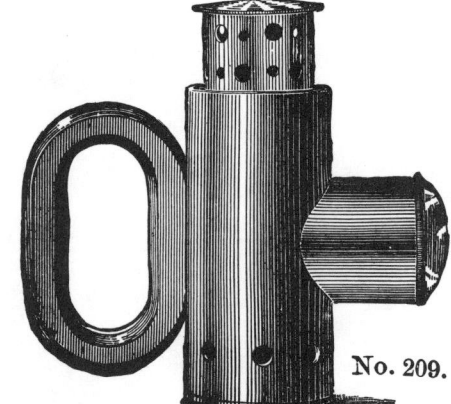

No. 209.

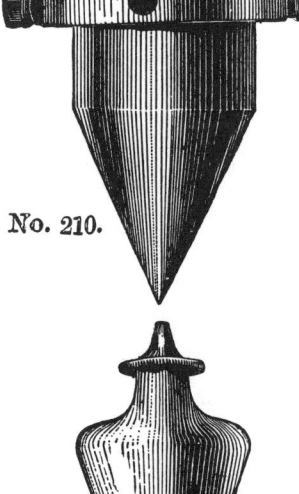

No. 210.

No. 209. Lamp for illuminating graduations, cross wires, etc., for use in underground work, of brass and nickel-plated, with ground lens, ... $7.00
" 210. Small Plummet Lamp, of brass, steel point, 16 oz., . 8.00
" 211. Large Plummet Lamp, of brass, steel point, 24 oz., . 10.00
Box with shoulder straps, for pair of Plummet Lamps, 3.00

Plumb Bobs.

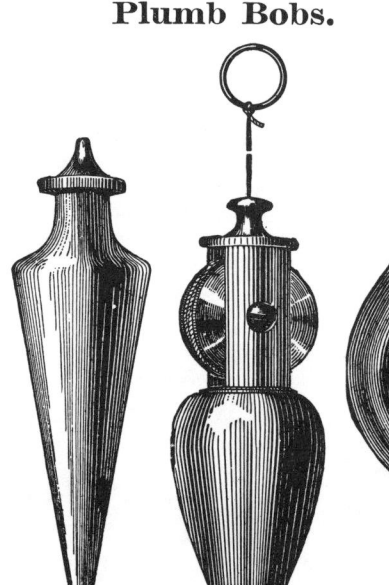

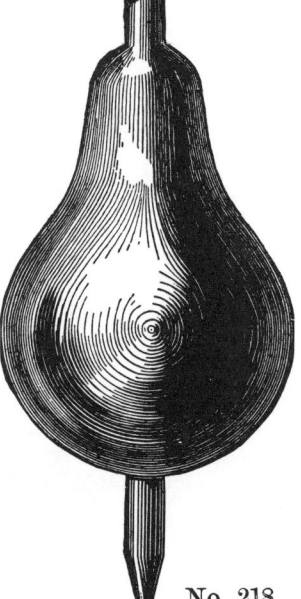

No. 218.

No. 212.

No. 214. No. 216.

No. 212. Plumb Bob of Brass, steel point, shape as in cut, 8 oz., . . $2.25
" 213.   "   "   "   " " No. 212, 11 oz., . . 2.50
" 214.   "   "   "   " " in cut. 6 oz., . . 2.00
" 215.   "   "   "   " " No. 214, 10 oz., . 2.50
" 216.   "   "   "   patent reel adjustment, 8 oz., 2.50
" 217.   "   "   "   "   "   "   14 oz., 2.80
" 218.   "   "   "   for use in shafts, 36 oz., . . 5.00

Pocket Magnifiers.

No. 221.   No. 224.

No. 221. Zylonite Case, as in cut, size of lens 1 inch diameter, . . . $0.60
" 222.   "   "   " 221, "   "   1¼ "   " 90
" 223.   "   "   " 221, "   "   1½ `   "   . . . 1.15
" 224.   "   "   " cut   " of lenses, 1⅛ and 1¼ in diameter, . 1.30

Gossamer and Silk Bags.

Gossamer or Water-proof Bag, to cover Transit or Level in case of rain or dust, $1.00
Silk Bag, to cover Transit, with solid silver graduations 1.00

Lubricants.

Bottle of Fine Watch Oil, for lubricating Transit Centers, etc. . . . **$0.25**

Utensils for Cleaning Instruments.

Camel's Hair Brush **$0.40**
Stiff Brush tor cleaning screw-threads 40
Chamois-skin for cleaning lenses, centers, etc. 50
Stick for cleaning centers 30

Spirit Levels.

Engineers' Spirit Levels, of all sizes and grades of sensitiveness, accurately
    ground and tested by us.
Per inch, according to length and diameter . . . . from $0.80 to $1.00

Portable Anemometers.

These instruments are extensively used in studying and controlling the ventilation of dwellings, public buildings, factories, mines, etc.

The velocity of the air current is measured by means of a very light fan wheel, whose revolutions are recorded on a dial.

This fan wheel is very delicate, the vanes being made of aluminum, and the axis of hard steel runs in jewel bearings.

The counting mechanism is enclosed in a dust-proof case, and can readily be thrown into or out of action by a disconnecting lever.

The instrument is provided with a thumb-screw for attaching it to a rod for use in measuring the velocity of air currents at any point on the surface of the earth, mine shafts, in pipes, conduits or narrow channels. In this case the counting mechanism is thrown in or out of gear by pulling on cords of different colors.

This Anemometer is carefully rated and supplied WITH A CORRECTION NUMBER.

Anemometer, Counting up to 10,000,000 ft.; diameter of fan, 3 in.; complete, packed in polished wooden box, **$25.00.**

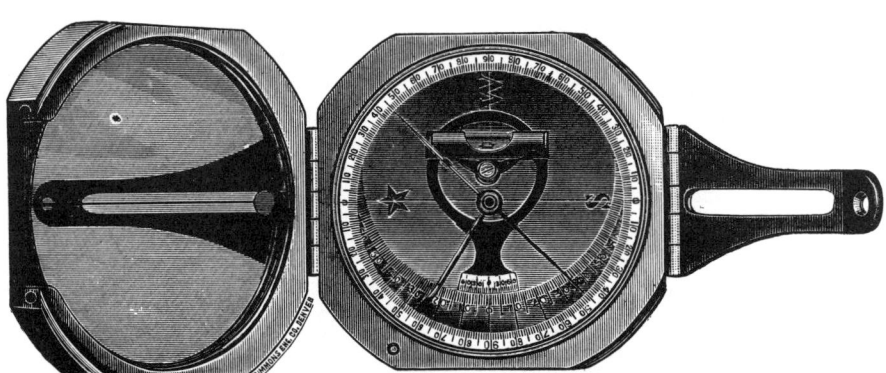

The Brunton Patent Pocket Mine Transit.

A pocket instrument which takes the place of a sighting compass, clinometer, prismatic compass, and an Abney level or Locke's level. Weight 8 ounces. Price, **$25.00**

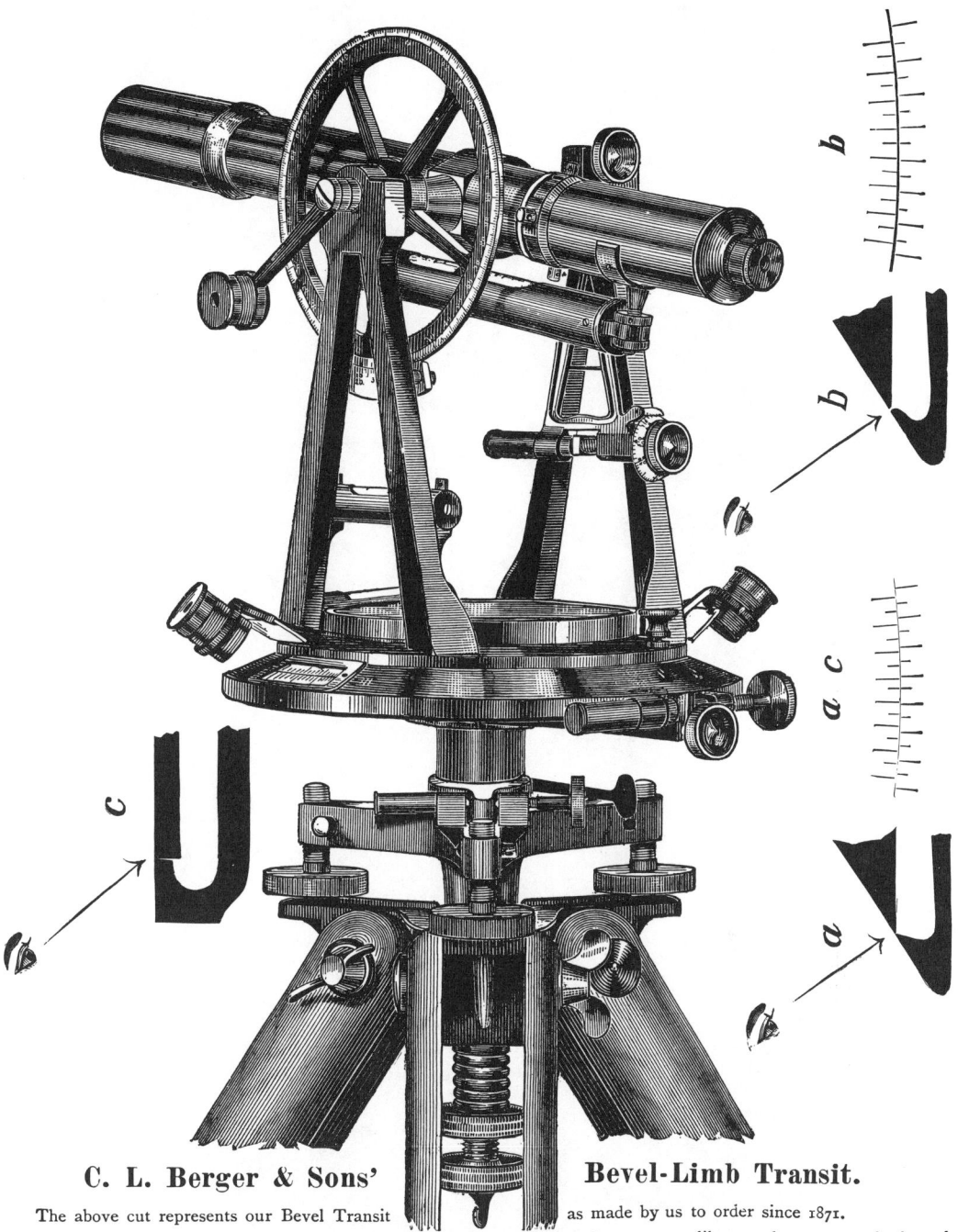

C. L. Berger & Sons'   Bevel-Limb Transit.

The above cut represents our Bevel Transit as made by us to order since 1871.

NOTE. While a bevel-limb graduation of a horizontal circle can be somewhat more readily seen than one on a horizontal surface, it is well to remember, before ordering instruments to be made so, that there are very serious objections to their general adoption for engineers' field-instruments. As will be seen in the annexed cuts, the sharper and therefore more delicate edges of the soft solid silver, necessitated by the bevel at the junction of limb and verniers, are much more liable to injury and wear in field use than the common horizontal graduation, where the same edges are carried down nearly rect-angularly below the graduated surface. Thus, while a bevel possesses some advantages when *new and well constructed*, it soon becomes impaired by slight dents and the edges rounded by brush or finger when dust and oxyd must be removed at certain times. It then can be read only with difficulty and becomes a source of great annoyance, particularly as the eye looks squarely at it, thus defeating the very object sought and rendering the instrument almost unfit for good work, although otherwise in good condition. We say this with an experience of twenty-five years to back up. To make it plain we must have recourse to the diagrams. Fig. *c* is the cross-section of a horizontal limb and vernier as commonly made. It is obvious that the fine silver edges at the junction of limb and verniers are in this form better protected from wear, and also that, when slightly rounded by wear, or when the graduated surfaces are not in the same plane, the eye, being stationed at an angle of about 45° to the limb, requiring an observer to glance along the graduated lines, will more readily see and esti-mate differences in the reading of limb and verniers, thereby enabling him to obtain closer results, *as verified by the superior results in triangulation obtained with horizontal graduations over bevel ones formerly in vogue*. Fig. *a* is a cross-section of a bevel limb, showing the sharper edges of the graduated surfaces. When new and properly made the edges of limb and verniers appear the same as those in Fig. *c* and *a c*, but when worn off they will leave a big circular space between them, making the reading of an angle all the more difficult and uncertain. There are other reasons, against their general adop-tion, such as that the standards for the telescope have in bevel instruments not the stability so important in angle-measuring instruments unless they are mounted on a special horizontal base provided for them, (instead of on the bevel surface of the upper plate) which means a great increase in weight, or that the distance between them be quite short, resulting in a shortening of transverse axis of the telescope, as in the cut above, thereby increasing the instability of the same, particularly if the tele-scope is of modern power and length, besides reducing the size of compass, length of plate levels, and with this the degree of sensitiveness of the latter. In both these instances the standards are apt to change their distance apart affecting the adjustment of the line of collimation for near distances. —The greater expense of repairing, in case of accident, is also an item that should be considered. It is often double or triple that of a horizontal graduation. To our mind bevel gradu-ation should be confined to exceptional cases and to the larger instruments read by micrometer microscopes.

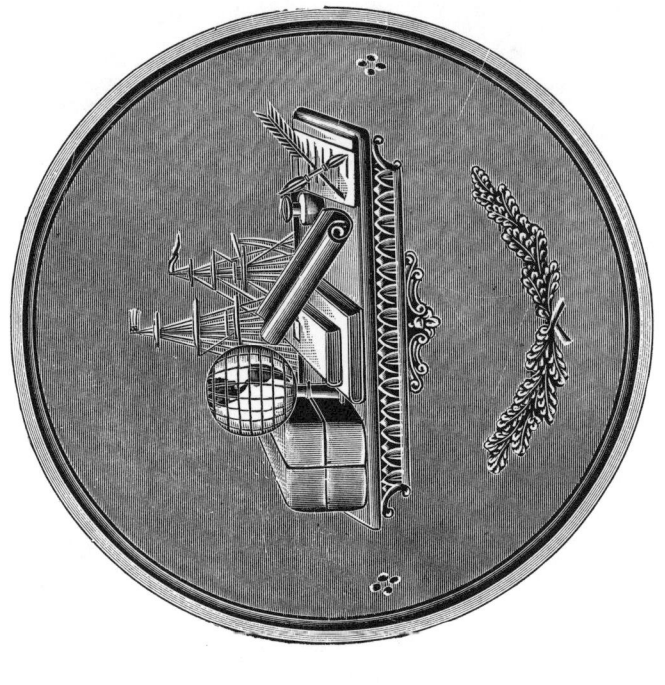

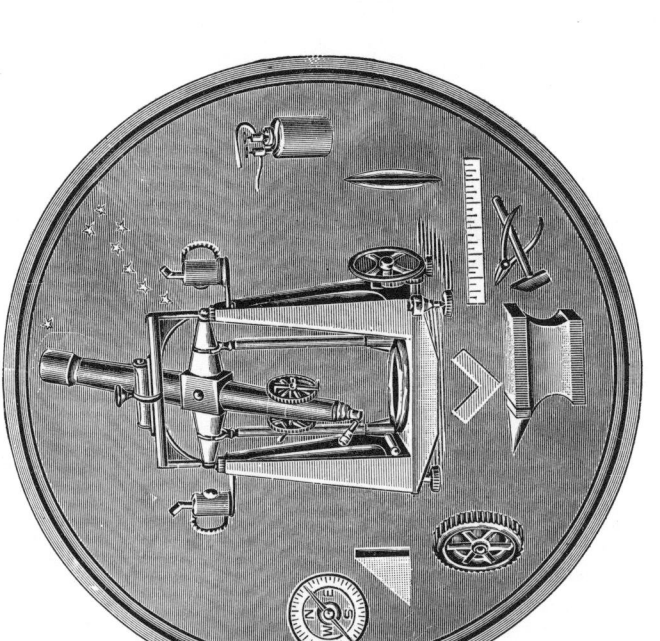

TRADE MARK.

MEDAL and DIPLOMA AWARDED for EXCELLENT WORK-
MANSHIP and NUMEROUS IMPORTANT IMPROVEMENTS in
DETAIL and CONSTRUCTION.

BOSTON, C. L. BERGER & SONS. MASS. U.S.A.
SURVEYING AND ENGINEERING INSTRUMENTS.

"THE STANDARD."

ALL ORDERS EXECUTED WITH FIDELITY AND DESPATCH

1. Complete 8-in. Transit-theodolite No. 12, for triangulation.
2. Complete 7-inch Transit-theodolite No. 12a, for triangulation.
3. Complete 6¼-in. Transit-theodolite No. 11a, for use in cities and triang'n.
4. Complete Tachymeter No. 1g, with Reversion Level.
5. Complete large Engineer's Transit No. 1b.
6. Complete small Engineer's Transit No. 2b; quick-leveling attachment.
7. Complete large Surveyor's Transit with Buff & Berger's Solar Attachment and Latitude Level. Variation plate.
8. Complete small Surveyor's Transit No. 2, with Davis' Solar Screen
   Attachment, Variation Plate and Reversion Level.
9. Complete Pat. Univ. Mining Transit No. 8; Duplex Telesc. Bear'gs.
10. Complete Mining Transit Side Telescope, Davis' Solar Attachment.
11. Mountain, Mining and Reconnaissance Transit, No. 4.
12. Interchangeable Lamp Targets, for Mining Transits Nos. 7 and 8,
13. Plane Table Alidade.
14. Inverting Dumpy Level.
15. 18-inch Engineer's Wye Level. Upright Telescope.
16. 18-in. Engineer's Y Level; with Reversion Level, Mirror, Upright Tel.
17. One New Engineer's Precise Level, with Mirror and one fixed and one
    Stride Level. Telescope Inverting.
18. Ellis Current Meter with Electric Register and Battery for measuring
    velocity of current in Rivers, Harbors, &c.
19. Electric Current Meter, after Fteeley and Stearns, for measuring velocity
    of current in conduits, flumes and shallow rivers.
20. Locke Hand-levels; Levels on Metal Base; Plumb-Bobs; Reading
    Glasses; Measuring Tapes; Targets, etc.
21. Two Tripods (One Extension and One Regular.)
22. Parts of instruments, made of aluminum, showing method of ribbing,
    etc., to make instruments light and stiff.

INDEX.